Biodiversity and Ecology of Organisms Associated with Woody Plants

Guest Editors

Katarína Pastirčáková
Rostislav Zemek

MDPI

Basel • Beijing • Wuhan • Barcelona • Belgrade • Novi Sad • Cluj • Manchester

Guest Editors
Katarína Pastirčáková
Department of Plant
Pathology and Mycology
Institute of Forest Ecology
Slovak Academy of Sciences
Nitra
Slovakia

Rostislav Zemek
Institute of Entomology
Biology Centre CAS
České Budějovice
Czech Republic

Editorial Office
MDPI AG
Grosspeteranlage 5
4052 Basel, Switzerland

This is a reprint of the Special Issue, published open access by the journal *Forests* (ISSN 1999-4907), freely accessible at: https://www.mdpi.com/journal/forests/special_issues/Woody_Plants_Organisms.

For citation purposes, cite each article independently as indicated on the article page online and as indicated below:

Lastname, A.A.; Lastname, B.B. Article Title. *Journal Name* **Year**, *Volume Number*, Page Range.

ISBN 978-3-7258-3423-5 (Hbk)
ISBN 978-3-7258-3424-2 (PDF)
https://doi.org/10.3390/books978-3-7258-3424-2

Cover image courtesy of Katarína Pastirčáková

Contents

About the Editors

Katarína Pastirčáková

Katarína Pastirčáková is a senior scientist at the Department of Plant Pathology and Mycology, Institute of Forest Ecology of the Slovak Academy of Sciences in Nitra, Slovakia. She is a plant pathologist and specializes in mycology. She earned her M.Sc. in Mathematics and Biology from the Faculty of Natural Sciences, Constantine the Philosopher University in Nitra in 1999 and her Ph.D. in Plant Protection from the Institute of Experimental Phytopathology and Entomology SAS in Ivanka pri Dunaji in 2003. Her main research interests include the taxonomy, species diversity, and ecology of fungal pathogens on herbs and woody plants, as well as fungi associated with bark beetles. She has been involved in national and international projects and has authored over 30 scientific papers in JRC-indexed journals. Dr. Pastirčáková is a member of the European Mycological Association, the Slovak Mycological Society, and the Slovak Plant Health Society. Since 2013, she has held a position as a curator of Plant Pathology Herbarium (NR) at IFE SAS.

Rostislav Zemek

Rostislav Zemek is a senior scientist at the Laboratory of Arthropod Ecology and Pest Management, Biology Centre of the Czech Academy of Sciences in České Budějovice, Czech Republic. From 2016 to 2020, he served as the head of the Arthropod Ecology and Biological Control Research Group at Ton Duc Thang University in Ho Chi Minh City, Vietnam. He earned his M.Sc. in Agronomy from the Agricultural University in České Budějovice in 1986 and his Ph.D. in Entomology from the Institute of Entomology, České Budějovice, in 1993. Dr. Zemek completed postdoctoral research at the Zoological Institute, University of Copenhagen (Denmark) from 1995 to 1996, with a fellowship from the Danish Natural Science Research Council. His research primarily focuses on the ecology and behavior of insects and mites, multitrophic interactions, and biological pest control. Currently, his work centers on the application of entomopathogenic fungi in crop pest management. Other research interests include plant pathogen–herbivore interactions and the non-consumptive effects of predators on their prey. Dr. Zemek has served as an editor for *Systematic and Applied Acarology* and was a member of the Executive Committee for the International Congress of Acarology and the European Association of Acarologists. He organized the IOBC working group meeting on "Integrated Control of Plant-Feeding Mites" in 2011.

forests

MDPI

Editorial

Biodiversity and Ecology of Organisms Associated with Woody Plants

Katarína Pastirčáková [1] and Rostislav Zemek [2,*]

1 Department of Plant Pathology and Mycology, Institute of Forest Ecology, Slovak Academy of Sciences, Akademická 2, 949 01 Nitra, Slovakia; uefezima@hotmail.com
2 Institute of Entomology, Biology Centre CAS, Branišovská 31, 370 05 Ceske Budejovice, Czech Republic
* Correspondence: rosta@entu.cas.cz

1. Introduction

Woody plants serve as hosts for a vast array of organisms, ranging from herbivores and pathogenic species—often considered detrimental—to beneficial organisms that contribute to plant health and ecosystem stability. Natural enemies of pests, for instance, play a crucial role in maintaining the health of wild woody plants, production forests, and ornamental tree vegetation in urban green spaces, including city parks, gardens, and urban forests. While interactions between woody plants and herbivores have been extensively studied, much less is known about the intricate relationships among the diverse organisms that inhabit these plants.

Recent research has highlighted the ecological significance of these interactions. For example, mycoparasitic fungi and endophytic entomopathogenic fungi offer promising avenues for the biological control of woody plant fungal pathogens and pests. Similarly, certain deciduous tree species serve as reservoirs for predatory mites (Acari: Phytoseiidae) [1,2], which contribute to the regulation of phytophagous mite populations in adjacent orchards and vineyards [3]. The presence of abandoned leaf mines has also been shown to enhance phytoseiid mite densities [4], further illustrating the complexity of these interactions. However, our understanding of such ecological networks remains incomplete, and further studies are needed to elucidate their roles in ecosystem functioning.

The impact of climate change on woody plant-inhabiting organisms is another pressing issue [5]. Rising temperatures, prolonged drought periods, shifts in hydrological conditions, and air pollution are contributing to the declining physiological activity of woody plants, often triggering outbreaks of insect pests and fungal pathogens. The widespread decline in forest trees due to bark beetles and their fungal associates continues to be a major research focus, yet significant knowledge gaps persist regarding their biology, ecology, epidemiology, and management [6,7]. Understanding the diverse types of bark beetle–fungus interactions is critical for developing effective mitigation strategies [8].

This Special Issue aims to present the recent advances in our knowledge of the biodiversity and ecological interactions of both beneficial and harmful organisms associated with woody plants. Relevant topics include, but are not limited to, the following: (1) above- and below-ground communities, (2) micro- and macro-organisms, (3) population dynamics and seasonal variations, (4) invasive pests and pathogens, (5) pollinators and natural enemies, (6) mutualistic interactions and competition, (7) effects of pollution and climate change, and (8) methods and modeling approaches for studying plant-associated organisms.

This Special Issue presented cutting-edge research on woody plant–pathogen and woody plant–arthropod interactions, as well as studies exploring their combined effects.

Received: 31 January 2025
Accepted: 6 February 2025
Published: 7 February 2025

Citation: Pastirčáková, K.; Zemek, R. Biodiversity and Ecology of Organisms Associated with Woody Plants. *Forests* **2025**, *16*, 283. https://doi.org/10.3390/f16020283

Contributions covered diverse ecosystems and geographical regions. While field studies were particularly emphasized, laboratory-based research was also included. By compiling the latest research in this field, this Special Issue will contribute to a deeper understanding of the ecological roles of the organisms associated with woody plants and their implications for forest health, conservation, and sustainable management.

2. Summary of Papers Included in the Special Issue

This Special Issue comprises 15 papers written by 72 authors from 11 countries, namely Croatia, Czech Republic, Japan, Latvia, Norway, Portugal, Russia, Serbia, Slovakia, Spain, and the USA. The collection represents a diverse array of studies on various organisms associated with woody plants ranging from microorganisms to invertebrates and mammals. As such, this Special Issue offers valuable examples of recent research activities in this global discipline.

Four studies published in this Special Issue highlight the significant role of fungal communities in forest health, resilience, and disease management. One study examined the diversity of culturable endophytic fungi in symptomatic and asymptomatic *Fraxinus excelsior*, identifying dominant species such as *Diaporthe eres* and *Alternaria alternata*. The inhibitory potential of certain endophytes against *Hymenoscyphus fraxineus*, the causative agent of ash dieback, was demonstrated, with *Fusarium lateritium* and *Didymella aliena* showing the strongest suppressive effects in planta [9].

In clonal forestry, mycorrhizal associations play a critical role in tree performance. Research on silver birch and hybrid aspen plantations revealed that mycorrhizal communities were moderately diverse, predominantly consisting of generalist ectomycorrhizal taxa. Clone-specific differences in mycorrhizal richness were observed, with hybrid aspen productivity showing an inverse relationship to fungal diversity, emphasizing the importance of host–genotype interactions in forest regeneration [10].

Efforts to mitigate *Fusarium circinatum*, the pathogen responsible for pine pitch canker, have focused on seed disinfection methods. A comparative study tested MennoFlorades, Captan, ethanol, and hot water treatments, with hot water emerging as the most effective and environmentally sustainable option for eliminating the pathogen without compromising seedling viability, providing a practical solution for nursery and forest management [11].

Finally, fungal biodiversity serves as a key indicator of forest habitat conservation. An analysis of macrofungi and lichenized fungi in Pyrenean oak and chestnut forests revealed greater species diversity in *Quercus pyrenaica* ecosystems. The presence of pollution-sensitive lichens in oak forests and a higher proportion of ectomycorrhizal fungi in chestnut groves highlight the ecological significance of these fungal communities and the need for finer-scale conservation assessments [12].

Recent research on invertebrate pests highlights emerging threats to forestry and agriculture, as well as the ecological complexities of pest population dynamics. A significant discovery was made with the first report of *Bursaphelenchus xylophilus*, the pinewood nematode, in *Pinus sylvestris* in Europe. Previously recorded in other *Pinus* species, its detection in Portugal raises serious phytosanitary concerns, reinforcing the need for stringent monitoring and management strategies to prevent further spread [13].

Understanding the dynamics of insect seed predators is crucial for predicting pest outbreaks in fruit crops. Long-term data from Norway on *Argyresthia conjugella* and *Megastigmus brevicaudis* revealed prolonged diapause as a key factor influencing their population cycles. Both seed predators and their associated parasitoids exhibited delayed emergence, sometimes spanning up to five years. These findings emphasize the need to account for diapause when studying insect guilds in rowanberry ecosystems, particularly given the occasional shift of *A. conjugella* to apple as an alternative host during poor rowan fruiting years [14].

Meanwhile, the increasing impact of *Ips sexdentatus*, a bark beetle species, has become evident in the Czech Republic. Population monitoring over two years revealed seasonal flight activity patterns, with peak swarming occurring in June and July. The study identified ACUMIPROTECT as the most effective pheromone lure for monitoring purposes and emphasized the potential influence of climate change on the voltinism and behavior of this pest. Such insights are essential for refining forest protection strategies [15].

The research conducted by Takagi and Kajimura [16] sheds light on the ecological traits of three species of *Xiphydria* woodwasps from Japan, emphasizing their host tree preferences and the intriguing relationships with symbiotic fungi and mucus in their bodies. As woodwasps play a significant role in forest ecosystems, understanding the ecology of Xiphydriidae is crucial, especially given the limited knowledge compared to their Siricidae counterparts. This study reveals new host records and highlights the complexity of their ecological interactions, including the presence of specialized mycangia for fungal associations and unique mucus characteristics. The findings contribute to the broader understanding of woodwasp biology and their role in forest dynamics, which is vital for effective pest management strategies.

Recent studies on natural enemies highlight their crucial role in regulating pest populations in diverse ecosystems, from urban trees to coniferous forests. A survey of predatory mites inhabiting *Aesculus hippocastanum* revealed a diverse community dominated by *Euseius finlandicus*, whose abundance was influenced by environmental factors such as urban greenery and air pollution. These findings underscore the potential of horse chestnut trees as reservoirs for phytoseiid mites, which could contribute to natural pest control in urban landscapes [17]. Similarly, research on mite communities associated with *Ips typographus* in Europe provided new insights into the diversity of these potential biological control agents. The study documented several mite species in Serbia for the first time, expanding the known distribution of key taxa and reinforcing the need for the further exploration of their ecological roles in bark beetle-infested forests [18].

Efforts to improve bark beetle management have also focused on reducing unintended impacts on natural enemies. A modification to pheromone traps used for monitoring *Orthotomicus erosus* in Croatia successfully reduced the capture of key predatory beetles, such as *Temnoscheila caerulea* and *Thanasimus formicarius*, without significantly affecting bark beetle captures. This approach offers a more balanced strategy for forest protection, ensuring that beneficial predators continue to contribute to pest suppression [19].

In agroforestry, parasitoids play a vital role in controlling invasive leafminers. A long-term study on *Cameraria ohridella* in the Czech Republic revealed fluctuations in parasitism rates and a shift in the dominant parasitoid species over time. The increasing prevalence of *Pediobius saulius* suggests an adaptive response of native parasitoid communities to the invasive host, highlighting the dynamic interactions shaping biological control success [20].

Ecosystem engineers play a critical role in shaping habitats and influencing biodiversity, often in ways that are underappreciated. Deadwood-dwelling beetles, particularly those in the Eucnemidae family, are key contributors to forest decomposition processes, yet their ecology remains poorly understood. A study conducted in a beech reserve in the Czech Republic provided new insights into the distribution and habitat preferences of these elusive beetles. Results indicated that lying logs offered a more stable microclimate, supporting a higher diversity of species, including several endangered ones. This underscores the importance of maintaining diverse deadwood structures in forest management practices to support saproxylic biodiversity [21].

Similarly, the underestimated ecological role of subdominant ants is highlighted in research on *Formica lemani*, a species capable of forming large nest complexes (NCs) that restructure behavioral interactions within forest ecosystems. Within these NCs, *F. lemani* exhibited higher population densities, increased aggressiveness, and more efficient trophobiotic relationships with aphids, particularly on birch trees. This behavior enhances the species' competitive ability, allowing it to play a significant role in plant protection against herbivores. The study challenges traditional views of ant hierarchy in forest systems, emphasizing that subdominant species can act as influential ecosystem engineers under the right ecological conditions [22].

Understanding the intricate relationships between species and their environment is key to predicting how ecosystems respond to change. One such complexity arises from the interaction between non-native plants, vegetative structure, and the foraging behavior of insectivorous bats. A study conducted in the forests of northern New Jersey [23] explored these dynamics, revealing that non-native vegetation significantly alters midstory clutter but does not directly impact insect abundance. Interestingly, bat activity was highest in areas with a high percentage of non-native vegetation and reduced midstory clutter, suggesting that structural openness is a stronger predictor of bat foraging than prey availability. However, the response varied by foraging guild—bats adapted to open spaces benefited from altered vegetation, while clutter-adapted species showed no clear preference. These findings highlight the multifaceted ways in which invasive species can reshape ecosystems, influencing not only plant communities but also higher trophic levels, such as insectivorous bats. The study underscores the need for further research to disentangle the indirect effects of non-native plants on predator–prey interactions and habitat use.

3. Concluding Remarks

This collection of studies provides critical insights into the multifaceted roles of invertebrates and microorganisms in forestry ecosystems, significantly enhancing our understanding of pest management and ecological interactions. The findings reveal how fungal communities contribute to disease suppression, optimize tree performance, and serve as bioindicators for forest health. By investigating pests like *B. xylophilus* and *I. sexdentatus*, as well as their natural enemies, these studies highlight the necessity of continuous surveillance and adaptive strategies in response to climate change and emerging pest threats. Furthermore, the research underscores the importance of preserving natural enemy communities, including predatory mites and parasitic wasps, as integral components of sustainable pest management in both forestry and urban environments. Recognizing the contributions of overlooked invertebrate groups, such as saproxylic beetles and ants, reinforces their significant influence on ecosystem dynamics. Together, these studies advocate for a holistic approach to biodiversity conservation and sustainable habitat management, emphasizing the need for further research into fungal-based biocontrol methods, mycorrhizal associations, and the complex interactions within these ecosystems.

Author Contributions: Conceptualization: K.P. and R.Z.; writing—original draft preparation: K.P. and R.Z.; writing—review and editing: K.P. and R.Z. All authors have read and agreed to the published version of the manuscript.

Funding: This research received no external funding.

Acknowledgments: We would like to thank all authors contributing their work to this Special Issue. We also appreciate all reviewers for their insightful comments and suggestions, which have significantly enhanced the quality of the papers.

Conflicts of Interest: The authors declare no conflicts of interest.

References

1. Kabíček, J. Broad Leaf Trees as Reservoirs for Phytoseiid Mites (Acari: Phytoseiidae). *Plant Prot. Sci.* **2003**, *39*, 65–69. [CrossRef]
2. Kopačka, M.; Stathakis, T.I.; Broufas, G.; Papadoulis, G.T.; Zemek, R. Diversity and Abundance of Phytoseiidae (Acari: Mesostigmata) on Horse Chestnut (*Aesculus hippocastanum* L.) in an Urban Environment: A Comparison between Greece and the Czech Republic. *Acarologia* **2018**, *58*, 83–90. [CrossRef]
3. Tuovinen, T. Influence of Surrounding Trees and Bushes on the Phytoseiid Mite Fauna on Apple Orchard Trees in Finland. *Agric. Ecosyst. Environ.* **1994**, *50*, 39–47. [CrossRef]
4. Villanueva, R.T.; Childers, C.C. Mine-Damaged Leaves by *Phyllocnistis citrella* Stainton Provide Refuge for Phytoseiids on Grapefruit in Florida and Texas. *Zoosymposia* **2011**, *6*, 118–123. [CrossRef]
5. Choi, W.I.; Park, Y.-S. Management of Forest Pests and Diseases. *Forests* **2022**, *13*, 1765. [CrossRef]
6. Netherer, S.; Kandasamy, D.; Jirosová, A.; Kalinová, B.; Schebeck, M.; Schlyter, F. Interactions among Norway Spruce, the Bark Beetle *Ips typographus* and Its Fungal Symbionts in Times of Drought. *J. Pest Sci.* **2021**, *94*, 591–614. [CrossRef]
7. Siegert, C.; Clay, N.; Pace, K.; Vissa, S.; Hofstetter, R.W.; Leverón, O.; Riggins, J.J. Bark Beetle-Driven Community and Biogeochemical Impacts in Forest Ecosystems: A Review. *Ann. Entomol. Soc. Am.* **2024**, *117*, 163–183. [CrossRef]
8. Reverchon, F.; Contreras-Ramos, S.M.; Eskalen, A.; Guerrero-Analco, J.A.; Quiñones-Aguilar, E.E.; Rios-Velasco, C.; Velázquez-Fernández, J.B. Microbial Biocontrol Strategies for Ambrosia Beetles and Their Associated Phytopathogenic Fungi. *Front. Sustain. Food Syst.* **2021**, *5*, 737977. [CrossRef]
9. Barta, M.; Pastirčáková, K.; Ostrovský, R.; Kobza, M.; Kádasi Horáková, M. Culturable Endophytic Fungi in *Fraxinus Excelsior* and Their Interactions with *Hymenoscyphus fraxineus*. *Forests* **2022**, *13*, 1098. [CrossRef]
10. Matisons, R.; Orlovskis, Z.; Blūms, K.T.; Ruņģis, D.; Baranova, M.; Krivmane, B.; Bitenieks, K.; Ramanenka, M.; Kļaviņa, D.; Zeps, M. Mycorrhizal Diversity on Roots of Silver Birch and Hybrid Aspen in Clonal Plantations in Northern Europe, Latvia. *Forests* **2024**, *15*, 2123. [CrossRef]
11. Fernandes, L.; Paiva, D.S.; Silva, A.C.; Fernandes, C.; Fernandes, A.R.; Ribeiro, D.; Martins, L.; Bragança, H.; Portugal, A. From Lab to Nursery: Novel Approaches of Seed Disinfection for Managing Pine Pitch Canker Propagation. *Forests* **2024**, *15*, 1154. [CrossRef]
12. Rodríguez-de La Cruz, D.; Perfecto-Arribas, S.; Delgado-Sánchez, L. Diversity Analysis of Macrofungi and Lichenised Fungi in Pyrenean Oak (*Quercus pyrenaica* Willd.) and Chestnut (*Castanea sativa* L.) Forests: Implications for the Conservation of Forest Habitats in Castilla y León (Central-Northwest Spain). *Forests* **2024**, *16*, 9. [CrossRef]
13. Fonseca, L.; Silva, H.; Cardoso, J.M.S.; Esteves, I.; Maleita, C.; Lopes, S.; Abrantes, I. *Bursaphelenchus xylophilus* in *Pinus sylvestris*—The First Report in Europe. *Forests* **2024**, *15*, 1556. [CrossRef]
14. Trandem, N.; Westrum, K.; Hofsvang, T.; Kobro, S. Distribution and Prolonged Diapause of the Rowan Seed Predators *Argyresthia conjugella* (Lepidoptera: Yponomeutidae) and *Megastigmus brevicaudis* (Hymenoptera: Torymidae) and Their Parasitoids in Norway. *Forests* **2023**, *14*, 847. [CrossRef]
15. Hlávková, D.; Davídková, M.; Koudelková, J.; Doležal, P. Population Dynamics of *Ips sexdentatus* (Börner) in the Czech Republic. *Forests* **2024**, *15*, 961. [CrossRef]
16. Takagi, R.; Kajimura, H. Ecological Traits of Three Species of *Xiphydria* Woodwasps from Japan: Host Tree Species and Eggs, Symbiotic Fungi and Mucus in Their Bodies. *Forests* **2025**, *16*, 264. [CrossRef]
17. Kopačka, M.; Zemek, R. Species Composition and Seasonal Abundance of Predatory Mites (Acari: Phytoseiidae) Inhabiting *Aesculus hippocastanum* (Sapindaceae). *Forests* **2023**, *14*, 942. [CrossRef]
18. Milosavljević, M.; Tabaković-Tošić, M.; Pernek, M.; Rakonjac, L.; Lučić, A.; Eremija, S.; Rindos, M. Mites Associated with the European Spruce Bark Beetle *Ips typographus* (Linnaeus, 1758) in Europe, with New Evidence for the Fauna of Serbia. *Forests* **2022**, *13*, 1586. [CrossRef]
19. Pernek, M.; Milas, T.; Kovač, M.; Lacković, N.; Koren, M.; Hrašovec, B. Effective Reduction in Natural Enemy Catches in Pheromone Traps Intended for Monitoring *Orthotomicus erosus* (Coleoptera, Curculionidae). *Forests* **2024**, *15*, 1298. [CrossRef]
20. Volter, L.; Prenerová, E.; Weyda, F.; Zemek, R. Changes in the Parasitism Rate and Parasitoid Community Structure of the Horse Chestnut Leafminer, *Cameraria ohridella* (Lepidoptera: Gracillariidae), in the Czech Republic. *Forests* **2022**, *13*, 885. [CrossRef]
21. Zumr, V.; Nakládal, O.; Remeš, J. Deadwood-Dwelling Beetles (Coleoptera: Eucnemidae) in a Beech Reserve: A Case Study from the Czech Republic. *Forests* **2024**, *15*, 469. [CrossRef]

22. Novgorodova, T.; Taranenko, D. Hidden Potential of the Subdominant Ant *Formica lemani* Bondroit (Hymenoptera: Formicidae): The Formation of Large Nest Complexes and Restructuring Behavioural Stereotypes. *Forests* **2024**, *15*, 1322. [CrossRef]
23. Mark, M.; Drake, E.; Kerwin, K.; Maslo, B. Non-Native Plants Influence Forest Vegetative Structure and the Activity of Eastern Temperate Insectivorous Bats. *Forests* **2024**, *15*, 711. [CrossRef]

forests

MDPI

Article

Culturable Endophytic Fungi in *Fraxinus excelsior* and Their Interactions with *Hymenoscyphus fraxineus*

Marek Barta, Katarína Pastirčáková *, Radovan Ostrovský, Marek Kobza and Miriam Kádasi Horáková

Department of Plant Pathology and Mycology, Institute of Forest Ecology, Slovak Academy of Sciences, Akademická 2, 94901 Nitra, Slovakia; marek.barta@savba.sk (M.B.); radovan.ostrovsky@ife.sk (R.O.); marek.kobza@ife.sk (M.K.); kadasi@ife.sk (M.K.H.)
* Correspondence: katarina.pastircakova@ife.sk or uefezima@hotmail.com

Abstract: The species diversity of culturable endophytic fungi was studied in the leaves and twigs of symptomatic and asymptomatic *Fraxinus excelsior* trees. Endophytic mycobiota was dominated by Ascomycota species, with Pleosporales (44.17%) and Diaporthales (23.79%) endophytes being the most frequently observed in the tree samples. The number of endophytic isolates and species richness varied depending on the sampling date (May and October) and tissue location. Of the 54 species identified based on ITS sequences, 14 were classified as dominant. The most frequently isolated species were *Diaporthe eres*, followed by *Alternaria alternata*, *Dothiorella gregaria*, and *Fraxinicola fraxini*. The inhibitory effect of 41 species (75 isolates) of endophytes on the radial growth of a *Hymenoscyphus fraxineus* isolate was studied under in vitro conditions (dual cultures). The radial growth of *H. fraxineus* was the most inhibited by four endophytic fungi from twigs (*Fusarium lateritium*, *Didymella aliena*, *Didymella macrostoma*, and *Dothiorella gregaria*). The inhibitory effect of the four isolates was also studied under in planta conditions. The isolates artificially inoculated into the trunks of ash trees reduced the length of necroses formed by *H. fraxineus* co-inoculated in the same trunks. This effect depended on the isolate, and the inhibition was most prominent only on trunks inoculated with *F. lateritium* and *D. aliena*. Although the total length of necrotic lesions formed by the *H. fraxineus* infection was shorter in the ash trunks co-inoculated with the endophytes, the difference was not significant.

Keywords: ash dieback; European ash; endophytic mycobiota; diversity; inhibitory effect

Citation: Barta, M.; Pastirčáková, K.; Ostrovský, R.; Kobza, M.; Kádasi Horáková, M. Culturable Endophytic Fungi in *Fraxinus excelsior* and Their Interactions with *Hymenoscyphus fraxineus*. *Forests* **2022**, *13*, 1098. https://doi.org/10.3390/f13071098

Academic Editor: Fred O. Asiegbu

Received: 6 June 2022
Accepted: 11 July 2022
Published: 13 July 2022

Publisher's Note: MDPI stays neutral with regard to jurisdictional claims in published maps and institutional affiliations.

1. Introduction

European ash (*Fraxinus excelsior*) plantations throughout Europe have been devastated by the fungal pathogen *Hymenoscyphus fraxineus*, causing ash dieback disease since the 1990s. Investigation into the etiology of *F. excelsior* decline in Europe has revealed a rich community of other fungal species. Few species are very common, and many occur only sporadically. In addition to *H. fraxineus*, ash branches and leaves are colonized by other parasitic and saprophytic fungi [1–5], which are secondary invaders of tissue weakened or dying from ash dieback. Frequently occurring fungi include *Alternaria alternata*, *Diaporthe eres*, *Diplodia mutila*, *Epicoccum nigrum*, *Fusarium* spp., and *Phomopsis* spp. The fungus *Phyllactinia fraxini*, which causes powdery mildew disease [6], is the most common foliage pathogen of ash trees. *Plagiostoma fraxini* (anamorph *Discula fraxinea*) causes anthracnose disease in ash trees grown in relatively cool regions [7,8]. Wood-inhabiting fungal communities in ash trees include species such as *Auricularia mesenterica*, *Bjerkandera adusta*, *Inonotus hispidus*, *Perenniporia fraxinea*, and *Ganoderma* spp., causing white rot and wood degradation [9,10]. *Armillaria* species (*Armillaria cepistipes*, *A. gallica*) attack the roots of ash trees and cause root and butt rot [11].

In Slovakia, Pastirčáková et al. [12] recorded a wide spectrum of parasitic and saprophytic species of fungi commonly colonizing ash trees, but *H. fraxineus* was the most widespread species in the country [13]. *Hymenoscyphus fraxineus* causes dieback of *F. excelsior* in Europe, but in its native East Asian range, it is typically a harmless endophyte in the leaves of several *Fraxinus* species [14–16].

Endophytes can be characterized in a variety of ways [17], but in general, they are microorganisms that spend at least part of their life cycle inside plant tissues without generating visible symptoms or damage to their hosts [18]. A single plant can harbor many endophytic organisms [19]. Fungal endophytes that asymptomatically colonize plants have the capacity to promote host plant growth and can play an important role in increasing host plant tolerance to abiotic stress, plant pests, and pathogens [20–24]. Endophytic fungi are also active against pathogenic fungi. Endophytes *Alternaria* sp., *Cladosporium* sp., *Fusarium* sp., and *Penicillium* sp. from *Aristotelia chilensis* and *Embothrium coccineum* significantly inhibited the growth of the common fungal pathogen *Botrytis cinerea* [25]. An endophytic isolate of *Trichoderma koningiopsis* from *Hevea guianensis* inhibited the causal agent of *Corynespora* leaf fall disease (*Corynespora cassiicola*), both in culture and in plants [26]. Several studies have already been published on the endophytic microbiome of ash in recent years, and the protective effects of some endophytic fungi against the ash dieback pathogen have also been studied [27–34]. Although endophytes that inhibit *H. fraxineus* in vitro (e.g., *Boeremia exigua, Botrytis cinerea, Clonostachys rosea, Epicoccum nigrum, Nemania diffusa, N. serpens, Peniophora cinerea, Phoma macrostoma, Rosellinia corticium, Setomelanomma holmii,* and *Xylaria polymorpha*) [28,29,34] are promising candidates for a biocontrol agent for ash dieback, their efficacy should be verified by in planta tests because the in planta situation is more complex [35]. Preliminary research has revealed that *Hypoxylon rubiginosum* has an antagonistic effect on *H. fraxineus* in planta trials [31]. Little is known about the endophytic mycobiome of ash trees in Slovakia. Only a few endophytes (*Alternaria alternata, Dothiorella sarmentorum,* and *Fusarium oxysporum*) colonizing *F. excelsior* branches and leaves have been recorded [12,36,37]. However, there is a need to further characterize and study endophytic fungi in ash trees to understand their potential role in the biocontrol of ash dieback disease.

The aims of this study were to (i) characterize species diversity of culturable fungal endophytes in leaves and twigs from symptomatic and asymptomatic *F. excelsior* trees in spring and autumn, (ii) evaluate interactions between ash dieback pathogen *H. fraxineus* and isolated endophytes in vitro (dual cultures), and (iii) evaluate the inhibitory effect of antagonistic endophytes under in planta conditions.

2. Materials and Methods

2.1. Fungal Cultures

The strain of *H. fraxineus* (D27) used in this study was isolated from necrotic lesions on twigs of *F. excelsior* collected in central Slovakia (locality of Duchonka; 48°42′53″ N, 18°02′33″ E) in 2015 [38,39]. Malt extract agar (MEA) supplemented with 50 g/L of frozen healthy ash (*F. excelsior*) leaflets removed after autoclaving [40], was used to isolate and grow the *H. fraxineus* strain and for a dual culture bioassay. MEA supplemented with Rose Bengal 10 mg/L and antibiotics (streptomycin sulphate 50 mg/L, penicillin G 50 mg/L; both added after autoclaving) was used for fungal endophyte isolation. All of the media were autoclaved at 120 °C for 20 min, and 20 mL per plate was poured into polystyrene Petri dishes (90 × 16 mm). The fungal cultures were incubated at 25 ± 1 °C in darkness.

2.2. Study Sites and Sampling

The collection of ash samples for fungal endophyte analyses was carried out in an *F. excelsior* forest (17.03 ha; with 5% of *Robinia pseudoacacia* as an admixture species) in the locality of Jarok (southwestern Slovakia, 48°16′38″ N, 17°57′51″ E). The forest was attacked by ash dieback, which was confirmed by direct observations of disease symptoms on trees (necrotic lesions on leaflets and petioles, dieback of branches, and sporadically whole tree crowns dead) and by *H. fraxineus* apothecia presence on ash leaf petioles from the previous year on the forest floor. In total, 20 trees showing symptoms of ash dieback disease and 20 asymptomatic trees were selected for sampling on 14 May and 14 October 2019. On each sampling date, 5 leaves and 5 twigs (only those that looked healthy) were collected from 10 symptomatic and 10 asymptomatic trees. Altogether, 100 leaves and 100 twigs (two years old) were collected on each sampling occasion. The individual samples were placed in polyethylene zipper bags and stored at 5 °C overnight.

2.3. Isolation of Fungal Endophytes

Collected leaves/twigs were processed for the isolation of fungal endophytes the next day. Under aseptic conditions, 5 leaflets of individual leaf samples and a 40 mm segment of each twig were surface-disinfected (96% ethanol for 1 min, followed by 2.5% sodium hypochlorite for 5 min and 3× rinses in sterile distilled water for 1 min) and dried shortly on sterilized tissue wipes (modified after Ibrahim et al. [41]). A single 10 mm diameter leaf disc was cut off from the central part of each surface-disinfected leaflet (including a midrib) by a cork borer. Five 5 mm sections were cut from the disinfected twigs. The leaf discs and twig sections were placed on the surface of the MEA plate and incubated in the dark at 25 ± 1 °C. In total, isolations were performed from 500 fragments of 100 leaves and 500 segments of 100 twigs. The plant fragments were checked daily, and all endophytic colonies were aseptically transferred to fresh MEA plates and cultivated for 10 days. Most cultures were transferred twice to obtain pure cultures. About 10% of cultures needed 3–4 transfers. Fungal colonies on agar plates were characterized as endophytic only if mycelia grew from internal plant tissue at the edge of plant sections. The quality of the surface-disinfection method was assessed by plating three replicates of the residual third rinse water (500 µL) on MEA plates. A three-week incubation at 25 ± 1 °C resulted in no fungal colonies on these plates, which confirmed the efficacy of the disinfection procedure. Pure cultures of endophytes were used for molecular species identification, and identified cultures were included in a dual culture bioassay. The colonization rate (*CR*) of the tissue samples by endophytic fungi was calculated using the formula by Kumar and Hyde [42]:

$$\%CR = EI/SS \times 100, \tag{1}$$

where *EI* is the number of endophytic isolates obtained from the particular tissue sample (leaf or twig), and *SS* is the number of tissue sections from the tissue sample tested for endophytes on the surface of the MEA plate.

2.4. Molecular Identification of Endophytes

DNA was extracted from two-week-old cultures using the EZ-10 Spin Column Fungal Genomic DNA Kit (Bio Basic Inc., Markham, ON, Canada) according to the manufacturer's instructions. The internal transcribed spacer (ITS) region of the ribosomal RNA gene was amplified using the primer combinations ITS1F/ITS4 [43,44] and ITS4/ITS5 [43]. PCR conditions for the primer pair ITS1F/ITS4 were set as reported by Kádasi Horáková et al. [45]. The amplification reaction conditions for primers ITS4/ITS5 included initial denaturation at 95 °C for 14 min, followed by 30 cycles of denaturation at 95 °C for 25 s, annealing at 56 °C for 50 s, elongation at 72 °C for 90 s, and final extension at 72 °C for 10 min. The PCR mix consisted of approximately 10 ng of template DNA, 10 pmol/µL of forward and reverse primers, 5x HOT FIREPol® Blend Master Mix (Solis Biodyne, Tartu, Estonia), and deionized water of molecular grade (Pro injection, B. Braun). All PCRs were performed in a total volume of 20 μL in Bio-Rad T100^{TM} Thermal Cycler (Bio-Rad Laboratories Inc., Hercules, CA, USA). The PCR products were visualized on 1% (*w*/*v*) TBE agarose gel stained with a SimplySafe stain (EURx, Gdansk, Poland). The target PCR fragments were purified using the QIAquick PCR Purification Kit (Qiagen, Hilden, Germany). Sequencing was performed in both directions using an ABI PRISM 3130 (Applied Biosystems, Waltham, MA, USA) by SEQme Ltd. (Dobříš, Czech Republic). The retrieved sequences were processed using SnapGene® Viewer 5.0.7 (GSL Biotech LLC, San Diego, CA, USA) and compared by BLASTN ver. 2.13.0+ [46] against ITS sequences deposited in the NCBI GenBank database.

2.5. Dual Culture Bioassay

In total, 75 fungal endophytes were screened for their capacity to suppress the growth of *H. fraxineus* (strain D27) in a dual culture bioassay on MEA (supplemented with ash leaflets) in Petri dishes (90 × 16 mm). Mycelial plugs (5 mm in diameter) of the actively growing margins of the *H. fraxineus* and endophyte colonies were placed at a fixed distance of 55 mm on the MEA plates. The plugs were placed at the opposing end of the same Petri dish. Due to the slower growth rate of *H. fraxineus* compared to the endophytes, *H. fraxineus* plugs were first cultured for four days at 25 °C, and on the fifth day, plugs of endophytes were placed on the MEA plates [47]. Dual culture tests of six slow-growing endophytic isolates (TA31-4M, TA55-5M, TA93-1M, TS94-4M, LS25-1M, and LS42-3M) were started simultaneously with the *H. fraxineus* strain on the MEA plates. Ten control plates were inoculated in a similar manner, but two plugs of *H. fraxineus* were placed in Petri dishes without endophytes. The plates for all combinations of *H. fraxineus* and endophytes were incubated at 25 ± 1 °C in the dark. The interactions between the dual culture partners were checked at three-day intervals, and the radii of colonies were measured using a digital caliper on day 27 post-inoculation. Each combination of the *H. fraxineus* strain and the endophytic isolates was performed in triplicate. Inhibition of radial growth of the *H. fraxineus* strain by the endophytes in dual cultures was evaluated by the inhibition index of radial growth (*IRG*), calculated using the following formula:

$$IRG = (R1 - R2)/R1, \tag{2}$$

where $R1$ is the radius of the *H. fraxineus* colony from the control Petri dishes, and $R2$ is the radius of the *H. fraxineus* colony measured on the line between the inoculation positions of the *H. fraxineus* strain and the endophyte in the dual culture plates. The radial growth of the *H. fraxineus* colony was inhibited by the co-culturing endophyte when the *IRG* value was greater than zero. The higher the *IRG* value within the range of 0–1, the greater the inhibition effect of the endophyte.

Interactions between colonies of *H. fraxineus* and endophytes were also visually assessed, and the following interaction types [48] were recognized: physical contact of mycelia in which neither isolate was able to overgrow the other (A), *H. fraxineus* colony partially overgrown by an endophyte after initial deadlock with mycelial contact (B1), *H. fraxineus* colony overgrown by an endophyte without initial deadlock (B2), an endophyte colony overgrown by *H. fraxineus* (C), an inhibition zone present between the colonies with the width of <2 mm (D1), the inhibition zone was >2 mm (D2).

2.6. Field Bioassay

Inoculation experiments were conducted on 5-year-old *F. excelsior* trees grown in the experimental area of the Institute of Forest Ecology SAS in Nitra. The experimental trees had an average trunk diameter of 15.4 mm at the inoculation point. Inoculations were performed with one isolate of *H. fraxineus* (strain D27) and four isolates of endophytes exhibiting inhibitory activity in a dual culture bioassay (TS105-4M—*Didymella aliena*, TA63-2O—*Didymella macrostoma*, TA52-5M—*Dothiorella gregaria*, and TS94-4M—*Fusarium lateritium*). For inoculum production, sterilized discs (4 mm in diameter) of *F. excelsior* sapwood were colonized with fungal isolates on MEA plates for 3 weeks [49]. Inoculations of the trees were carried out in the tree trunk (ca. 40 cm from the base of the trunk) on 21 April

2021. Two holes (4 mm in diameter) were drilled by a cork borer in the bark tissue at a distance of 2 cm. The holes were patched with the colonized discs, one colonized with an endophyte isolate, and one colonized with *H. fraxineus*. The endophyte was inoculated upward. Combinations of *H. fraxineus* strain with each endophytic isolate were inoculated in four replicates. As a negative control, a single sterile non-colonized disc was applied to the trunk, whereas an *H. fraxineus*-colonized disc was applied to the trunk as a positive control. In the case of positive and negative controls, three replicate inoculations were performed. The discs were then covered with parafilm. Parafilm wrapping was removed after 30 days. Together, 34 trees were inoculated. The host's response to the fungal isolates was measured by the formation of a callus or necrotic tissue at 30-day intervals. The length of superficial necrosis on the bark was measured in the acropetal and basipetal directions from the inoculation point. All experimental trees were cut down 180 days after inoculation (on 18 October 2021) and transported to the laboratory for analysis. After the bark tissue was removed, the length of the cambial necrosis was measured. The tree trunk was cut at the inoculation point, and the depth of necrosis was measured.

Two samples of wood tissue, each approximately 5 × 5 mm, were taken from the margin of necrosis that formed after inoculation of *H. fraxineus* on all experimental trees. The samples were used to confirm the presence of *H. fraxineus* in the tissue. The samples were homogenized in liquid nitrogen, and the total genomic DNA was extracted according to the manufacturer's protocol using the EZ-10 Spin Column Fungal Genomic DNA Kit. DNA was suspended in 50 μL of elution buffer and stored at −20 °C. The identity of *H. fraxineus* was confirmed by species-specific primers targeting the 18S gene and the ITS-2 region of the rDNA operon [50]. The PCR components and conditions were in accordance with Pastirčáková et al. [51].

2.7. Data Analysis

Chi-square tests were performed to determine if the colonization rate by endophytic fungi was affected by the tissue type (leaf and twig), the health condition of the studied trees (symptomatic and asymptomatic), and from the sampling time (May and October). The Shannon and Simpson diversity indices [52,53] were used to determine the diversity of fungal endophytes found in ash tree samples. Species dominance was estimated for each species of fungal endophyte according to Camargo [54]. A species was considered dominant if its relative abundance was higher than $1/S$, where S denotes species richness. *IRG* data from the dual culture bioassay were arcsine transformed ($n' = arcsin\sqrt{n}$) before analysis of variance (ANOVA) was used to determine differences among fungal endophytes. If significant differences were detected, the post hoc Tukey HSD test ($p = 0.05$) was performed. All statistical analyses were performed using Minitab 17® (© 2013 Minitab Inc., State College, PA, USA).

3. Results

3.1. Endophytic Mycobiota in Ash Leaves and Twigs

Altogether, 400 tissue samples (2000 tissue sections) were cultured from the leaves and twigs of 40 ash trees. Endophytic fungi grew from all of the examined trees, and 799 tissue sections (39.95%) yielded fungal colonies. The colonization rate (*CR*) of samples by endophytes varied depending on the tissue type, the health status of the trees, and from the date of sampling (Table 1). The mean *CR* of the different tissue types ranged from 13.6 to 72.0%. The lowest rate of colonization was observed in leaves sampled in May, whereas the highest *CR* was detected in twigs collected in May. Chi-square analysis identified a significant difference in *CR* between leaves and twigs collected from both symptomatic ($X^2_{(1,N=500)} = 63.51, p < 0.001$) and asymptomatic trees ($X^2_{(1,N=500)} = 30.29, p < 0.001$). This was observed regardless of the sampling date, and twigs were colonized by endophytes at a significantly higher rate than leaves in May ($X^2_{(1,N=500)} = 392.71, p < 0.001$), but more endophytes were retrieved from leaves than twigs in October ($X^2_{(1,N=500)} = 40.14, p < 0.001$). However, the health status of the sampled trees had no significant effect on the *CR* of leaf

($X^2_{(1,N=500)}$ = 3.51, p = 0.061) and twig samples ($X^2_{(1,N=500)}$ = 0.42, p = 0.517). The date of sampling was a significant factor in retrieving endophytes from the trees. A significantly higher *CR* was detected for leaf samples collected in October from both symptomatic ($X^2_{(1,N=500)}$ = 66.59, $p < 0.001$) and asymptomatic trees ($X^2_{(1,N=500)}$ = 171.96, $p < 0.001$), but the twig samples were significantly more colonized in May ($X^2_{(1,N=500)}$ = 73.63, $p < 0.001$ for asymptomatic trees and $X^2_{(1,N=500)}$ = 62.78, $p < 0.001$ for symptomatic trees).

Table 1. Isolation of endophytic fungi on malt extract agar plates from different types of tissue samples collected from *Fraxinus excelsior* in Slovakia in 2019.

Date of Sampling	Health Status of Trees *	Tissue Location on Trees	Mean Colonization Rate of Tissue Samples (%) **	Total Number of Isolates	Mean Number of Isolates per Tissue Sample **
May 2019	Asymptomatic	Leaf	13.6 ± 2.0	34	0.7 ± 0.1
		Twig	68.4 ± 4.8	197	4.0 ± 0.3
	Symptomatic	Leaf	19.2 ± 2.9	48	1.0 ± 0.2
		Twig	72.0 ± 4.7	198	4.0 ± 0.3
October 2019	Asymptomatic	Leaf	68.0 ± 4.1	179	3.6 ± 0.3
		Twig	41.2 ± 3.3	103	2.0 ± 0.2
	Symptomatic	Leaf	54.4 ± 3.2	136	2.7 ± 0.2
		Twig	44.8 ± 2.9	112	2.2 ± 0.2

* Trees displaying symptoms of ash dieback diseases were categorized "symptomatic", and trees without symptoms of the diseases were considered "asymptomatic" in this study; ** mean with standard error (±SE).

Altogether, 1007 isolates of endophytic fungi were obtained from the analyzed ash trees. The number of isolates obtained was not uniformly distributed across the collected samples and was dependent on the sampling date, tree health, and tissue type (Table 1). The highest number of isolates was retrieved from the twigs collected in May (197 and 198 isolates from asymptomatic and symptomatic trees, respectively) and the asymptomatic leaves sampled in October (179 isolates). The lowest number of cultures was isolated from the leaves in May. The mean number of isolates per tissue sample (twig or leaf) varied from 0.7 to 4.0 (Table 1) with a significant difference ($F_{(7,72)}$ = 15.57, $p < 0.001$) among sample types, sampling dates, and the health status of trees. For example, leaf samples collected in May yielded significantly fewer isolates than twig samples on the same collection date or leaf samples in October (Figure 1). The sampling date was an important factor in obtaining endophytes from a particular tissue type. In October, a significantly higher mean number of isolates was recovered from leaves than in May, whereas more isolates were obtained from twigs in May than in October.

Based on macromorphological characteristics, 206 endophytic isolates were selected for molecular identification, 199 isolates were identified at the species level, 5 isolates were determined at the genus level, and 2 isolates could only be placed in the order rank (Table 2). Endophytic mycobiota isolated from the tissue samples comprised 54 species from 42 genera and 13 orders. Sequences of four isolates provided a BLAST match with the GenBank database of less than 98% and might therefore be different taxa than those listed in Table 2. Interestingly, *H. fraxineus* was not isolated from the tissue samples of symptomatic trees using the isolation method employed in this study. The numbers of species varied depending on the sampling dates (May and October) and tissue locations (leaf and twig) (Figure 2). As many as 43 endophytes were detected in either from leaves or twigs, and 16 endophytes were identified in both tissue types. On both sampling dates, 20 endophytes were detected. Twelve endophytes were detected in both tissue types and on both sampling dates.

Table 2. Endophytic fungi identified from leaves and twigs of ash trees displaying (symptomatic) or not displaying (asymptomatic) symptoms of ash dieback disease collected in Slovakia in May and October 2019.

Order	Taxon	Number of Fungal Isolates					GenBank Acc. No. ITS	BLASTn	Identities %
		AL (May/Oct)	SL (May/Oct)	AT (May/Oct)	ST (May/Oct)	Total Number (May/Oct)			
Agaricales	*Hygrophorus* sp. *	0/1				0/1		LT716040	78.49
Amphisphaeriales	*Lepteutypa fuckelii*			2/0	1/0	3/0	OM950730	MZ045855	100
Botryosphaeriales	*Diplodia fraxini* **		2/0	1/0	2/0	5/0	OM950731	MT587349	99.82
	Dothiorella gregaria **		0/1	6/1	2/2	8/4	OM950732	MN685280	99.44
	Microdiplodia sp.			1/0		1/0	OM950733	FJ228194	99.11
Cladosporiales	*Cladosporium allicinum*	0/1				0/1	OM950734	MT573471	100
	Cladosporium cladosporioides	1/0			1/0	2/0	OM950735	MT635286	100
	Cladosporium tenuissimum		0/1		0/1	0/2	OM950736	LT603045	100
Diaporthales	*Cytospora pruinosa*				1/0	1/0	OM950737	MW447045	99.83
	Diaporthe eres **	0/1		8/0	17/3	25/4	OM950738	OM442980	100
	Diaporthe nobilis			1/0	2/0	3/0	OM950739	KJ609011	99.65
	Diaporthe oncostoma **	1/2	1/2	0/1	0/1	2/6	OM950740	LN714541	99.82
	Diaporthe rudis **		0/1	2/0	3/0	5/1	OM950741	MW032267	99.66
	Diaporthe vacuae				2/0	2/0	OM950742	MZ127189	99.66
Dothideales	*Aureobasidium pullulans*	1/1			1/0	2/1	OM950743	MW560221	100
Eurotiales	*Aspergillus pseudoglaucus*				1/0	1/0	OM950744	KX696361	100
Helotiales	*Neofabraea vagabunda*		0/1			0/1	OM950745	KT923785	99.63
Hypocreales	*Fusarium avenaceum*				1/1	1/1	OM950746	MW016661	100
	Fusarium lateritium **			4/1	2/1	6/2	OM950747	JQ693397	100
Mycosphaerellales	*Ramularia endophylla*	0/1				0/1	OM950748	MH859364	100

Table 2. *Cont.*

Order	Taxon	Number of Fungal Isolates					GenBank Acc. No. ITS	BLASTn	Identities %
		AL (May/Oct)	SL (May/Oct)	AT (May/Oct)	ST (May/Oct)	Total Number (May/Oct)			
Pleosporales	*Alternaria alternata* **	0/3	1/8	0/7	5/2	6/20	OM950749	MK183818	100
	Alternaria infectoria **	0/2	0/2	0/1		0/5	OM950750	MK461063	100
	Alternaria longipes		0/1			0/1	OM950751	MH712187	99.82
	Aposphaeria corallinolutea	0/1				0/1	OM950752	MT177916	99.80
	Ascochyta medicaginicola			1/2		1/2	OM950753	KF293990	100
	Ascochyta pisi			0/1		0/1	OM950754	MH854853	99.61
	Comoclathris incompta **			1/1	1/1	2/2	OM950755	KU973715	99.83
	Coniothyrium ferrarisianum *				1/0	1/0		MH860854	97.09
	Didymella aliena **		1/1		2/3	3/4	OM950756	KC311486	99.81
	Didymella glomerata			0/1	2/0	2/1	OM950757	MN075513	100
	Didymella macrostoma			0/1		0/1	OM950758	MH858090	99.81
	Epicoccum nigrum **		0/1	0/2	2/1	2/4	OM950759	KX664321	100
	Epicoccum thailandicum *				1/0	1/0		MG975626	80.79
	Foliophoma camporesii				1/1	1/1	OM950760	MN244200	99.65
	Lophiostoma corticola			0/1	1/0	1/1	OM950761	KT004559	99.62
	Muriphaeosphaeria viburni				1/0	1/0	OM950762	MW446984	99.48
	Neodidymelliopsis camporesii		0/1		1/1	1/2	OM950763	MN244199	99.81
	Neosetophoma aseptata			2/0		2/0	OM950764	NR164449	99.12
	Nothophoma spiraeae				2/0	2/0	OM950765	OM287410	99.43
	Paracucurbitaria corni			1/0		1/0	OM950766	MT547826	98.70
	Phaeosphaeria sp.			0/1	0/1	0/2	OM950767	LC171698	99.81
	Phoma herbarum			0/1		0/1	OM950768	KP739881	99.62
	Phoma sp.				0/1	0/1	OM950769	MG098303	99.31
	Pleosporales sp.		2/0			2/0	OM950770	MT777338	100

Table 2. *Cont.*

Order	Taxon	Number of Fungal Isolates					GenBank Acc. No. ITS	BLASTn	Identities %
		AL (May/Oct)	SL (May/Oct)	AT (May/Oct)	ST (May/Oct)	Total Number (May/Oct)			
	Praetumpfia obducens				1/0	1/0	OM950771	NR147688	98.30
	Pseudocamarosporium brabeji			0/1		0/1	OM950772	KR909143	99.32
	Pyrenophora triseptata			0/1		0/1	OM950773	MT548680	99.84
	Sporormiella minima			2/0		2/0	OM950774	MG098329	99.43
	Stemphylium vesicarium **	1/2	0/2	0/1	1/0	2/5	OM950775	MZ452063	100
Sordariales	*Dichotomopilus erectus*	0/1				0/1	OM950776	MN956887	99.64
	Chaetomium globosum *	0/1				0/1		MH858130	74.45
	Sordaria fimicola		0/1	1/0	1/0	2/1	OM950777	MN341410	99.82
	Sordaria lappae	0/1				0/1	OM950778	MH858210	100
Venturiales	*Fraxinicola fraxini* **	2/1	6/0			8/1	OM950779	MW447009	99.63
Xylariales	*Anthostoma amoenum*	0/1				0/1	OM950780	KC774569	98.67
	Hypoxylon fragiforme			0/1		0/1	OM950781	EF155508	100
	Nemania diffusa	0/1				0/1	OM950782	MZ078701	99.65
	Nemania serpens				1/0	1/0	OM950783	MF161316	99.48
	Rosellinia corticium **	1/1	0/1		1/0	2/2	OM950784	KY593990	99.81
	Total number of isolates	7/22	13/24	33/26	61/20	114/92			
	Species richness (S)	6/17	6/14	14/17	30/14	38/41			
	Simpson's index of diversity (D)	0.816/0.934	0.722/0.852	0.872/0.895	0.899/0.910	0.925/0.929			
	Shannon's index of diversity (H)	1.748/2.471	1.525/2.179	2.331/2.502	2.931/2.528	3.152/3.232			

Abbreviations: AL—leaf from asymptomatic trees, SL—leaf from symptomatic trees, AT—twig from asymptomatic treeKPs, ST—twig from symptomatic trees. * Indicates a BLAST match below 98% and likely a different species than listed; the sequences were not submitted to GenBank. ** Indicates the dominant species of endophytes, determined according to Camargo [54].

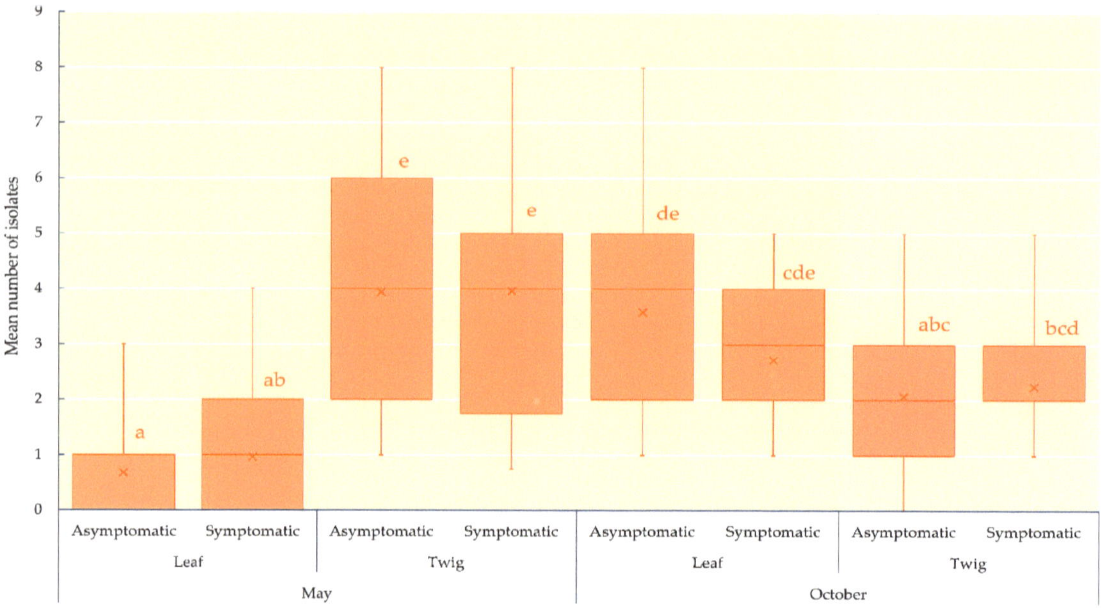

Figure 1. Box and whisker plot showing the mean number of isolates of endophytic fungi per tissue sample (leaves or twigs) of ash trees displaying (symptomatic) or not displaying (asymptomatic) symptoms of ash dieback disease collected in May and October 2019; mean values indicated by the same letter are not significantly different (Tukey HSD test; $p = 0.05$).

The mycobiota of the endophytes from the ash tissue were significantly dominated by Ascomycota (205 isolates, 99.51%) and included 13 orders. One isolate was identified as *Hygrophorus* sp. (Agaricales) from Basidiomycota, however, with a low percent identity (78.49%) after BLAST in the GenBank database. Therefore, the identity of this isolate must be considered doubtful. Pleosporales endophytes were the most frequently identified (91 isolates, 44.17%) from the tissue samples, and this order was the most species-rich (29 species, 49.15%), followed by Diaporthales, with 23.79% of relative isolate frequency (49 isolates) and six species (10.17%) (Figure 3c). The remaining fungal orders were represented by $\leq$5 species and $\leq$18 isolates. Relative frequencies of isolates by taxonomic order and percentage proportions of species numbers in particular orders were similar on both sampling dates (Figure 3a,b). Although the number of fungal orders discovered in May and October was comparable (11 orders in May vs. 12 orders in October), the structure of the identified orders varied. On both sampling dates, Pleosporales was the most species-rich order (19 species in May and 20 species in October). Diaporthales was the second most abundant order. Although only six Diaporthales species were detected in May, the relative frequency of isolates reached 33.33% (38 isolates), which was higher than the relative frequency of the Pleosporales isolates (29.82%, 34 isolates). Diaporthales were less abundant in October (11 isolates, 11.58%) than in May. While Amphisphaeriales and Eurotiales (each represented by a single species) were only detected in May, Helotiales, Mycosphaerellales, and Agaricales (each represented by a single species) were detected only in October. The species richness in the genera was variable. The most species-rich genus, *Diaporthe*, was represented by five species; *Cladosporium*, *Alternaria*, and *Didymella* were each represented by three species (Table 2). Most genera (29) were represented by a single species. Of the 54 identified species, 14 were classified as dominant. *Diaporthe eres* was the most frequently isolated species (25 isolates in May, and 4 isolates in October), followed by *Alternaria alternata* (6 isolates in May, and 20 isolates in October), *Dothiorella gregaria* (8 isolates in May, and 4 isolates in October), and *Fraxinicola fraxini* (8 isolates in May, and 1 in October).

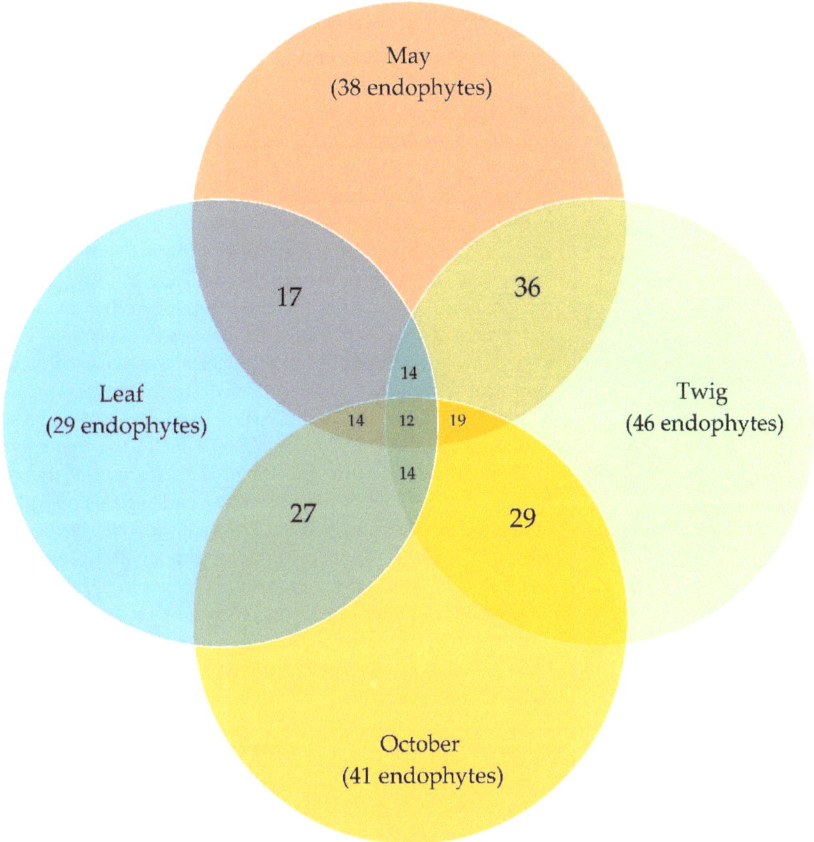

Figure 2. Venn diagram showing the number of identified endophytes shared among different sampling dates (May and October) and tissue locations (leaf and twig) on *Fraxinus excelsior*.

The species diversity indices calculated for all samples in May reached 0.925 and 3.152 for Simpson's (D) and Shannon's (H) indices, respectively, and they were similar to the indices for samples collected in October (D = 0.929; H = 3.232) (Table 2). The highest diversity indices were recorded for endophyte populations in asymptomatic leaves in May (D = 0.934; H = 2.471), and the lowest values were recorded for symptomatic leaves in May (D = 0.722; H = 1.525).

3.2. Inhibitory Effect of Endophytes against *H. fraxineus* on Artificial Medium

As many as 75 isolates of 41 species of endophytic fungi were tested for their inhibitory effect on the radial growth of *H. fraxineus* in dual cultures. Various types of mycelial interactions were observed between the *H. fraxineus* isolate and the endophytic fungi after 27 days of co-incubation on MEA (Table 3, Figure 4). The most frequent type of interaction was the formation of an inhibition zone between the co-partners in the dual cultures. This interaction type was observed for 49 endophyte isolates (31 species). The width of the inhibition zone depended on the endophyte isolates. It was greater than 2 mm (D1 interaction type) for 25 isolates and less than 2 mm (D2 interaction type) for 24 isolates. The largest mean zone of growth inhibition was formed around the colonies of *Phoma herbarum* (9.33 mm) and *Phaeosphaeria* sp. (8.67 mm), both isolated from asymptomatic twigs collected in October. The *H. fraxineus* colony was overgrown by 20 isolates (13 species) of endophytic fungi, 14 isolates (9 species) partially overgrew the *H. fraxineus* colony after an initial deadlock with mycelial contact (B1 interaction type), and 6 isolates (4 species) overgrew *H. fraxineus* without an initial deadlock (B2 interaction type). The capacity of *H. fraxineus* to grow over endophyte colonies was also observed (C interaction type). Such a situation was observed for one isolate of *Sporormiella minima* (Figure 4w) when the *H. fraxineus* isolate partly overgrew the edge of the endophyte colony. The interaction of five isolates of endophytic fungi from leaves resulted in physical contact with the colony of *H. fraxineus* (A interaction type). In this case, neither the endophyte nor *H. fraxineus* could overgrow the co-partner after mutual contact of mycelia. There was no observed variation in the interaction type between replicates of the same isolate; however, isolates of the same species could form different types of interactions with *H. fraxineus*, e.g., *Didymella aliena* isolates: LS25-3O type B1, TS105-4M type D1, and TS43-1M and LS102-1M type D2. In the dual culture experiment, the inhibitory effect of endophytes on the radial growth of *H. fraxineus* was observed for 57 isolates (35 species). The radius of *H. fraxineus* colonies adjacent to their endophytic co-partners was reduced compared to that in the control. The mean inhibition index (*IRG*) varied significantly ($F_{(56,114)} = 6.37$, $p < 0.001$) among the endophytes (Table 3). The *IRG* values for most endophytes (52 isolates) did not exceed 0.25. The highest value of *IRG* was observed for *Fusarium lateritium* isolate TS94-4M, reaching 0.575 ± 0.017. The radial growth of *H. fraxineus* was the most reduced by four endophytic isolates from twigs (TS94-4M *Fusarium lateritium*, TS105-4M *Didymella aliena*, TA63-2O *Didymella macrostoma*, and TA52-5M *Dothiorella gregaria*) (Table 3). These isolates were selected for the in planta bioassay. Eighteen endophytic isolates (17 species) did not show an inhibitory effect on the growth of *H. fraxineus*. In this case, the radius of the *H. fraxineus* colonies was greater in the dual cultures than in the control.

Table 3. Inhibitory effect of endophytic fungi on the radial growth of *Hymenoscyphus fraxineus* in a dual culture experiment.

Isolate Name	Endophytic Fungus	Interaction Type *	Mean Inhibition Index ± SE **
TS94-4M	*Fusarium lateritium*	D2 (4.00)	0.575 ± 0.017 a
TS105-4M	*Didymella aliena*	D1	0.326 ± 0.019 b
TA63-2O	*Didymella macrostoma*	D2 (3.33)	0.318 ± 0.020 bc
TA52-5M	*Dothiorella gregaria*	D2 (3.33)	0.280 ± 0.041 bcd
LS92-2O	*Alternaria alternata*	D1	0.267 ± 0.026 bcde
TA93-2O	*Dothiorella gregaria*	D2 (3.00)	0.212 ± 0.010 bcdef
TA35-3O	*Ascochyta medicaginicola*	D1	0.202 ± 0.050 bcdef
LS25-3O	*Didymella aliena*	B1	0.197 ± 0.007 bcdef
TA31-4M	*Sporormiella minima*	C	0.196 ± 0.018 bcdef
TA31-3O	*Pyrenophora triseptata*	D1	0.195 ± 0.001 bcdef
LA45-4O	*Rosellinia corticium*	B2	0.191 ± 0.043 bcdef
LA74-1O	*Sordaria lappae*	D1	0.189 ± 0.029 bcdef
TS85-2M	*Diaporthe rudis*	D2 (2.00)	0.188 ± 0.038 bcdef
LS62-2O	*Alternaria alternata*	D1	0.176 ± 0.020 bcdef
TS102-1O	*Diaporthe eres*	B1	0.175 ± 0.007 bcdef
TA63-1O	*Ascochyta medicaginicola*	D1	0.174 ± 0.020 bcdef
TA34-1M	*Sordaria fimicola*	B1	0.153 ± 0.015 bcdef
TA23-1O	*Phoma herbarum*	D2 (9.33)	0.150 ± 0.019 bcdef
TA102-2O	*Didymella glomerata*	D1	0.139 ± 0.007 bcdef
TA63-1M	*Ascochyta medicaginicola*	D1	0.134 ± 0.024 bcdef
TA93-1M	*Microdiplodia* sp.	D2 (6.00)	0.128 ± 0.051 bcdef
TS31-1M	*Epicoccum nigrum*	B1	0.126 ± 0.007 bcdef
TS91-2M	*Rosellinia corticium*	D2 (5.00)	0.119 ± 0.014 bcdef
TA52-2O	*Hypoxylon fragiforme*	B1	0.117 ± 0.072 bcdef
LS21-1O	*Stemphylium vesicarium*	A	0.114 ± 0.028 bcdef

Table 3. *Cont.*

Isolate Name	Endophytic Fungus	Interaction Type *	Mean Inhibition Index ± SE **
LA44-5O	*Diaporthe eres*	B1	0.111 ± 0.031 bcdef
LA25-3O	*Stemphylium vesicarium*	A	0.106 ± 0.044 bcdef
LS31-2O	*Diaporthe rudis*	D1	0.104 ± 0.022 bcdef
TS43-1M	*Didymella aliena*	D2 (4.00)	0.101 ± 0.035 bcdef
LS102-1M	*Didymella aliena*	D2 (3.33)	0.100 ± 0.020 bcdef
LA94-1M	*Rosellinia corticium*	B2	0.100 ± 0.023 bcdef
LS61-2O	*Neodidymelliopsis camporesii*	D1	0.100 ± 0.028 bcdef
TS41-1M	*Diaporthe eres*	D2 (3.33)	0.098 ± 0.027 cdef
LA13-3O	*Alternaria alternata*	B1	0.086 ± 0.055 def
LS73-3O	*Sordaria fimicola*	B1	0.083 ± 0.032 def
LA52-1O	*Fraxinicola fraxini*	D1	0.077 ± 0.140 def
TS64-1O	*Cladosporium tenuissimum*	D1	0.077 ± 0.050 def
LA72-1O	*Anthostoma amoenum*	A	0.074 ± 0.017 def
LS22-1M	*Diplodia fraxini*	D1	0.072 ± 0.028 def
TA61-1M	*Dothiorella gregaria*	B1	0.071 ± 0.024 def
TS35-2O	*Diaporthe oncostoma*	D1	0.071 ± 0.021 def
LS103-3O	*Alternaria infectoria*	D1	0.070 ± 0.012 def
LA25-4O	*Nemania diffusa*	B2	0.068 ± 0.021 def
TS64-4O	*Epicoccum nigrum*	D2 (6.67)	0.057 ± 0.093 def
TA82-1O	*Stemphylium vesicarium*	D1	0.052 ± 0.011 ef
TA94-1O	*Phaeosphaeria* sp.	D2 (8.67)	0.052 ± 0.054 ef
TA91-3O	*Ascochyta pisi*	D2 (3.00)	0.052 ± 0.026 ef
TA55-5M	*Lepteutypa fuckelii*	B2	0.042 ± 0.022 ef
TA75-3M	*Diaporthe nobilis*	B1	0.041 ± 0.017 ef
LS44-1O	*Cladosporium tenuissimum*	D1	0.035 ± 0.006 f
TS52-2M	*Diaporthe eres*	B1	0.031 ± 0.044 f
TS43-1O	*Comoclathris incompta*	D2 (2.33)	0.030 ± 0.030 f
TS35-3O	*Foliophoma camporesii*	D1	0.018 ± 0.054 f
TS12-3M	*Diaporthe eres*	B1	0.017 ± 0.030 f
TA35-3M	*Neosetophoma aseptata*	B1	0.012 ± 0.022 f
TA55-2M	*Comoclathris incompta*	B2	0.010 ± 0.072 f
LA65-2O	*Aposphaeria corallinolutea*	D1	0.009 ± 0.034 f
TS62-1O	*Neodidymelliopsis camporesii*	D2 (4.33)	−0.005 ± 0.045
LS15-1O	*Diaporthe oncostoma*	D1	−0.006 ± 0.015
TS12-1O	*Phoma* sp.	D2 (8.00)	−0.015 ± 0.029
TS101-2O	*Fusarium lateritium*	D2 (3.33)	−0.017 ± 0.045
LA21-5O	*Ramularia endophylla*	D2 (6.00)	−0.019 ± 0.028
TS31-5M	*Diaporthe rudis*	D1	−0.022 ± 0.016
LS72-2O	*Rosellinia corticium*	B2	−0.027 ± 0.024
TA63-2M	*Dothiorella gregaria*	D2 (3.33)	−0.027 ± 0.024
LA51-2O	*Aureobasidium pullulans*	D2 (3.00)	−0.038 ± 0.008
TS24-1M	*Diaporthe eres*	B1	−0.039 ± 0.079
LS11-1O	*Neofabraea vagabunda*	D2 (8.00)	−0.042 ± 0.040
TA92-1O	*Lophiostoma corticola*	A	−0.053 ± 0.038
LS25-1M	*Fraxinicola fraxini*	D2 (3.33)	−0.102 ± 0.029
LS42-3M	*Fraxinicola fraxini*	D1	−0.106 ± 0.007
TA71-6M	*Diaporthe rudis*	D1	−0.109 ± 0.030
LA93-1O	*Cladosporium allicinum*	A	−0.111 ± 0.045
TS44-2M	*Diaporthe eres*	D1	−0.140 ± 0.009
LS85-2M	*Diaporthe oncostoma*	D2 (5.33)	−0.201 ± 0.010

* Observed interaction types: A—physical contact of mycelia; B1—*H. fraxineus* colony partially overgrown by an endophyte after initial deadlock with mycelial contact; B2—*H. fraxineus* colony overgrown by an endophyte without initial deadlock; C—an endophyte colony overgrown by *H. fraxineus*; D1—an inhibition zone present with a width of <2 mm; D2—an inhibition zone of >2 mm; the average width of the inhibition zone (mm) in parenthesis; ** mean inhibition indices followed by the same letter are not significantly different ($p = 0.05$).

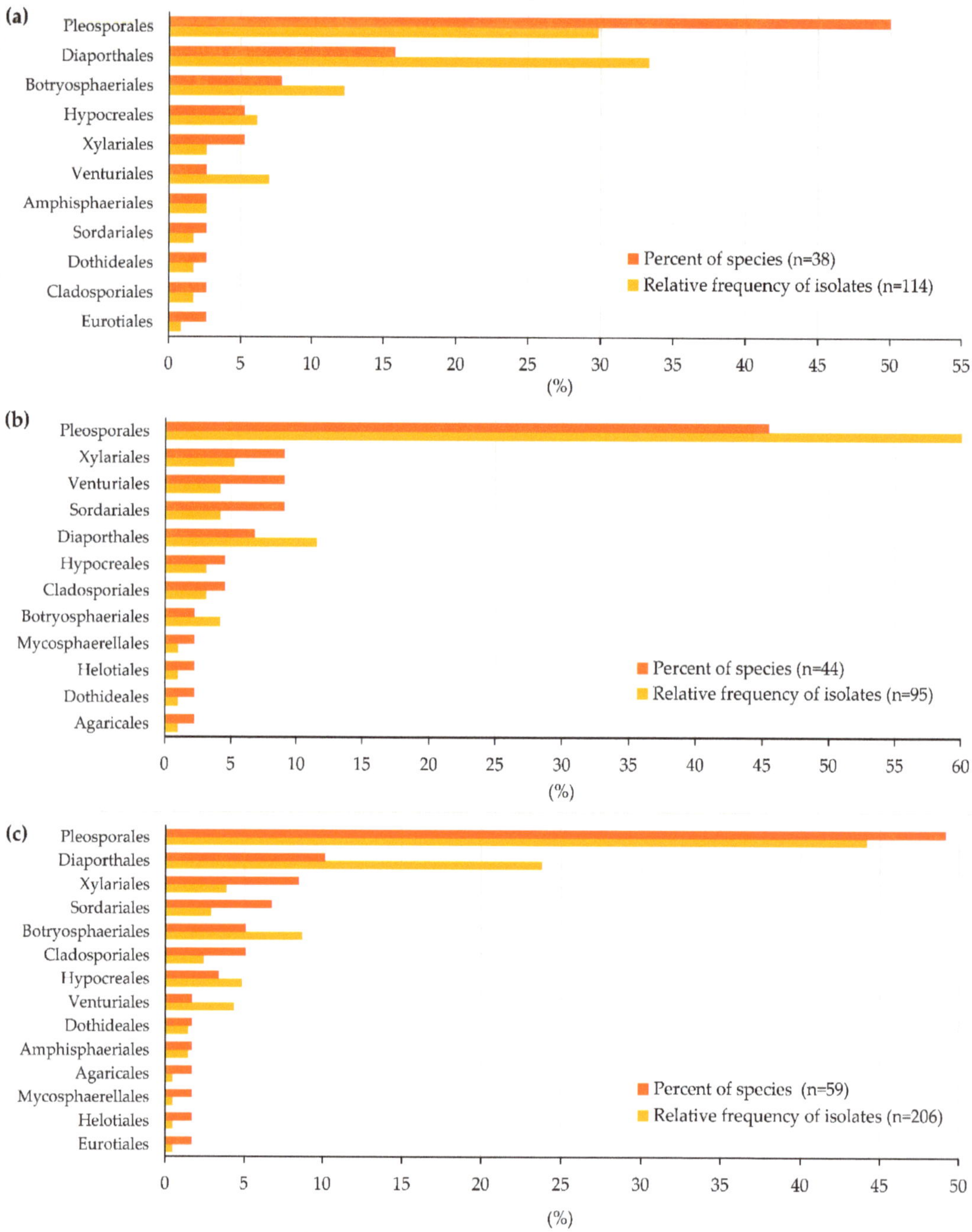

Figure 3. Relative frequency of isolates by taxonomic order and percentage proportions of species number in a particular order to the overall number of isolated species; (**a**) analysis of isolates from May 2019; (**b**) analysis of isolates from October 2019; (**c**) total number of isolates in 2019.

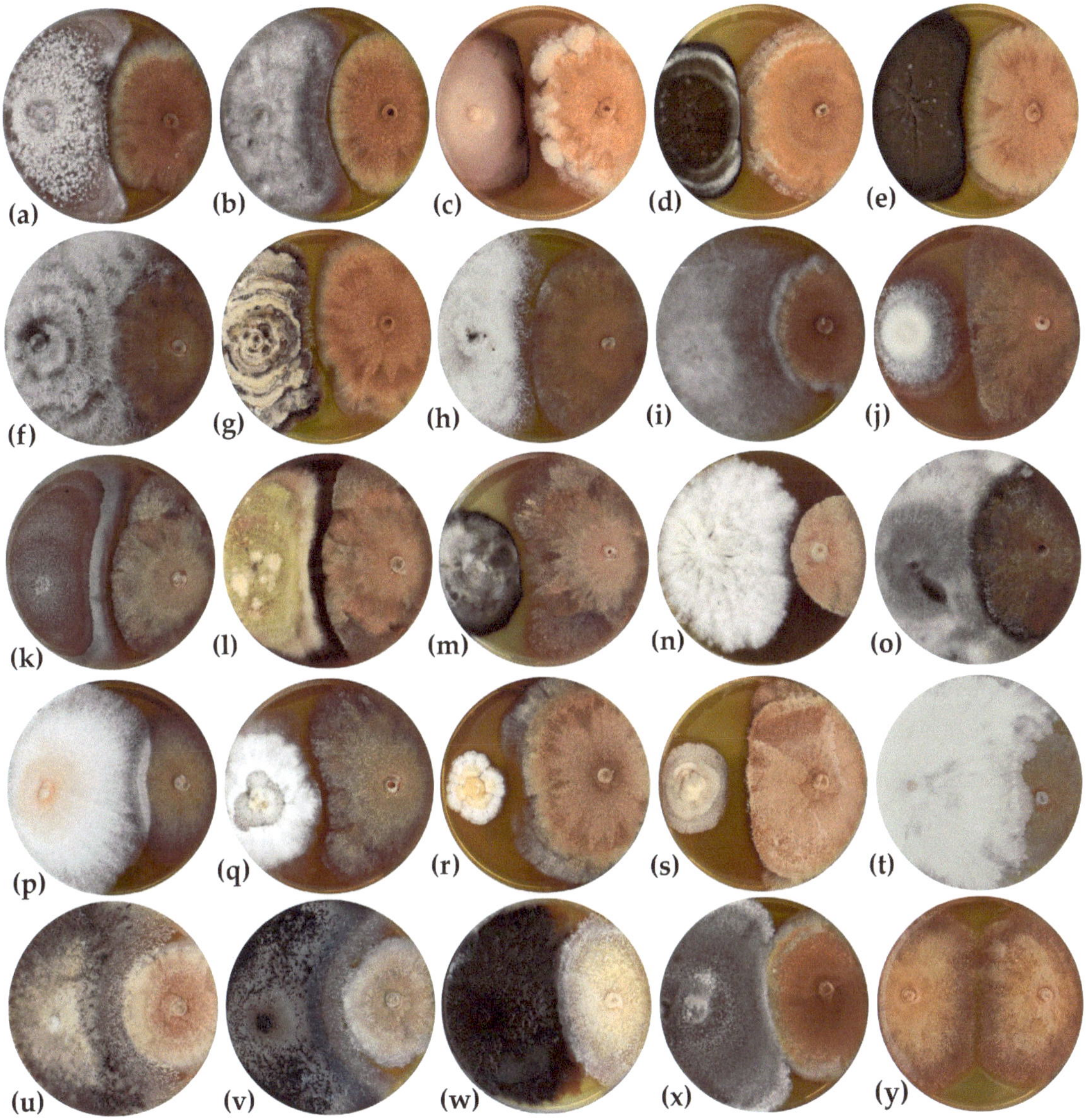

Figure 4. Mycelial interactions after 27 days of co-incubation between the isolate of *Hymenoscyphus fraxineus* (right) and selected isolates of endophytic fungi (left): (**a**) *Alternaria infectoria*, (**b**) *Ascochyta medicaginicola*, (**c**) *Aureobasidium pullulans*, (**d**) *Cladosporium allicinum*, (**e**) *Cladosporium tenuissimum*, (**f**) *Diaporthe eres*, (**g**) *Diaporthe oncostoma*, (**h**) *Diaporthe rudis*, (**i**) *Didymella aliena*, (**j**) *Didymella macrostoma*, (**k**) *Dothiorella gregaria*, (**l**) *Epicoccum nigrum*, (**m**) *Fraxinicola fraxini*, (**n**) *Fusarium lateritium*, (**o**) *Hypoxylon fragiforme*, (**p**) *Lepteutypa fuckelii*, (**q**) *Microdiplodia* sp., (**r**) *Neofabraea vagabunda*, (**s**) *Phaeosphaeria* sp., (**t**) *Rosellinia corticium*, (**u**) *Sordaria fimicola*, (**v**) *Sordaria lappae*, (**w**) *Sporormiella minima*, (**x**) *Stemphylium vesicarium*, (**y**) control—two plugs of *H. fraxineus* taken from the same colony.

3.3. In Planta Evaluation of the Inhibitory Effect of Endophytes

Four endophytic isolates from ash twigs of symptomatic (TS94-4M *Fusarium lateritium* and TS105-4M *Didymella aliena*) and asymptomatic (TA63-2O *Didymella macrostoma* and TA52-5M *Dothiorella gregaria*) trees that inhibited the *H. fraxineus* strain on MEA plates were selected for the in planta bioassay. The endophytic isolates used to artificially inoculate the trees formed no necroses, and the wounds made on the bark during inoculation were covered by protective callus tissue and were almost totally regenerated 180 days after inoculation (Figure 5b). The same situation was observed in the negative control when the trees were treated with sterile, non-colonized discs. Following *H. fraxineus* inoculation, typical ash dieback lesions appeared surrounding the inoculation points (Figure 5a). The necrotic lesions developed on the surface of the bark and in the wood tissue. The length of superficial necroses developed on the bark by the artificially inoculated *H. fraxineus* isolate (Figure 6a) varied depending on whether (the positive control) endophytic fungal isolates were co-inoculated on the experimental trees. Necroses were shorter on trees co-inoculated with endophytes TS94-4M and TS105-4M ($\overline{x}$ = 24.25 ± 4.77 and 17.50 ± 4.29, respectively) than on trees inoculated only with *H. fraxineus* ($\overline{x}$ = 34.00 ± 12.50 mm). This indicates a suppressive effect of the endophytes on *H. fraxineus* growth and corresponds with the results of the in vitro bioassay. However, the difference in necrosis length was not significant ($F_{(2,8)} = 1.305, p > 0.05$). No suppressive effect on *H. fraxineus* was observed in trees inoculated with the endophytic isolates TA63-2O and TA52-5M. The lesions were of the same length ($\overline{x}$ = 34.00 ± 12.29 mm) or longer ($\overline{x}$ = 37.25 ± 8.41 mm) than lesions in the positive control.

The lengths of superficial necroses (on the bark) were shorter than the lengths of cambial necroses (Figure 6a,b), but the difference was not significant ($p > 0.05$). The average length of cambial necroses in the positive control reached 39.83 ± 14.61 mm, and the necroses on the experimental trees co-inoculated with the endophytes TS94-4M and TS105-4M were shorter (25.75 ± 4.46 mm and 17.75 ± 4.27 mm, respectively) than the control necrosis.

The necroses did not develop evenly above or below the inoculation points. On average, the length of necrosis in the acropetal direction (to the endophyte inoculation point) was shorter by 79.26 ± 18.37% (superficial necroses) and 75.25 ± 14.69% (cambial necroses) than the length in the basipetal direction (Figure 5a). However, the difference was not significant ($F_{(1,30)} = 3.734$, $p > 0.05$ for superficial necroses, and $F_{(1,30)} = 3.304$, $p > 0.05$ for cambial necroses). The growth of necroses depended on co-inoculated endophytes. Necroses on experimental trees co-inoculated with isolates TA52-5M and TS94-4M grew evenly in both directions, while necroses on trees co-inoculated with TA63-2O and TS105-4M grew predominantly in the basipetal direction. Necrotic wood depth at the inoculation point reached the pith of the tree trunk (Figure 5a), and the mean depth varied from 5.25 ± 0.25 mm to 7.50 ± 2.25 mm, depending on the co-inoculated endophytes.

The presence of *H. fraxineus* was confirmed in all necrotic lesions by species-specific PCR, which produced an amplicon of 456 bp.

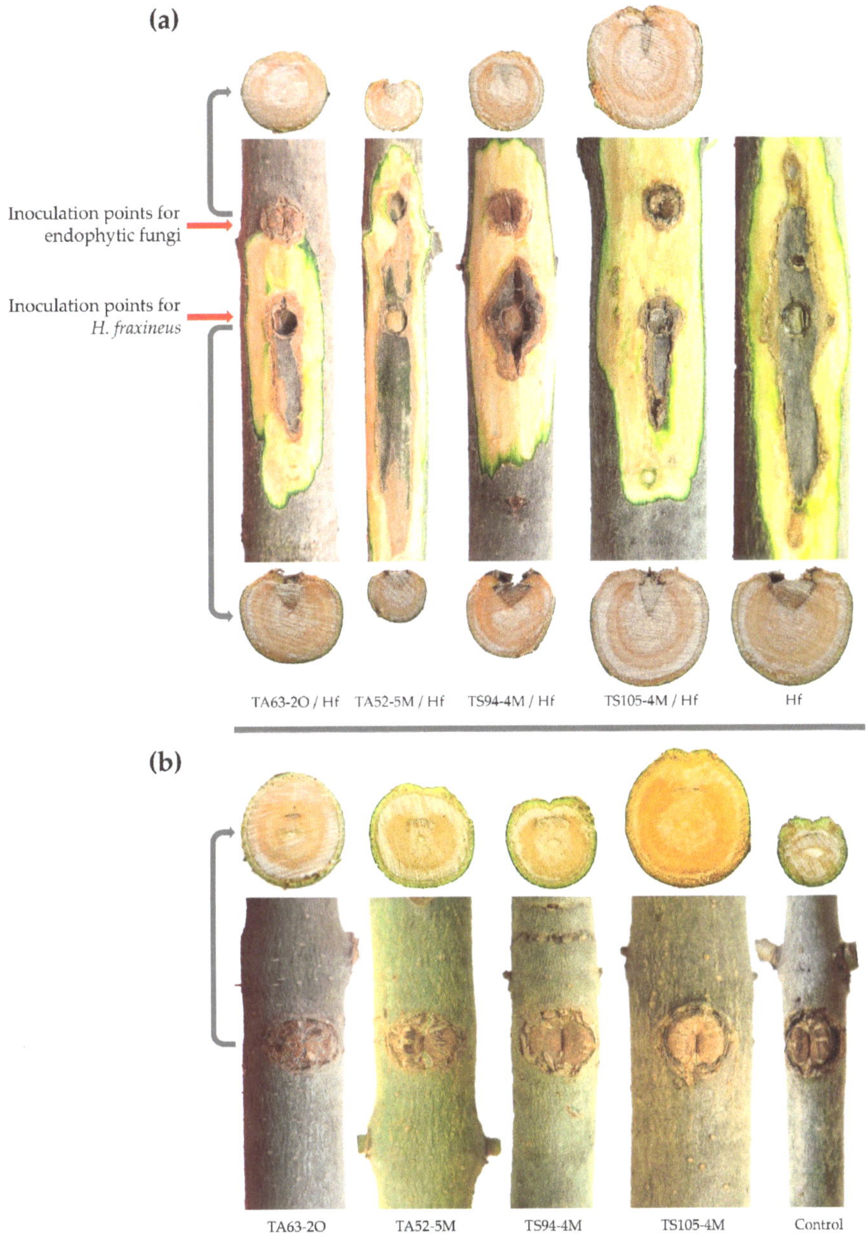

Figure 5. Necrotic lesions on the stems of five-year-old *Fraxinus excelsior* trees developed 180 days after their artificial inoculation with *Hymenoscyphus fraxineus* (Hf) and endophytic fungi *Didymella macrostoma* (TA63-2O), *Dothiorella gregaria* (TA52-5M), *Fusarium lateritium* (TS94-4M), and *Didymella aliena* (TS105-4M): (**a**) necroses developed after inoculation of the bark with *H. fraxineus* alone (the positive control) or in combination with endophytic fungal isolates including cross-sections through the ash stems at the inoculation points; (**b**) trees recovered from wounds performed on the bark of ash trees during inoculation with endophytic fungal isolates, including non-inoculated tree (the negative control).

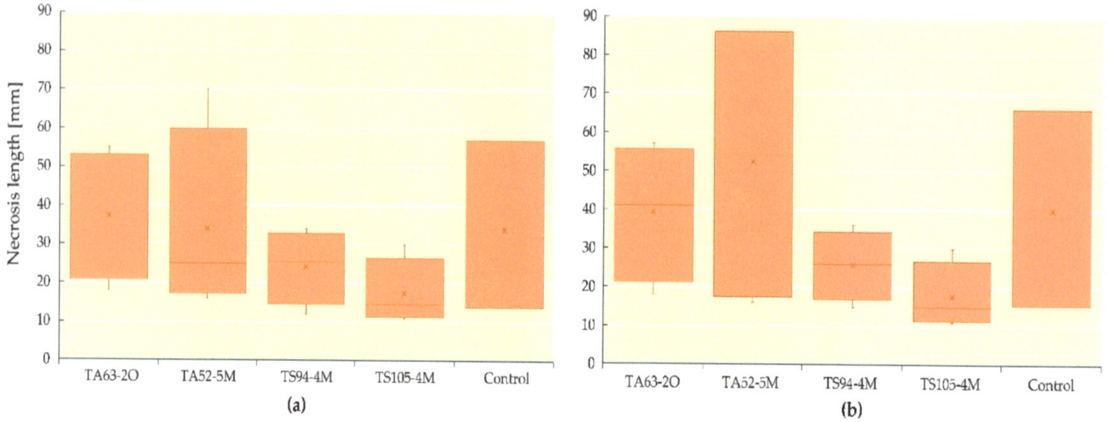

Figure 6. Box and whisker plots showing the length of superficial (**a**) and cambial (**b**) necroses developed by artificially inoculated *Hymenoscyphus fraxineus* on the bark of ash trees colonized or not colonized (control) by selected endophytic fungal isolates (TA63-2O—*Didymella macrostoma*, TA52-5M—*Dothiorella gregaria*, TS94-4M—*Fusarium lateritium*, TS105-4M—*Didymella aliena*).

4. Discussion

The results of this study revealed high species variability of culturable fungal endophytes in leaves and twigs from asymptomatic and symptomatic ash trees. Species from Pleosporales and Diaporthales were the most common endophytes, which corresponds to the findings of earlier research on endophyte diversity in ash trees in Europe [4,29,55]. Only limited information is available on fungal endophytes in ash trees in Slovakia [12,36,37], and this work considerably expanded the existing data. Most endophytic fungi identified in this study are not host-specific, e.g., *Alternaria alternata*, *Cladosporium cladosporioides*, and *Epicoccum nigrum*, and only two are strictly host-specific fungi, namely, *Diplodia fraxini* and *Fraxinicola fraxini*. The generalists were anticipated to predominate in the analyzed samples because the method of tissue examination for endophytes favored this group of fungi. *Nemania serpens*, *Diaporthe eres*, *Venturia fraxini* (a synonym for *Fraxinicola fraxini*), and *Diaporthe* sp. were the most frequent endophytes found in the leaf petioles of *F. excelsior* in Poland, as shown in a recent study [34]. Regardless of tissue type, the species structure of the most frequent endophytes identified from the ash trees in this study was different. Although *D. eres* was also the most frequently isolated species, *F. fraxini* and *N. serpens* were recovered from only nine leaf samples or a single twig sample, respectively. This discrepancy is possible because endophyte communities are not equal throughout the leaf tissue. For example, Schlegel et al. [56] discovered considerable variations in endophyte communities between leaf petioles and leaf laminae of ash and maple trees. It was evident that *D. eres* dominated in twig samples in May, whereas the second most frequent endophyte, *A. alternata*, was more prevalent in leaf and twig samples collected in October. The colonization frequency and species richness of endophytes increased with the age of leaf tissue but slightly decreased with the age of twig tissue. These differences may reflect the short life span of the leaves, and the nutrient composition and physiochemical variations between these two environments. We presume that leaf endophytes are transmitted horizontally, and that leaf colonization exists via infections directly from the environment, particularly in the case of generalist species that predominate in the analyzed samples. Different species of fungi require different lengths of time to produce spores, colonize tissue, and establish themselves in the hosts. Therefore, the colonization rate and species richness were lower in the samples collected in the spring (May) than in the fall season (October). It is generally accepted that the colonization rate of plant tissues by endophytic fungi increases with tissue age. Numerous studies have observed this finding [57–60]. In contrast to the leaf samples, the rate of colonization in twigs did not increase during the season but decreased. The decrease was more prominent in twigs from symptomatic trees, and we presume that ash dieback disease has a negative effect on endophyte colonization. The species diversity indices calculated for both sampling dates were similar to those observed by Bilański and Kowalski [34] for endophytes in the leaf petioles of *F. excelsior*. The diversity increased slightly with the age of the leaves, but not with the age of the twigs.

Factors influencing the abundance and diversity of endophyte communities in woody plants have recently been discussed in detail by Sieber [61]. Because the identification of endophytes in this study was limited to culturable fungi, the true diversity and abundance of the endophytic community in the ash trees remained undiscovered. Many nonculturable and obligate biotrophic species could not be detected by the method used to examine endophytes in ash tree samples. Another important selection factor was the use of only one type of agar medium for endophyte isolation. However, we expect that incorporating ash leaf extract into the cultivating medium might facilitate the isolation of species that have closer interactions with the host trees. Slower-growing endophytic fungi could be outcompeted or inhibited in the medium by faster-growing species, and to eliminate this effect, Rose Bengal was used to retard the growth of fast-growing species [62], while allowing slower-growing fungi to emerge from ash tissues. Culture-dependent techniques tend to favor dominant endophytic fungi [63], and rarer species with irregular occurrences in ash trees might be missed with only two sampling occasions. Due to the limitations of the methodology used in this study, it is possible that a range of potential candidate endophytes with biocontrol capabilities was overlooked.

Fungal endophytes may interact with host plants in manifold ways, and a potential protective effect against plant pathogens is one of them [63,64]. The use of endophytic microorganisms to control plant pathogens is receiving increasing attention as a sustainable alternative to synthetic pesticides. In the last decade, the mutual relationships between endophytes and plant hosts have been studied to understand the effects of endophytes on plant pathogens and their potential use for biological control. The endophyte-based biocontrol strategy is not a novel idea and has been studied in many agricultural and horticultural crops [65–67]. Treatment of trees with endophytes with the purpose of inhibiting the development of plant-pathogenic fungi is also under consideration [21,63,68–70]. Recently, several studies have been published that have determined the antagonistic potential of fungal endophytes against *H. fraxineus* [27,29,34,71]. This study confirmed the antagonistic activity of several local endophytic isolates against *H. fraxineus* in a dual culture bioassay. The strongest antagonistic effect was observed for *F. lateritium*, *D. aliena*, *D. macrostoma*, and *D. gregaria*. The antagonistic effect was coupled with the production of a wide inhibition zone in the cases of *F. lateritium*, *D. macrostoma*, and *D. gregaria*, which indicated a release of metabolites into the culture medium with an inhibitory effect against *H. fraxineus* growth. In a similar dual culture experiment, *F. lateritium* was listed among five endophytes with the strongest inhibition effect against *H. fraxineus* [34]. This fungus is also known as a natural antagonist of the plant pathogen *Eutypa armeniacae* (synonym of *Eutypa lata*), causing sapwood necrosis in fruit trees, grapevines, and ornamental plants [72]. However, *F. lateritium* is known as a globally distributed plant pathogen and has been reported in approximately 180 hosts, mainly woody plants, where it causes wilt, tip, or branch dieback, and cankers [73,74]. *Dothiorella gregaria* is also linked to branch and trunk canker in several plants [75]. However, the *Didymella* genus mainly includes saprobes commonly found in living or dead parts of plants [76]. In a recent study, another endophyte from the ash tree *Hypoxylon rubiginosum* appeared to be a promising biocontrol endophyte with strong fungitoxic properties and an antagonistic effect on *H. fraxineus* in planta [31]; however, it is also a mild pathogen, sometimes causing cankers on plants, including ash trees [77]. In other in vitro bioassays, different endophytes demonstrated inhibitory effects against isolates of *H. fraxineus*. For example, *Sclerostagonospora* sp., *Setomelanomma holmii*, *Epicoccum nigrum*, *Boeremia exigua*, and *Fusarium* sp. [29] or *Plenodomus biglobosus* and *Paracucurbitaria corni* [71] inhibited the growth of *H. fraxineus* in dual cultures. The inhibitory effect on the germination of *H. fraxineus* ascospores has also been documented for several leaf endophytes isolated from *F. excelsior*, e.g., *Fraxinicola fraxini*, *Paraconiothyrium* sp., *Boeremia exigua*, *Kretzschmaria deusta*, *Pezicula* sp., *Neofabraea alba* (synonym of *Neofabraea vagabunda*), and *Ampelomyces quisqualis* [27].

Although several studies have confirmed the antagonistic effect of fungal endophytes against *H. fraxineus* under in vitro conditions, their effect under in planta conditions has not yet been verified. Laboratory studies may not be good predictors of biocontrol agents' protective capacity, and, unfortunately, most research on tree pathogen–endophyte interactions has been conducted in the laboratory. It is uncertain how the interactions would change in the face of changing environmental conditions and existing competition with other species in the tree ecosystem; therefore, in planta bioassays are necessary. Moreover, the modes of action of most endophytes as biocontrol agents are still unknown, and bioassays on host plants may explain them. For example, the antagonistic effect of fungal endophytes on trunk necrosis development has already been documented in plants. Endophytic *Trichoderma aureoviride* used to inoculate the trunk was able to significantly reduce the necrosis size compared to the control on 30-year-old beech trees artificially inoculated with *Phytophthora plurivora* [69]. We evaluated the trunk inoculations of the four endophytic isolates, showing a strong inhibitory effect in the laboratory against inoculated *H. fraxineus*. Although the total length of necrotic lesions formed by the *H. fraxineus* infection was shorter in the ash trunks co-inoculated with the endophytes than in the trunks without the endophytes, the difference was not significant. The presence of *H. fraxineus* was confirmed in the necrotic lesions on all trunks (inoculated and non-inoculated with endophytes) six months after inoculation, which demonstrated that the endophytes could not eliminate the pathogen. The effect of endophytes on the development of necrosis was most prominent in the direction of the inoculation points of *Didymella macrostoma* and *Didymella aliena*. The weakest effect on *H. fraxineus* was *Dothiorella gregaria*. Although trunk inoculations did not produce optimistic results in this research, trunk inoculation with endophytes against phytopathogens has potential. It is likely that some fungal species can stimulate the ash's immune system against *H. fraxineus* infection, and this supports the importance of further research in the fight against this pernicious pathogen.

Author Contributions: Conceptualization and methodology, K.P., M.K.H. and M.B.; formal analysis, K.P., M.K.H. and M.B.; investigation, K.P., M.K.H., M.B., R.O. and M.K.; data curation, K.P. and M.K.H.; writing—original draft preparation, K.P., M.K.H. and M.B.; writing—review and editing, K.P., M.K.H., M.B., R.O. and M.K.; visualization, K.P., M.B. and R.O.; supervision, K.P. and M.K.H.; project administration and funding acquisition, M.K.H. All authors have read and agreed to the published version of the manuscript.

Funding: This research was funded by the Scientific Grant Agency of the Ministry of Education, Science, Research and Sport of the Slovak Republic and the Slovak Academy of Sciences, grant number VEGA 2/0062/18.

Data Availability Statement: Sequences generated during this study were deposited in GenBank.

Acknowledgments: We would like to thank the anonymous reviewers for their useful and insightful comments on the manuscript.

Conflicts of Interest: The authors declare no conflict of interest. The funders had no role in the design of the study; in the collection, analyses, or interpretation of data; in the writing of the manuscript; or in the decision to publish the results.

References

1. Przybyl, K. Fungi associated with necrotic apical parts of *Fraxinus excelsior* shoots. *For. Pathol.* **2002**, *32*, 387–394. [CrossRef]
2. Bakys, R.; Vasaitis, R.; Barklund, P.; Thomsen, I.M.; Stenlid, J. Occurrence and pathogenicity of fungi in necrotic and non-symptomatic shoots of declining common ash (*Fraxinus excelsior*) in Sweden. *Eur. J. For. Res.* **2009**, *128*, 51–60. [CrossRef]
3. Kowalski, T.; Czekaj, A. Symptomy chorobowe i grzyby na zamierających jesionach (*Fraxinus excelsior* L.) w drzewostanach Nadleśnictwa Staszów. *Leśne Pract. Badaw.* **2010**, *71*, 357–368. [CrossRef]
4. Kowalski, T.; Kraj, W.; Bednarz, B. Fungi on stems and twigs in initial and advanced stages of dieback of European ash (*Fraxinus excelsior*) in Poland. *Eur. J. For. Res.* **2016**, *135*, 565–579. [CrossRef]

5. Cross, H.; Sønstebø, J.H.; Nagy, N.E.; Timmermann, V.; Solheim, H.; Børja, I.; Kauserud, H.; Carlsen, T.; Rzepka, B.; Wasak, K.; et al. Fungal diversity and seasonal succession in ash leaves infected by the invasive ascomycete *Hymenoscyphus fraxineus*. *New Phytol.* **2017**, *213*, 1405–1417. [CrossRef] [PubMed]
6. Braun, U.; Cook, R.T.A. *Taxonomic Manual of Erysiphales (Powdery Mildews)*, 1st ed.; CBS Biodiversity Series 11; CBS: Utrecht, The Netherlands, 2012; 707p.
7. Sinclair, W.A.; Lyon, H.H.; Johnson, W.T. *Diseases of Trees and Shrubs*, 1st ed.; Cornell University Press: Ithaca, NY, USA, 1987; 575p.
8. Mejía, L.C.; Castlebury, L.A.; Rossman, A.Y.; Sogonov, M.V.; White, J.F. A systematic account of the genus *Plagiostoma* (*Gnomoniaceae, Diaporthales*) based on morphology, host-associations, and a four-gene phylogeny. *Stud. Mycol.* **2011**, *68*, 211–235. [CrossRef]
9. Ryvarden, L.; Melo, I. *Poroid Fungi of Europe*, 2nd ed.; Synopsis Fungorum 37; Fungiflora A/S: Oslo, Norway, 2017; 430p.
10. Zúbrik, M.; Kunca, A.; Vakula, J.; Galko, J.; Leontovyč, R.; Konôpka, B.; Gubka, A.; Nikolov, C.H.; Rell, S.; Longauerová, V.; et al. *Hmyz a Huby: Atlas Poškodení Lesných Drevín*, 2nd ed.; National Forest Centre, Forest Research Institute: Zvolen, Slovakia, 2019; 243p.
11. Longauerová, V.; Maľová, M.; Kunca, A. Poznatky z hynutia jaseňov spôsobovaného hubou *Hymenoscyphus pseudoalbidus* (ana. *Chalara fraxinea*). In *Aktuálne Problémy v Ochrane Lesa 2013*; Kunca, A., Ed.; National Forest Centre: Zvolen, Slovakia, 2013; pp. 77–81.
12. Pastirčáková, K.; Ivanová, H.; Pastirčák, M. Species diversity of fungi on damaged branches and leaves of ashes (*Fraxinus* spp.) in different types of stands in Slovakia. *Cent. Eur. For. J.* **2018**, *64*, 133–139. [CrossRef]
13. Kádasi Horáková, M.; Barta, M.; Adamčíková, K.; Ostrovský, R.; Pastirčáková, K. *Hymenoscyphus fraxineus* on *Fraxinus excelsior* in Slovakia: Distribution and mating types. *Biologia* **2022**. [CrossRef]
14. Zhao, Y.-J.; Hosoya, T.; Baral, H.-O.; Hosaka, K.; Kakishima, M. *Hymenoscyphus pseudoalbidus*, the correct name for *Lambertella albida* reported from Japan. *Mycotaxon* **2012**, *122*, 25–41. [CrossRef]
15. Gross, A.; Holdenrieder, O.; Pautasso, M.; Queloz, V.; Sieber, T.N. *Hymenoscyphus pseudoalbidus*, the causal agent of European ash dieback. *Mol. Plant Pathol.* **2014**, *15*, 5–21. [CrossRef]
16. Cleary, M.; Nguyen, D.; Marčiulynienė, D.; Berlin, A.; Vasaitis, R.; Stenlid, J. Friend or foe? Biological and ecological traits of the European ash dieback pathogen *Hymenoscyphus fraxineus* in its native environment. *Sci. Rep.* **2016**, *6*, 21895. [CrossRef] [PubMed]
17. Hyde, K.D.; Soytong, K. The Fungal Endophyte Dilemma. *Fungal Divers* **2008**, *33*, 163–173.
18. Wilson, D. Endophyte: The Evolution of a Term, and Clarification of Its Use and Definition. *Oikos* **1995**, *73*, 274–276. [CrossRef]
19. Saikkonen, K.; Faeth, S.H.; Helander, M.; Sullivan, T.J. Fungal Endophytes: A Continuum of Interactions with Host Plants. *Annu. Rev. Ecol. Syst.* **1998**, *29*, 319–343. [CrossRef]
20. Redman, R.S.; Sheehan, K.B.; Stout, R.G.; Rodriguez, R.J.; Henson, J.M. Thermotolerance Generated by Plant/Fungal Symbiosis. *Science* **2002**, *298*, 1581. [CrossRef] [PubMed]
21. Arnold, A.E.; Mejía, L.C.; Kyllo, D.; Rojas, E.I.; Maynard, Z.; Robbins, N.; Herre, E.A. Fungal endophytes limit pathogen damage in a tropical tree. *Proc. Natl. Acad. Sci. USA* **2003**, *100*, 15649–15654. [CrossRef]
22. Rodriguez, R.J.; Redman, R.S.; Henson, J.M. The Role of Fungal Symbioses in the Adaptation of Plants to High Stress Environments. *Mitig. Adapt. Strateg. Glob. Chang.* **2004**, *9*, 261–272. [CrossRef]
23. Gao, F.K.; Dai, C.C.; Liu, X.Z. Mechanisms of fungal endophytes in plant protection against pathogens. *Afr. J. Microbiol. Res.* **2010**, *4*, 1346–1351.
24. Mengistu, A.A. Endophytes: Colonization, Behaviour, and Their Role in Defense Mechanism. *Int. J. Microbiol.* **2020**, *2020*, 6927219. [CrossRef]
25. González-Teuber, M.; Vilo, C.; Guevara-Araya, M.J.; Salgado-Luarte, C.; Gianoli, E. Leaf resistance traits influence endophytic fungi colonization and community composition in a South American temperate rainforest. *J. Ecol.* **2020**, *108*, 1019–1029. [CrossRef]
26. Pujade-Renaud, V.; Déon, M.; Gazis, R.; Ribeiro, S.; Dessailly, F.; Granet, F.; Chaverri, P. Endophytes from Wild Rubber Trees as Antagonists of the Pathogen *Corynespora cassiicola*. *Phytopathology* **2019**, *109*, 1888–1899. [CrossRef] [PubMed]
27. Schlegel, M.; Dubach, V.; Von Buol, L.; Sieber, T.N. Effects of endophytic fungi on the ash dieback pathogen. *FEMS Microbiol. Ecol.* **2016**, *92*, fiw142. [CrossRef] [PubMed]
28. Haňáčková, Z.; Havrdová, L.; Černý, L.; Zahradník, D.; Koukol, O. Fungal endophytes in ash shoots—Diversity and inhibition of *Hymenoscyphus fraxineus*. *Balt. For.* **2017**, *23*, 89–106.
29. Kosawang, C.; Amby, D.B.; Bussaban, B.; McKinney, L.V.; Xu, J.; Kjær, E.D.; Collinge, D.B.; Nielsen, L.R. Fungal communities associated with species of *Fraxinus* tolerant to ash dieback, and their potential for biological control. *Fungal Biol.* **2018**, *122*, 110–120. [CrossRef] [PubMed]
30. Becker, R.; Ulrich, K.; Behrendt, U.; Kube, M.; Ulrich, A. Analyzing Ash Leaf-Colonizing Fungal Communities for Their Biological Control of *Hymenoscyphus fraxineus*. *Front. Microbiol.* **2020**, *11*, 590944. [CrossRef]
31. Halecker, S.; Wennrich, J.-P.; Rodrigo, S.; Andrée, N.; Rabsch, L.; Baschien, C.; Steinert, M.; Stadler, M.; Surup, F.; Schulz, B. Fungal endophytes for biocontrol of ash dieback: The antagonistic potential of *Hypoxylon rubiginosum*. *Fungal Ecol.* **2020**, *45*, 100918. [CrossRef]
32. Kowalski, T.; Bilański, P. Fungi Detected in the Previous Year's Leaf Petioles of *Fraxinus excelsior* and Their Antagonistic Potential against *Hymenoscyphus fraxineus*. *Forests* **2021**, *12*, 1412. [CrossRef]

33. Nawrot-Chorabik, K.; Marcol-Rumak, N.; Latowski, D. Investigation of the Biocontrol Potential of Two Ash Endophytes against *Hymenoscyphus fraxineus* Using In Vitro Plant–Fungus Dual Cultures. *Forests* **2021**, *12*, 1750. [CrossRef]
34. Bilański, P.; Kowalski, T. Fungal endophytes in *Fraxinus excelsior* petioles and their in vitro antagonistic potential against the ash dieback pathogen *Hymenoscyphus fraxineus*. *Microbiol. Res.* **2022**, *257*, 126961. [CrossRef]
35. Collinge, D.B.; Jørgensen, H.J.L.; Latz, M.A.C.; Manzotti, A.; Ntana, F.; Rojas, E.C.; Jensen, B. Searching for novel fungal biological control agents for plant disease control among endophytes. In *Endophytes for a Growing World*, 1st ed.; Hodkinson, T., Doohan, F., Saunders, M., Murphy, B., Eds.; Cambridge University Press: Cambridge, UK, 2019; pp. 25–51. [CrossRef]
36. Ivanová, H. Identification and characterization of the fungus *Dothiorella sarmentorum* on necrotic shoots of declining ash in Slovakia. *Folia Oecol.* **2018**, *45*, 53–57. [CrossRef]
37. Ivanova, H.; Malinicova, L.; Piknova, M.; Pristas, P. New Endophytic *Fusarium* spp. from *Fraxinus excelsior* Leaves in Slovakia. *Planta Med.* **2020**, *86*, 1025–1031. [CrossRef] [PubMed]
38. Adamčíková, K.; Kádasi-Horáková, M.; Jankovský, L.; Havrdová, L. Identification of *Hymenoscyphus fraxineus*, the causal agent of ash dieback in Slovakia. *Biologia* **2015**, *70*, 559–564. [CrossRef]
39. Adamčíková, K.; Pažitný, J.; Pastirčáková, K. Individual resistance of *Fraxinus angustifolia* and *F. excelsior* clones to *Hymenoscyphus fraxineus*. *J. Plant Protect. Res.* **2018**, *58*, 227–233.
40. Kirisits, T.; Dämpfle, L.; Kräutler, K. *Hymenoscyphus albidus* is not associated with an anamorphic stage and displays slower growth than *Hymenoscyphus pseudoalbidus* on agar media. *For. Pathol.* **2013**, *43*, 386–389. [CrossRef]
41. Ibrahim, M.; Sieber, T.N.; Schlegel, M. Communities of fungal endophytes in leaves of *Fraxinus ornus* are highly diverse. *Fungal Ecol.* **2017**, *29*, 10–19. [CrossRef]
42. Kumar, D.S.S.; Hyde, K.D. Biodiversity and tissue-recurrence of endophytic fungi in *Tripterygium wilfordii*. *Fungal Divers.* **2004**, *17*, 69–90.
43. White, T.J.; Bruns, T.; Lee, S.; Taylor, J. Amplification and direct sequencing of fungal ribosomal RNA genes for phylogenetics. In *PCR Protocols: A Guide to Methods and Applications*, 1st ed.; Innis, M.A., Gelfand, D.H., Sninsky, J.J., White, T.J., Eds.; Academic Press: San Diego, CA, USA, 1990; pp. 315–322. [CrossRef]
44. Gardes, M.; Bruns, T.D. ITS primers with enhanced specificity for basidiomycetes—Application to the identification of mycorrhizae and rusts. *Mol. Ecol.* **1993**, *2*, 113–118. [CrossRef]
45. Kádasi-Horáková, M.; Adamčíková, K.; Pastirčáková, K.; Longauerová, V.; Maľová, M. Natural infection of *Fraxinus angustifolia* by *Hymenoscyphus fraxineus* in Slovakia. *Balt. For.* **2017**, *23*, 52–55.
46. Zhang, Z.; Schwartz, S.; Wagner, L.; Miller, W. A Greedy Algorithm for Aligning DNA Sequences. *J. Comput. Biol.* **2000**, *7*, 203–214. [CrossRef]
47. Abdallah, M.F.; De Boevre, M.; Landschoot, S.; De Saeger, S.; Haesaert, G.; Audenaert, K. Fungal Endophytes Control *Fusarium graminearum* and Reduce Trichothecenes and Zearalenone in Maize. *Toxins* **2018**, *10*, 493. [CrossRef]
48. Skidmore, A.M.; Dickinson, C.H. Colony interactions and hyphal interference between *Septoria nodorum* and phylloplane fungi. *Trans. Br. Mycol. Soc.* **1976**, *66*, 57–64. [CrossRef]
49. Gross, A.; Sieber, T.N. Virulence of *Hymenoscyphus albidus* and native and introduced *Hymenoscyphus fraxineus* on *Fraxinus excelsior* and *Fraxinus pennsylvanica*. *Plant Pathol.* **2016**, *65*, 655–663. [CrossRef]
50. Johansson, S.B.K.; Vasaitis, R.; Ihrmark, K.; Barklund, P.; Stenlid, J. Detection of *Chalara fraxinea* from tissue of *Fraxinus excelsior* using species-specific ITS primers. *For. Pathol.* **2010**, *40*, 111–115. [CrossRef]
51. Pastirčáková, K.; Adamčíková, K.; Barta, M.; Pažitný, J.; Hoťka, P.; Sarvašová, I.; Kádasi Horáková, M. Host Range of *Hymenoscyphus fraxineus* in Slovak Arboreta. *Forests* **2020**, *11*, 596. [CrossRef]
52. Shannon, C.E. A Mathematical Theory of Communication. *Bell Syst. Tech. J.* **1948**, *27*, 379–423. [CrossRef]
53. Simpson, E.H. Measurement of Diversity. *Nature* **1949**, *163*, 668. [CrossRef]
54. Camargo, J.A. On Measuring Species Evenness and Other Associated Parameters of Community Structure. *Oikos* **1995**, *74*, 538–542. [CrossRef]
55. Scholtysik, A.; Unterseher, M.; Otto, P.; Wirth, C. Spatio-temporal dynamics of endophyte diversity in the canopy of European ash (*Fraxinus excelsior*). *Mycol. Prog.* **2013**, *12*, 291–304. [CrossRef]
56. Schlegel, M.; Queloz, V.; Sieber, T.N. The Endophytic Mycobiome of European Ash and Sycamore Maple Leaves—Geographic Patterns, Host Specificity and Influence of Ash Dieback. *Front. Microbiol.* **2018**, *9*, 2345. [CrossRef]
57. Osono, T.; Mori, A. Seasonal and leaf age-dependent changes in occurrence of phyllosphere fungi of giant dogwood. *Mycoscience* **2005**, *46*, 273–279. [CrossRef]
58. Guo, L.-D.; Huang, G.-R.; Wang, Y. Seasonal and Tissue Age Influences on Endophytic Fungi of *Pinus tabulae formis* (Pinaceae) in the Dongling Mountains, Beijing. *J. Integr. Plant Biol.* **2008**, *50*, 997–1003. [CrossRef] [PubMed]
59. Juybari, H.Z.; Ghanbary, M.A.T.; Rahimian, H.; Karimi, K.; Arzanlou, M. Seasonal, tissue and age influences on frequency and biodiversity of endophytic fungi of *Citrus sinensis* in Iran. *For. Pathol.* **2019**, *49*, e12559. [CrossRef]
60. Sadeghi, F.; Samsampour, D.; Seyahooei, M.A.; Bagheri, A.; Soltani, J. Diversity and Spatiotemporal Distribution of Fungal Endophytes Associated with *Citrus reticulata* cv. Siyahoo. *Curr. Microbiol.* **2019**, *76*, 279–289. [CrossRef] [PubMed]
61. Sieber, T.N. The phyllosphere mycobiome of woody plants. In *Forest Microbiology*, 1st ed.; Tree Microbiome: Phyllosphere, Endosphere and Rhizosphere; Asiegbu, F.O., Kovalchuk, A., Eds.; Academic Press: London, UK, 2021; Volume 1, pp. 111–132. [CrossRef]

62. Pan, J.J.; May, G. Fungal-Fungal Associations Affect the Assembly of Endophyte Communities in Maize (*Zea mays*). *Microb. Ecol.* **2009**, *58*, 668–678. [CrossRef] [PubMed]
63. Rabiey, M.; Hailey, L.E.; Roy, S.R.; Grenz, K.; Al-Zadjali, M.A.S.; Barrett, G.A.; Jackson, R.W. Endophytes vs tree pathogens and pests: Can they be used as biological control agents to improve tree health? *Eur. J. Plant Pathol.* **2019**, *155*, 711–729. [CrossRef]
64. Ownley, B.H.; Pereira, R.M.; Klingeman, W.E.; Quigley, N.B.; Leckie, B.M. *Beauveria bassiana*, a dual purpose biological control with activity against insect pests and plant pathogens. In *Emerging Concepts in Plant Health Management*, 1st ed.; Lartey, R.T., Caesar, A.J., Eds.; Research Signpost: Thiruvananthapuram, India, 2004; pp. 255–269.
65. Anjum, R.; Afzal, M.; Baber, R.; Khan, M.A.J.; Kanwal, W.; Sajid, W.; Raheel, A. Endophytes: As Potential Biocontrol Agent—Review and Future Prospects. *J. Agric. Sci.* **2019**, *11*, 113. [CrossRef]
66. De Silva, N.I.; Brooks, S.; Lumyong, S.; Hyde, K.D. Use of endophytes as biocontrol agents. *Fungal Biol. Rev.* **2019**, *33*, 133–148. [CrossRef]
67. Bamisile, B.S.; Siddiqui, J.A.; Akutse, K.S.; Ramos Aguila, L.C.; Xu, Y. General Limitations to Endophytic Entomopathogenic Fungi Use as Plant Growth Promoters, Pests and Pathogens Biocontrol Agents. *Plants* **2021**, *10*, 2119. [CrossRef]
68. Ganley, R.J.; Sniezko, R.A.; Newcombe, G. Endophyte-mediated resistance against white pine blister rust in *Pinus monticola*. *For. Ecol. Manag.* **2008**, *255*, 2751–2760. [CrossRef]
69. Berger, G.; Czarnocka, K.; Cochard, B.; Oszako, T.; Lefort, F. Biocontrol Endotherapy with *Trichoderma* spp. and *Bacillus amyloliquefaciens* against *Phytophthora* spp.: A Comparative Study with Phosphite Treatment on *Quercus robur* and *Fagus sylvatica*. *J. Agric. Sci. Technol.* **2015**, *5*, 428–439. [CrossRef]
70. Poveda, J.; Baptista, P. Filamentous fungi as biocontrol agents in olive (*Olea europaea* L.) diseases: Mycorrhizal and endophytic fungi. *Crop Prot.* **2021**, *146*, 105672. [CrossRef]
71. Marcol, N.; Pukalski, J.; NawrotChorabik, K.; Kowalski, T.; Latowski, D. Will black pigments save European ash (*Fraxinus excelsior* L.)?—EPR analysis of melanin production in dual-culture of endophytic and pathogenic fungi. In Proceedings of the 45th FEBS Congress, Molecules of Life: Towards New Horizons, Ljubljana, Slovenia, 3–8 July 2021.
72. Tsantrizos, Y.S.; Xu, X.-J.; Sauriol, F.; Hynes, R.C. Novel quinazolinones and enniatins from *Fusarium lateritium* Nees. *Can. J. Chem.* **1993**, *71*, 1362–1367. [CrossRef]
73. Gerlach, W.; Nirenberg, H. *The Genus Fusarium, a Pictorial Atlas*, 1st ed.; Mitt. Biol. Bundesanst. Land-Forstwirsch. Berlin-Dahlem; Kommissionsverlag P. Parey: Berlin, Germany, 1982; Volume 209, 406p. [CrossRef]
74. Vitale, S.; Santori, A.; Wajnberg, E.; Castagnone-Sereno, P.; Luongo, L.; Belisario, A. Morphological and Molecular Analysis of *Fusarium lateritium*, the Cause of Gray Necrosis of Hazelnut Fruit in Italy. *Phytopathology* **2011**, *101*, 679–686. [CrossRef] [PubMed]
75. Yu, Z.D.; Gao, A.Q. Pathogenicity of *Dothiorella gregaria* to Rosaceae and other host trees. *For. Pest Dis.* **2004**, *23*, 21–23.
76. Chen, Q.; Jiang, J.; Zhang, G.; Cai, L.; Crous, P. Resolving the *Phoma* enigma. *Stud. Mycol.* **2015**, *82*, 137–217. [CrossRef]
77. Farr, D.F.; Rossman, A.Y. Fungal Databases, U.S. National Fungus Collections, ARS, USDA. Available online: https://nt.ars-grin.gov/fungaldatabases/ (accessed on 15 May 2022).

forests

MDPI

Article

Mycorrhizal Diversity on Roots of Silver Birch and Hybrid Aspen in Clonal Plantations in Northern Europe, Latvia

Roberts Matisons [1], Zigmunds Orlovskis [2,*], Kārlis Trevors Blūms [2], Dainis Ruņģis [1], Margarita Baranova [1], Baiba Krivmane [1], Krišs Bitenieks [1], Maryna Ramanenka [1], Dārta Kļaviņa [1] and Mārtiņš Zeps [1]

1 Latvian State Forests Research Institute "Silava", Rīgas Str. 111, LV-2169 Salaspils, Latvia
2 Latvian Biomedical Research and Study Centre, Rātsupītes Str. 1 k-1, LV-1067 Riga, Latvia
* Correspondence: zigmunds.orlovskis@biomed.lu.lv

Abstract: Mycorrhizal fungi contribute to crop growth, yields, and stress tolerance. In forests, common mycorrhizal networks are suggested to function as carbon storage and to transfer substances and signals between trees, thus likely contributing to their resilience. Such properties are crucial under increasing environmental stresses, particularly for clonal forestry. However, mycorrhizal communities in relation to tree field performances have been scarcely studied. In this study, mycorrhizal communities on the roots of clones of silver birch and hybrid aspen growing in distinct trials in deep automorphous mineral soils (podzolic and fluvic) under hemiboreal conditions were assessed using internal transcribed spacer sequencing, bioinformatics, and community analysis. The mycorrhizal communities were moderately rich/diverse and were mostly formed by generalist taxa (prevailingly ectomycorrhizal) common for the region. The differences in communities among the tree clones were estimated for silver birch, while for hybrid aspen, the productivity of clones was inversely related to the richness and diversity of the communities, suggesting a top-down effect of the host. Accordingly, some mycorrhizal taxa (e.g., *Hyaloscypha* sp.) showed clone-specific abundances indicating a preference for a specific host. These findings prompt further functional studies and highlight the need to consider genetic differences of forest regenerative material for maximizing mycorrhizal diversity, as well as for more effective inoculation.

Keywords: clonal plantations; community composition; forest productivity; ITS sequencing; mycorrhizal diversity

Citation: Matisons, R.; Orlovskis, Z.; Blūms, K.T.; Ruņģis, D.; Baranova, M.; Krivmane, B.; Bitenieks, K.; Ramanenka, M.; Kļaviņa, D.; Zeps, M. Mycorrhizal Diversity on Roots of Silver Birch and Hybrid Aspen in Clonal Plantations in Northern Europe, Latvia. *Forests* **2024**, *15*, 2123. https://doi.org/10.3390/f15122123

Academic Editors: Katarína Pastirčáková and Rostislav Zemek

Received: 7 November 2024
Revised: 24 November 2024
Accepted: 28 November 2024
Published: 1 December 2024

1. Introduction

Due to warming in Northern Europe, the abundance of deciduous trees is increasing [1], expanding their commercial application throughout the region [2,3]. Still, the ameliorating thermal conditions and extending vegetation season are counteracted by the increasing frequency and severity of other climatic events, e.g., drought [4], and subsequent legacy biotic risks, particularly pest outbreaks [5]. Given the large-scale nature and the economic importance of such effects, long-lasting and self-perpetuating solutions are desirable for sustaining forest productivity [3,6].

In the eastern Baltic region, which is important for timber production in Europe, silver birch (*Betula* spp.) and aspen (*Populus* spp.) are increasingly economically important species [7]. The importance of birch within the region is highlighted by the advances plywood industry [2] and subordinate breeding programs [3,8]. Hence, stands are being partially regenerated by improved material [8], which can be more homogeneous genetically [9]. Given the weather sensitivity of birch [10], increasing its tolerance to environmental risks is crucial [2].

Due to abundance, productivity, and efficient self-regeneration, trembling aspen (*P. tremula*) [7], and particularly its hybrids, which have superior productivity, are a promising alternative, allowing a notable reduction in the rotation cycle [11]. The reproductive material of hybrid aspen (e.g., *P. tremuloides* × *tremula*) is vegetatively propagated and is genetically homogeneous [11,12], which also increases its susceptibility to disturbances [13]. The superior productivity comes with the cost of reduced plant defense [14] and susceptibility to herbivores [11,15]. In this regard, timely mobilization of the defense mechanisms can potentially alleviate such risks [16–18].

As a strategy to mitigate environmental and, particularly, biotic risks, plants have evolved the ability to eavesdrop on neighbor danger signals, allowing the proactive mobilization of defense mechanisms [19,20]. Inter-plant signal exchange can occur by several mechanisms, such as root exudates or airborne chemical signals [20–22]. In addition, common mycorrhizal networks have been suggested as a much more efficient network of substance and information exchange between plants, functioning as an "underground internet" [17,19,23,24]. This highlights the potential to improve the stress tolerance of crops and forest stands by the initial inoculation of planting material with relevant mycorrhizal fungi (MF) to enable the formation of a mycorrhizal network [18,25,26]. However, the underlying mechanisms of mycorrhizal-network-mediated inter-plant signal exchange are not fully explored, and there are still large uncertainties regarding the involvement of different species of ecto- and endomycorrhizal fungi [27,28]. Accordingly, attempts to implement such biotic solutions emerge as highly knowledge-intensive endeavors and require the characterization of the endemic MF communities [18,27].

The genetic background of various clones of silver birch and hybrid aspen is associated with accelerated growth, increased productivity, and stress tolerance [3,8,11,29–33]. Nevertheless, clone-specific differences in MF communities may constitute functionally different mycelial networks [18,34–37]. Therefore, as the initial step for the further assessment of the role of mycorrhizas in inter-tree signal exchange and substance transfer regarding the commercial forest reproductive material, an assessment of in situ MF communities is needed [38,39], particularly accounting for spatial specifics [18,34–36,40].

The MF communities associated with birch and aspen in mesotrophic sites with mineral soils in the Eastern Baltic region are scarcely studied [41,42]. Still, the genera *Cenococcum*, *Geopora*, *Helotiales*, *Scleroderma*, *Helobema*, and *Russula* have been described to occur on birch [43–47]. On aspen, *Cenococcum*, *Helobema*, *Inocybe*, *Tomentella*, *Hymenogaster*, and *Russula* have been considered as common [37,48], while the communities on the hybrids can be specific [37]. Nevertheless, due to spatial specifics, local data on MF and their relationships with host genotypes are needed for further research and development of practical applications. This study aimed to characterize MF communities on roots of clones of silver birch and hybrid aspen, which are intended for wide commercial applications, in trials in the hemiboreal forest zone in Latvia. Hence, it can be hypothesized that clones may bear specific MF communities, which contribute to field performance [37,45].

2. Material and Methods

2.1. Trials

The study sites were located in the central part of Latvia (56.451° N, 22.874° E; 56.713° N, 23.840° E), which is situated in lowland conditions in the mid-part of the hemiboreal forest zone. The studied trials were located in lowland conditions with a flat topography on mineral soils. The site fertility was higher in the hybrid aspen trial, which was established on former agricultural land (eutrophic fluvisoil with neutral reaction, fertility grade 40) and had deep silty bedrock. The birch trial was situated on forest land with fresh mesotrophic (*Hylocomiosa*; [49]) podzolic soil (pH ~3.5–4.5) with deep sandy bedrock. The trial was established after the previous stand (mixed coniferous stand with birch admixture) was removed in a clear-cut; hence, the soil had locally explicit podzolic characteristics.

The climate at the sites was cold moist continental with cool summers [50]. According to the gridded meteorological data (CRU TS4; [51]), the mean annual temperature (±standard deviation) was 7.4 ± 0.7 °C with monthly temperature ranging from −2.5 ± 2.6 to 18.1 ± 1.7 °C in January and July, respectively (Supplementary Material, Figure S1). The annual amount of precipitation was 667 ± 81 mm; the highest precipitation fell during the summer months (June–September), when the monthly precipitation sums ranged from 65 ± 21 to 83 ± 33 mm in June and July, respectively. Summer precipitation generally comprises half of the annual total. Climatic changes have manifested as warming, particularly during the dormancy period, thus extending the vegetation season by approximately two weeks during the last century [52]. Concomitantly, summer precipitation has been becoming more heterogeneous with moist periods shifting with extending hot droughts [4,52].

The hybrid aspen trial was established in 2008 to test the field performance of clones of the locally bred genotypes (hybrids) of *P. tremuoides* and *P. tremula* [12]. The clones were progenies from a mother tree, growing in the National Botanical Garden (in the central part of Latvia), and father trees (plus trees) collected in stands scattered across the territory of Latvia [12]. One-year-old containerized microclonally propagated seedlings were planted. Prior to planting, the soil was prepared by plowing. In total, six clones with contrasting field performance were tested in the studied trial [12,53]. The design of the trial was randomized blocks; each block had four to six replications. Initially, the blocks consisted of 5 × 5 trees with the initial spacing of 3 × 3 m (1100 trees ha^{-1}). In 2020, the trial was thinned diagonally. After the thinning, the density of the trial decreased to 529 trees ha^{-1} (Supplementary Material, Figure S2). However, considering recent disturbance, the formation of root suckers was quite common across the trial. The trial was fenced.

The studied trial of silver birch was established in 2017 to test the geographic variation in the field performance of clones of 34 best-performing individuals of half-sib progenies of open-pollinated plus trees collected across Latvia. One-year-old containerized microclonally propagated seedlings, representing clonal material from a 14-year-old trial, were planted. The trial design was randomized blocks of 5 × 5 trees with the initial spacing of 3 × 2 m (1667 trees ha^{-1}). Blocks had four or five replications. Prior to planting, the soil was prepared by mechanized mound preparation. Planting was performed manually. No management except mechanical weed control was implemented since the establishment (Supplementary Material, Figure S2). The tested genotypes were also maintained in in vitro collections, thus enabling further laboratory experimentation with MF inoculation of the same genotypes of silver birch.

2.2. Clone Selection and Sampling

To evaluate the potential relationships between MF communities and tree growth, clones with differing/contrasting field performance according to consolidated rankings among other trials [12,53], which were also consistent in the studied ones, were selected. For the hybrid aspen, clones "43" and "44", which showed moderate and superior field performance, respectively [53], were selected. According to the last inventory conducted in 2020, when trees were 13 years old, the mean (±standard. error) height and diameter at the breast height for the selected clones differed, being 21.2 ± 0.3 vs. 19.1 ± 0.5 m and 18.3 ± 0.4 vs. 14.5 ± 0.6 cm for the clones "44" and "43", respectively. The mean stem diameter of the clones was 117 and 92% of the trial mean, respectively.

For silver birch, clones "54-146-143" and "55-875", which show above-average field performance and exceed wild populations by 7%–30% in volume growth [3,8], were selected. According to the last inventory in 2021, when trees were four years old, height and diameter at the breast height for the selected clones were similar: 2.48 ± 0.06 vs. 2.54 ± 0.07 m (101.1% and 103.8% of the plantation mean) and 1.37 ± 0.05 vs. 1.53 ± 0.06 cm (82.3% and 92.4% of the plantation mean), respectively.

To represent fungal diversity associated with the clones within the plantations, stratified tree selection was performed. Three trees from the interior part (avoiding the edge trees) of four randomized blocks, which were visually healthy and represented the diameter distribution of the clone (dominant/co-dominant and intermediate) were selected for sampling. From each tree, a fine root sample was collected at a 60–90 cm distance from the stem up to the depth of 25 cm. The size of the sample was approximately 5–10 g of fine roots, which were cleaned from the bulk of the soil. Samples were kept in sealed plastic bags at 5–10 °C during the transfer to the lab, where they were kept in a freezer (−20 °C). The trials were sampled in June 2023 (a month after the onset of vegetation season), when hybrid aspen and birch explants were 16 and 7 years old, respectively.

2.3. Preparations, Sequencing, and Bioinformatic Treatments

A random subsample of 1–3 g of fine roots was taken from each sample for DNA extraction. The material was washed and rinsed with water. The DNA was extracted using a modified CTAB–chloroform method [54]. The extracted DNA was diluted to 20 ng/μL; 40 ng of DNA per sample was used for amplification with the 5× HOT FIREPol Blend Master Mix (Solis Biodyne, Tallin, Estonia) solution, with added 0.2 μM ITS9mun and ITS4ngsUni primers [55]. The primers amplify 850 bp long fungal DNA fragments, which contain ITS1 and ITS2 regions. The program used was 15 min at 95 °C, with a total of 30 cycles consisting of 30 s at 95 °C, 30 s at 57 °C, and 60 s at 72 °C. The duration of the terminal extension was 600 s at 72 °C. The amplification was performed in two technical replications.

To prepare the amplified material for sequencing (purification, reparation, and barcode ligation), Native Barcoding Kit 96V14 (SQK-NBD114.96, Oxford Nanopore, UK) with the NEB Blunt/TALigase Master Mix (M0367, NEB, USA) and NEBNext Ultra II End repair/dA-tailingModule (E7546, NEB, USA) set of enzymes was used. The Qubit® dsDNA HS Assay and Agilent HS DNA Kit were used to determine library concentration and length prior to pooling. R10.4.1 flow cells for long reads (FLO-MIN114, Oxford Nanopore, UK) and a MinION (Oxford Nanopore, UK) sequencer were used. The session was approximately 72 h long.

The raw sequencing data were initially analyzed with a super accuracy base calling model (SUP) and demultiplexed using the open source basecaller Dorado (v0.5.0, Oxford Nanopore Technologies), which was run on a high-performance computing cluster (RTU High-Performance Computing Centre). By using the program Seed2 [56], the sequences were reordered properly, allowing up to two mismatches on each of the primers, and filtered, leaving sequences of 300–1200 bp in length, with an average read quality above Q25. Chimeric sequences were removed using the uchime de novo algorithm in the program vsearch [57]. Full fungal ITS regions (ITS1 and ITS2) were extracted by the program *ITS*x (v1.1.3; [58]). The clustering of the sequences was performed in the program vsearch at the 86% similarity level [59]. The threshold was empirically lowered to avoid the overinflating of genetic diversity, likely due to adopting the method for the Oxford Nanopore platform. Taxonomy was assigned to operational taxonomic units (OTUs) using BLAST and the UNITE fungal sequence database (v. 9.0, issued on 18 July 2023; https://doi.org/10.15156/BIO/2938067).

2.4. Data Analysis

The estimated OTU (community) table was filtered for spurious records first by omitting the entries with less than two reads (singletons). Further, OTUs showing occurrences with less than 10 sequences per clone were omitted. The OTUs identified as mycorrhiza according to FunGuild classification [60] were then selected for the analysis, thus reducing and focusing the MF community matrix. The diversity of OTUs for clones and individual trees (ramets) was estimated by the Shannon diversity index (H'). The differences in the per-tree diversity and richness of OTUs according to clone for each species, as well as between the species, were compared by ANOVA.

To assess the similarity of MF communities inhabiting roots of the studied clones of silver birch and hybrid aspen, nonmetric multidimensional scaling, which is an unsupervised ordination technique, was used. Prior to the analysis, the community matrix was square-root-transformed and normalized by scaling the sum of squares of trees to one, as the scales of the OTU intensities were highly heterogeneous. The ordination was based on the Bray–Curtis dissimilarity matrix allowing 20 starts. The analysis was conducted for silver birch and hybrid aspen together, as well as separately. The significance of the differences in

MF communities between the "species"/trials and between clones for each tree "species" was estimated by the permutational multivariate analysis of variance (PerMANOVA, 1000 iterations). To account for spatial variation, replication (block) was included in the model as a fixed effect.

To identify OTUs with the strongest clonal dependence, sparse partial least squares discriminant analysis [61,62] based on the transformed and normalized community matrix was used. This method allows one to distinguish the individual contributions of variables in multidimensional space in pre-defined (fixed) groups. Considering the limitations of the dataset, the analysis was restricted to distinguishing up to five principal components with up to 40 variables (OTUs). The number of meaningful components was estimated based on multiple error estimates according to M-fold validation, maximum distance, and balanced error rate as the misclassification measures. The data analysis was performed in R v. 4.4.1 [63] using the libraries "vegan" [64] and "mixOmics" [62].

3. Results

3.1. Fungal Taxonomic and Functional Diversity

The amplification of the fungal ITS region was successful (Supplementary Material, Figure S3). The sequencing run yielded 204 ± 80 thousand (±st. dev.) reads per sample, and the median number of raw reads was higher for silver birch compared to hybrid aspen (Supplementary Material, Figure S4). The negative (no template) controls displayed a small number of reads (raw) corresponding to OTUs compared to reads from root samples. During the data decontamination, these OTUs were removed from further fungal diversity estimates. In total, 1081 fungal OTUs were detected from the root samples, but 20 taxa constituted >75% of the total fungal community in around 73% of all sequenced samples (Figure 1A). The taxonomic diversity of the fungal communities from roots of hybrid aspen was higher than that of silver birch, but without explicit uniform dominance structure. In contrast, fungal communities from the roots of silver birch showed high dominance of *Scolicosporum* irrespective of clone. *Mycena* and *Tomentella* were dominant in the roots of a few scattered birch and aspen individuals.

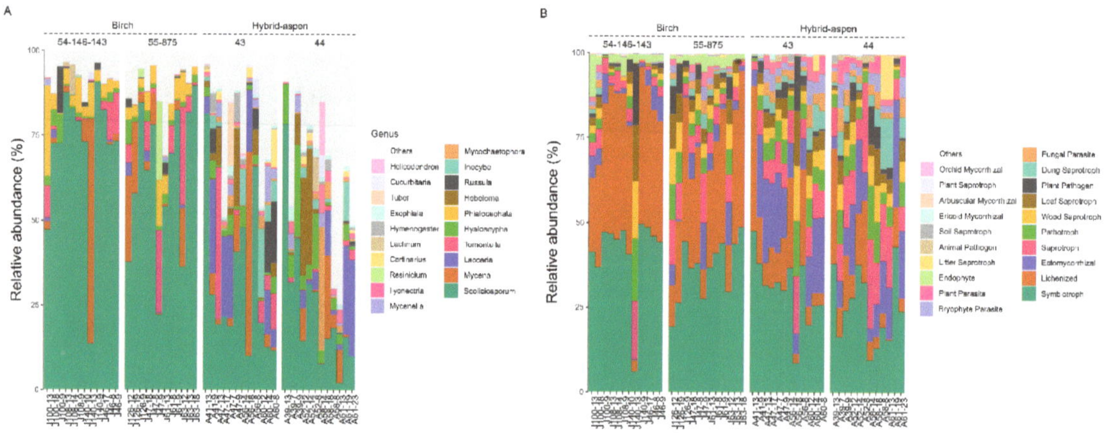

Figure 1. The relative abundance of the most common fungal genera (**A**) and *FunGuilds* functional groups (**B**) from the roots of hybrid aspen (clones 43 and 44) and silver birch (clones 54-146-143 and 55-875) in trial plantations in the central part of Latvia on deep automorphous mineral soils (podzolic and fluvic soils).

The diversity and evenness of functional guilds were high, particularly for hybrid aspen (Figure 1B). Symbiotrophs were the most abundant functional group in the samples from all clones of hybrid aspen and silver birch (Figure 1B). Root samples from silver birch contained a high proportion of taxa that correspond to fungi associated with lichens, whereas hybrid aspen roots displayed a high abundance of ectomycorrhizal species. Interestingly, the root fungal community of birch clone 54 contained a smaller proportion of pathogenic and parasitic fungi compared to birch clone 55 or both aspen clones. Noteworthily, fungal communities from both birch clones contained around 5%–10% of endophytic fungi (Figure 1B), indicating them as a considerable component of tree root microbiomes.

3.2. *Mycorrhizal Diversity and Community Structure in Birch and Aspen Clones*

The MF communities in the studied clones of silver birch and hybrid aspen consisted of 81 OTUs in total (Supplementary Material, Table S1), most of which were identified to the genus or species level. The number of OTUs, and hence the richness of MF communities, however, was somewhat higher for the clones of silver birch compared to hybrid aspen (mean of 47 and 35, respectively; Table 1). The diversity (Shannon index) of the detected OTUs of MF ranged from 1.71 to 2.47 for clones "55-875" (silver birch) and "43" (hybrid aspen), respectively, yet the differences were larger between the clones than tree "species". The individual-tree richness of MF OTUs was significantly higher on silver birch (p-value = 0.02) than on hybrid aspen (13.0 ± 1.4 and 15.4 ± 1.3; means ± 95% confidence interval), yet the diversity, as indicated by Shannon index, was similar (p-value = 0.16). The clonal differences in richness (number) of OTUs lacked statistical significance (p-value > 0.30) for both silver birch and hybrid aspen; yet significant clonal differences in per-tree diversity of OTUs were estimated only for hybrid aspen (p-value < 0.01; 1.08 and 1.45 for clones 44 and 43, respectively).

Table 1. Richness and diversity (Shannon diversity index H´) of the mycorrhizal fungi communities detected in the roots of silver birch and hybrid aspen clones. Numbers represent pooled data for each clone. The lowercase letters indicate the results of the post-hoc comparison (within-clone); similar letters depict a lack of significant differences at α = 0.05.

Species	**Hybrid Aspen**		**Birch**	
Clone	**43**	**44**	**54-146-143**	**55-875**
Number of taxonomic units detected	38	32	44	50
Shannon diversity index of taxonomic units detected	2.47	1.99	2.20	1.71
The mean (±95% confidence interval) per-tree number of taxonomic units detected	13.8 ± 1.9 a	12.2 ± 2.0 a	15.2 ± 2.1 a	15.6 ± 2.0 a
The mean (±95% confidence interval) per-tree Shannon diversity of taxonomic units detected	1.45 ± 0.22 a	1.08 ± 0.21 b	1.48 ± 0.23 a	1.38 ± 0.21 a

The dominance structure of OTUs of MF (Figure 2) was intermediate to low, as the relative abundance of single entities reached 0.44 for silver birch (*Tomentella stipitobasidia*), indicating generally high competitiveness among several ectomycorrhiza taxa. For hybrid aspen, the dominance of any particular MF was not observed, as the relative abundances of the most common OTUs were relatively similar (10%–23%). Among the most abundant MF taxa were several ectomycorrhizal fungi—*Laccaria populina*, *Agricales* spp., and *Russula emetica*—which were found in the roots of both tree "species" relative abundances (Figure 2). In comparison, endomycorrhizal fungi were less abundant and constituted 16% of the MF taxa on both tree hosts. The most frequently found endomycorrhizal species in both hybrid aspen and silver birch was *Hymenoschypha finlandica*. The total endomycorrhizal richness appeared higher on roots of silver birch than on hybrid aspen (nine and six taxa, respectively; 16% of the MF taxa). Most MF taxa identified were found in root samples from both tree hosts across different trial sites, suggesting that MF communities mostly consisted of generalist species.

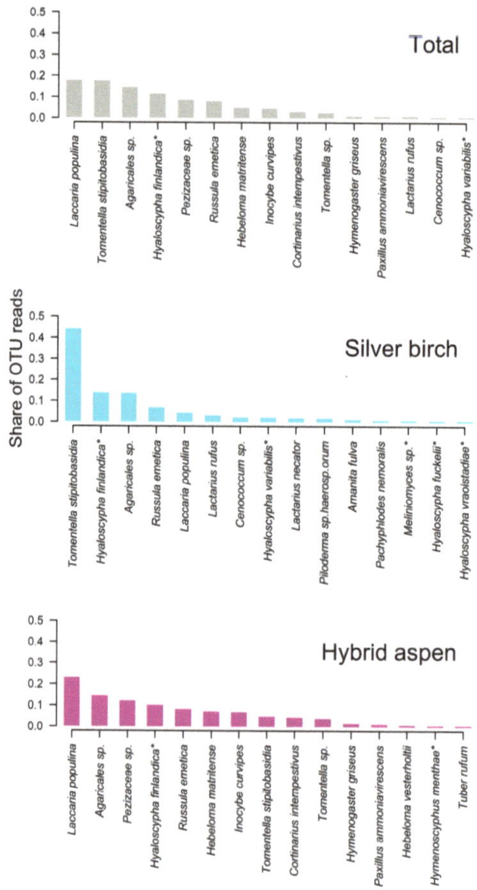

Figure 2. Relative abundance of the most common taxonomic units of mycorrhizal fungi detected on roots of silver birch and hybrid aspen clones. The 15 most abundant taxa are shown. Asterisks denote taxa associated with the endomycorrhizal guild. The other taxa are classified as ectomycorrhizal.

The communities of MF significantly differed between the tree "species"/trials (Table 2), as highlighted by the distinct grouping in the ordination of trees (Figure 3). The differences in the communities between the tree "species"/trials were highlighted by the abundance of *Peziza*, *Russula*, *Imleria*, *Hyaloschypha*, and *Pezoloma* species. Marginal clonal differences in MF communities were observed only for silver birch (Table 2), and these differences were primarily related to the abundance of *Inocybe*, *Russula*, *Lactarius*, *Hebeloma*, *Pezoloma*, and *Hyaloscypha* species (Figure 3). Nevertheless, the partial discriminant analysis suggested that nine of the studied MF OTUs, which represented the sole meaningful component (explaining 12% of variance), had the strongest contribution to the clonal differences in birch (Figure 4). Among these, two OTUs were endomycorrhizal (*Hyaloscypha variabilis* and *H. vraolstadiae*). Generally, these nine taxa were congruent with the taxa identified by species loadings from the ordination (Figure 3). In both trials, there was explicit local variability in MF communities, as indicated by the significant effects of replication (block), particularly for silver birch (Table 2). In contrast, local variability was weaker for hybrid aspen, which was older and had considerably larger trees than silver birch.

Table 2. The significance of the differences (MANOVA table) in mycorrhizal fungal community composition from roots of the studied clones of silver birch and hybrid aspen.

	R^2	*F*-Value	*p*-Value
Between tree species			
Species	0.24	17.26	0.001
Replication (block)	0.31	1.66	0.002
Residual	0.46		
Total	1.00		
Between clones of silver birch			
Clone	0.08	2.24	0.037
Replication (block)	0.38	1.86	0.004
Residual	0.54		
Total	1.00		
Between clones of hybrid aspen			
Clone	0.05	1.37	0.223
Replication (block)	0.33	1.45	0.049
Residual	0.61		
Total	1.00		

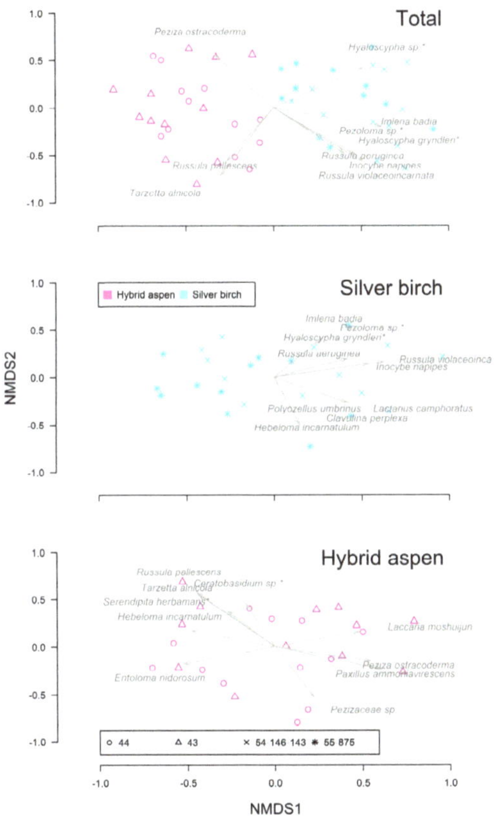

Figure 3. The ordination (nonmetric multidimensional scaling) of mycorrhizal fungal communities on the roots of the studied clones of silver birch and hybrid aspen. Ten taxa showing the strongest correlation with axes are shown in each panel. The asterisks next to the species names denote taxa associated with the endomycorrhizal guild.

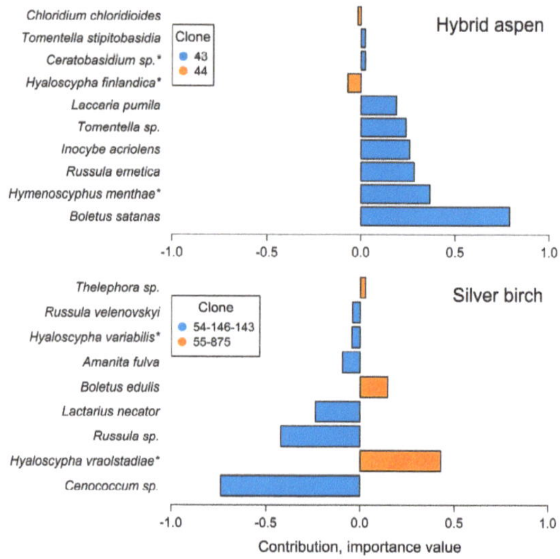

Figure 4. Contribution of species abundances to differences in mycorrhizal fungal community structure between clones of silver birch and hybrid aspen. The partial least squares discriminant analysis identified 9 OTUs explaining clonal variation in silver birch and 10 OTUs in hybrid aspen. The length of the bar indicates the relative contribution of an OTU to clonal differences in community structure. The asterisks next to the species names denote endomycorrhizal taxa.

Clones of hybrid aspen bared highly similar MF communities, as highlighted by the overlap in the ordination plots (Figure 3) and lack of significant differences (Table 2). The variability of the communities on trees was particularly related to the abundance of *Peziza*, *Laccaria*, *Entoloma*, and *Russula* species. The discriminant analysis estimated a sole meaningful component explaining only 8% of the variance, which highlighted ten OTUs as having the strongest contribution to possible clonal differences for hybrid aspen (Figure 4). Among those, the strongest contributions to the clonal differences were estimated for *Boletus satanas*, *Hymenoscyphus menthae*, *Russula emerica*, *Inocybe acriolens*, *Tomentell*a sp., and *Hyaloscypha finlandica*. Among these, three were associated with endomycorrhiza—*Hyaloscypha finlandica*, *Ceratobasidium* sp., *Hymenoscyphus menthae*.

4. Discussion

4.1. Host and Trial Site Effects on Fungal Diversity

The dominant fungal and MF taxa (Figures 1–3) were similar to those observed in other studies within the region [37,45], though with differing community structures, thus highlighting spatial (local and regional) variability in MF communities [38]. The total number of OTUs detected was relatively high, probably due to contamination from soil or misidentification of the sequences [57,59]. While the root samples were washed in sterile water prior to DNA extraction, fungal species from the tree rhizosphere are likely to be captured in study data and contribute to some of the less abundant functional guilds, such as soil saprotrophs or pathogens (Figure 1B). Rähn [37] suggests that saprotrophs were more abundant in forests than plantations and are associated with a higher abundance of arbuscular MF, as well as a lower abundance of pathogenic fungi. However, there was a somewhat lower relative composition of saprotrophs and pathogens in birch clone "54-146-143" compared to the other birch clone grown on forest soil or aspen plantations on former agricultural land (Figure 1B). Such discrepancies might be related to the differences in sampling, as the communities directly recruited by roots might deviate from those in soil in general [65].

The dominance structure in the entire community of OTUs on the roots of silver birch (Figure 1A) can be related to plant-host-specificity of fungal taxa [43,45], as well as the younger age of trees, as the colonization of later succession taxa could still be expected to occur [40,60]. The dominance structure of MF was more expressed in younger birch (Figure 2) with a high prevalence of the cosmopolitan ectomycorrhizal species *T. stipitobasidia* [66], which may demonstrate relatively higher competitiveness in nutrient-poor forest soils. Nevertheless, the diversity of functional guilds of the studied trees (Figure 1B) was high, suggesting that the communities were balanced [67]. In addition to the host-specificity and tree age, the soil preparation also differed between aspen and birch trials and could have potentially contributed to the differences in fungal community structure. Plowing was used in hybrid aspen trials, while mound preparation was used in silver birch trials. Since mycorrhizal networks are sensitive to mechanical damage, this may suggest that fungal communities were persistent or able to recover within the 7 (birch) to 16 (aspen) years after plantation [43,68]. The higher MF OTU richness in silver birch compared to hybrid aspen might be related to the younger age and design of the trials (Table 1), which had more open canopies with less shading, thus favoring herbs as additional plant hosts for MF [37,69,70]. The generally low dominance of specific MF in hybrid aspen (Figure 2) may indicate their even competitiveness in reclaimed agricultural soils and explain the rapid recovery of MF after the disturbance (establishment and planting of trees), indicating the good environmental quality of the sites [44,71].

4.2. Mycorrhizal Diversity in Silver Birch and Hybrid Aspen

The MF formed only a fraction of the fungal communities in the roots of deciduous trees (Figure 2), similar to other studies within the region [37,44,45]. However, hybrid aspen displayed typical ectomycorrhizal species associated with the *Populus rhizobiome*, e.g., *Laccaria, Inocybe, Hebeloma, Cortinarius,* and *Tuber*, as found in several studies [68,72–75]. Considering the abundance of common MF in remote stands (Figure 2), the most abundant OTUs likely represented regional specifics of root fungal communities [18,36,37,39,45]. Among the MF (Figure 2), arbuscular (endo)mycorrhizas were less abundant compared to ectomycorrhizas. This was also found in other studies [37,45,69] and is typically related to differences in biomass and available space [38,39,42]. Although MF communities in a plantation (represented by hybrid aspen) tend to be more specific compared to forest habitats (silver birch) [37,45], several MF species were actually associated with both tree hosts—*Tormentella, Agaricales, Rissula, Hyaloscypha* (Figure 2). Even though the occurrences of most common taxa were comparable (Figure 2), the species and local dependence of MF communities [38,44,45] was explicitly highlighted by the ordination (Figures 3 and 4) and indicated shifts in certain ecto- and endomycorrhizal species abundances in a host- and clone-dependent manner.

4.3. Clonal Effects on Mycorrhizal Communities

The abundance of endomycorrhiza, as well as the complexity of MF communities, has been positively related to productivity in intensively managed ecosystems [35,36,70], suggesting synergic interactions [18,69]. The clonal differences in the per-tree diversity (H′) of MF for hybrid aspen (Table 1) partially complied with the hypothesis; however, the relationships were inverse, with the superior clone showing a less diverse MF community. Probably, the higher productivity of a host resulted in an advantage for specific MF taxa, allowing them to outcompete others [34,39,76,77].

Given that the birch clones "54-146-143" and "55-875" both showed equally high field performance, differences in MF community diversity were not expected. Interestingly, the clonal differences in MF communities for silver birch were significant (Figure 3, Table 2) and might be related to the genetic differences of the hosts, while the hybrids of aspen were progenies of the sole mother tree [12]. This might also explain the lower local variability of MF communities, as indicated by the effect of the block (Table 2). Since the within-stand diversity of MF communities has been related to the productivity and sustainability of stands [36], this points to the positive effects of genetic diversity of the tree clones within a plantation on mycorrhizal diversity [13].

Furthermore, the discriminant analysis suggested that the clonal differences in mycorrhizal communities were small ($\leq$12% in data variation; Figure 4) and were related to the abundance of ecto- and endomycorrhizal taxa. The MF taxa, which differentiated the clones (Figures 3 and 4), are commonly found within the Baltic region [37,45,78]; nevertheless, such specifics suggested that studied MF exhibit some preference for hosts, likely resulting in spatial heterogeneity of distribution [38]. Although some clone-specific taxa were identified for hybrid aspen (Figure 4), the abundance of those taxa was generally low, suggesting a lack of clear relationships between specific MF and the productivity of the host [34]. Nevertheless, the rare *Boletus satanas* [79], despite showing a low number of reads (Figures 3 and 4), was specifically present on most of the ramets of the moderately productive clone "43", suggesting the conservation potential of hybrid aspen [37].

Generally, the study hypothesis that differences in MF communities are strongly associated with different field performances of tree clones was partially rejected, rather suggesting a top-down effect of host genetic differences on mycorrhizal community composition. Nevertheless, tree yield may be associated with particular fungal genotypes rather than genus- or species-level OTUs detected by universal fungal primers *ITS9mun* and *ITS4ngsUni* [55]. For example, AMS1 and AML2 primers may be better at resolving intra-specific differences among mycorrhizal fungi ecotypes [80,81]. Therefore, comparing different amplicon data and benchmarking long-read platforms such as Oxford Nanopore with traditionally more robust shorter-read technologies like Illumina MiSeq may be useful for accurately assessing MF community diversity in finer detail.

4.4. Implications for Sustainability and Functional Studies of Common Mycorrhizal Networks

The effects of mycorrhizal communities may not be limited to yields only, but likely extend to environmental resilience and resistance to pests and diseases [39,77]. The observed specialization of the MF communities with host genotypes (Figures 3 and 4) appears to increase the overall genetic diversity across forest stands and could potentially counteract the increasing environmental hazards [5], while also highlighting the risks for large-scale clonal forestry [9]. Therefore, for maximizing positive effects on soil microbial diversity, genetic differences of forest regenerative material need to be considered [34]. The clonal specifics in MF communities appear to be generalists (Figure 3), suggesting their adaptability [39,77] and the potential of further exploring the clonal genetic diversity within trees, such as silver birch.

Moreover, the ecto- and, especially, endomycorrhizal species are likely to be involved in inter-plant communication networks [19,20] and signal transfer between host plants [17,23,24]. Even though their abundance was relatively low (Figures 1 and 2), the presence of endomycorrhizal and endophytic genera *Hymenoscyphys, Hyalosscypha, Ceratobasidium, Serendipita,* and *Pezoloma* in birch and aspen rhyzobiomes was confirmed. Furthermore, in vitro propagated isogenic material of silver birch and hybrid aspen clones can be an excellent host system for future functional studies on inter-plant defense signals mediated by common mycorrhizal networks of endemic field-isolated fungi.

5. Conclusions

The MF communities on the roots of clones of silver birch and hybrid aspen prevailingly consisted of the generalist taxa, among which the majority were ectomycorrhizal. The clonal differences in the composition of MF communities were weak, yet significant for the genetically distinct silver birch clones. In contrast, hybrid aspen clones, which were progenies of a common mother tree, did not display differences in MF communities, but showed higher species richness in the clone with a moderate field performance. The estimated clonal differences suggested that MF communities display some preferences for host genotypes. Accordingly, clonal preferences of MF communities should be considered to maximize the positive effects of inoculation of forest reproductive material, particularly as a higher number of clones is recommended for the establishment of more resilient tree plantations. Still, the level of the detected MF taxa varied due to the application of universal primers, and some specific relationships might be overlooked, suggesting that the detection of MF taxa using more specific primers might be needed. Also, an assessment of clonal differences across edaphic and age gradients based on orthogonal design is needed.

Supplementary Materials: The following supporting information can be downloaded at https://www.mdpi.com/article/10.3390/f15122123/s1: Figure S1: Climatic diagram depicting mean (±standard deviation) monthly temperature (lines) and precipitation sums (bars) for the trial sites between 1992 and 2022; Figure S2: Amplification of 900 bp region between ribosomal SSU and LSU genes using ITS4ungsUNI and ITS9Mun primer pair. Amplification performed with 20–30 PCR cycles on DNA extracted from root samples. The amplicon includes ITS1, 5.8S, and ITS2 regions. NK—negative (no template) control; Figure S3: The sequencing run summary: number of reads for the studied clones of silver birch and hybrid aspen. The lines represent the median, the boxes represent the quartiles (first and third), and the whiskers show the range of the subsets (up to 1.5 times the interquartile distance). The raw read number before data decontamination is shown; Figure S4. The sequencing run summary: number of reads for the studied clones of silver birch and hybrid aspen. The lines represent median, the boxes represent the quartiles (first and third) and the whiskers show the range of the subsets (up to 1.5 times the interquartile distance). The raw read number before data decontamination is shown; Table S1: List of mycorrhizal OTUs detected on roots of silver birch and hybrid aspen.

Author Contributions: Conceptualization: R.M., Z.O. and D.R.; methodology: R.M., Z.O., D.R., K.B. and D.K.; investigation: R.M., Z.O., K.T.B., M.B., B.K., K.B., M.R., B.K. and D.K.; formal analysis: R.M., Z.O., D.R. and K.B.; data curation: K.B., M.R., B.K. and M.B. ; writing—original draft preparation: R.M. and Z.O.; writing—review and editing: R.M., Z.O., D.R., K.B. and M.Z.; visualization: R.M. and K.B.; supervision: Z.O., D.R. and M.Z.; project administration: Z.O. and M.Z.; funding acquisition: Z.O. All authors have read and agreed to the published version of the manuscript.

Funding: This work was supported by the Latvian Council of Science Fundamental and Applied Research grant lzp-2022/1-0283 and the University of Latvia Foundation (SIA MikroTik donation).

Data Availability Statement: Data are available with the NCBI Bioproject No. PRJNA1182978.

Conflicts of Interest: The authors declare no conflicts of interest.

References

1. Buras, A.; Menzel, A. Projecting tree species composition changes of European forests for 2061–2090 under RCP 4.5 and RCP 8.5 scenarios. *Front. Plant Sci.* **2019**, *9*, N1986. [CrossRef] [PubMed]
2. Dubois, H.; Verkasalo, E.; Claessens, H. Potential of birch (*Betula pendula* Roth and *B. pubescens* Ehrh.) for forestry and forest-based industry sector within the changing climatic and socio-economic context of Western Europe. *Forests* **2020**, *11*, N336. [CrossRef]
3. Jansson, G.; Hansen, J.K.; Haapanen, M.; Kvaalen, H.; Steffenrem, A. The genetic and economic gains from forest tree breeding programmes in Scandinavia and Finland. *Scand. J. For. Res.* **2017**, *32*, 273–286. [CrossRef]
4. Allen, C.D.; Breshears, D.D.; McDowell, N.G. On underestimation of global vulnerability to tree mortality and forest die-off from hotter drought in the Anthropocene. *Ecosphere* **2015**, *6*, 1–55. [CrossRef]
5. Seidl, R.; Thom, D.; Kautz, M.; Martin-Benito, D.; Peltoniemi, M.; Vacchiano, G.; Wild, J.; Ascoli, D.; Petr, M.; Honkaniemi, J.; et al. Forest disturbances under climate change. *Nat. Clim. Change* **2017**, *7*, 395–402. [CrossRef]
6. Aitken, S.N.; Bemmels, J.B. Time to get moving: Assisted gene flow of forest trees. *Evol. Appl.* **2016**, *9*, 271–290. [CrossRef] [PubMed]
7. Girdziušas, S.; Löf, M.; Hanssen, K.H.; Lazdiņa, D.; Madsen, P.; Saksa, T.; Liepiņš, K.; Fløistad, I.S.; Metslaid, M. Forest regeneration management and policy in the Nordic–Baltic region since 1900. *Scand. J. For. Res.* **2021**, *36*, 513–523. [CrossRef]
8. Gailis, A.; Kārkliņa, A.; Purviņš, A.; Matisons, R.; Zeltiņš, P.; Jansons, Ā. Effect of breeding on income at first commercial thinning in silver birch plantations. *Forests* **2020**, *11*, N327. [CrossRef]
9. Galeano, E.; Bousquet, J.; Thomas, B.R. SNP-based analysis reveals unexpected features of genetic diversity, parental contributions and pollen contamination in a white spruce breeding program. *Sci. Rep.* **2021**, *11*, 4990. [CrossRef]
10. Matisons, R.; Jansone, D.; Elferts, D.; Schneck, V.; Kowalczyk, J.; Wojda, T.; Jansons, Ā. Silver birch shows nonlinear responses to moisture availability and temperature in the eastern Baltic Sea region. *Dendrochronologia* **2022**, *76*, 126003. [CrossRef]
11. Tullus, A.; Lukason, O.; Vares, A.; Padari, A.; Lutter, R.; Tullus, T.; Karoles, K.; Tullus, H. Economics of hybrid aspen (*Populus tremula* L. × *P. tremuloides* Michx.) and silver birch (*Betula pendula* Roth.) plantations on abandoned agricultural lands in Estonia. *Balt. For.* **2012**, *18*, 288–298.

12. Zeps, M.; Adamovics, A.; Smilga, J.; Sisenis, L. Productivity and quality of hybrid aspen at the age of 18 years. *Res. Rural. Dev.* **2016**, *2*, 55–61.
13. Kelblerova, R.; Dvořák, J.; Korecký, J. Genetic diversity maximization as a strategy for resilient forest ecosystems: A case study on Norway spruce. *Forests* **2022**, *13*, 489. [CrossRef]
14. Monson, R.K.; Trowbridge, A.M.; Lindroth, R.L.; Lerdau, M.T. Coordinated resource allocation to plant growth–defense tradeoffs. *New Phytol.* **2022**, *233*, 1051–1066. [CrossRef] [PubMed]
15. Smilga, J.; Zeps, M.; Sisenis, L.; Kalnins, J.; Adamovics, A.; Donis, J. Profitability of hybrid aspen breeding in Latvia. *Agron. Res.* **2015**, *13*, 430–435.
16. Li, T.; Holopainen, J.K.; Kokko, H.; Tervahauta, A.I.; Blande, J.D. Herbivore-induced aspen volatiles temporally regulate two different indirect defences in neighbouring plants. *Funct. Ecol.* **2012**, *26*, 1176–1185. [CrossRef]
17. Rasheed, M.U.; Brosset, A.; Blande, J.D. Tree communication: The effects of "wired" and "wireless" channels on interactions with herbivores. *Curr. For. Rep.* **2023**, *9*, 33–47. [CrossRef]
18. Yu, L.; Zhang, W.; Geng, Y.; Liu, K.; Shao, X. Cooperation with arbuscular mycorrhizal fungi increases plant nutrient uptake and improves defences against insects. *Front. Ecol. Evol.* **2022**, *10*, 833389. [CrossRef]
19. Bais, H.P.; Park, S.W.; Weir, T.L.; Callaway, R.M.; Vivanco, J.M. How plants communicate using the underground information superhighway. *Trends Plant Sci.* **2004**, *9*, 26–32. [CrossRef]
20. Karban, R. Plant communication. *Annu. Rev. Ecol. Evol. Syst.* **2021**, *52*, 1–24. [CrossRef]
21. Ninkovic, V.; Markovic, D.; Rensing, M. Plant volatiles as cues and signals in plant communication. *Plant. Cell Environ.* **2021**, *44*, 1030–1043. [CrossRef] [PubMed]
22. Wang, N.Q.; Kong, C.H.; Wang, P.; Meiners, S.J. Root exudate signals in plant–plant interactions. Plant. *Cell Environ.* **2021**, *44*, 1044–1058. [CrossRef]
23. Gorzelak, M.A.; Asay, A.K.; Pickles, B.J.; Simard, S.W. Inter-plant communication through mycorrhizal networks mediates complex adaptive behaviour in plant communities. *AoB Plants* **2015**, *7*, plv050. [CrossRef] [PubMed]
24. Simard, S.W.; Beiler, K.J.; Bingham, M.A.; Deslippe, J.R.; Philip, L.J.; Teste, F.P. Mycorrhizal networks: Mechanisms, ecology and modelling. *Fungal Biol. Rev.* **2012**, *26*, 39–60. [CrossRef]
25. Babikova, Z.; Gilbert, L.; Bruce, T.J.; Birkett, M.; Caulfield, J.C.; Woodcock, C.; Pickett, J.A.; Johnson, D. Underground signals carried through common mycelial networks warn neighbouring plants of aphid attack. *Ecol. Lett.* **2013**, *16*, 835–843. [CrossRef] [PubMed]
26. Boyno, G.; Demir, S. Plant-mycorrhiza communication and mycorrhizae in inter-plant communication. *Symbiosis* **2022**, *86*, 155–168. [CrossRef]
27. Figueiredo, A.F.; Boy, J.; Guggenberger, G. Common mycorrhizae network: A review of the theories and mechanisms behind underground interactions. *Front. Fungal Biol.* **2021**, *2*, 735299. [CrossRef]
28. Karst, J.; Jones, M.D.; Hoeksema, J.D. Positive citation bias and overinterpreted results lead to misinformation on common mycorrhizal networks in forests. *Nat. Ecol. Evol.* **2023**, *7*, 501–511. [CrossRef]
29. Kasurinen, A.; Koikkalainen, K.; Anttonen, M.J.; Possen, B.; Oksanen, E.; Rousi, M.; Vapaavuori, E.; Holopainen, T. Root morphology, mycorrhizal roots and extramatrical mycelium growth in silver birch (*Betula pendula* Roth) genotypes exposed to experimental warming and soil moisture manipulations. *Plant Soil* **2016**, *407*, 341–353. [CrossRef]
30. Liziniewicz, M.; Barbeito, I.; Zvirgzdins, A.; Stener, L.-G.; Niemistö, P.; Fahlvik, N.; Johansson, U.; Karlsson, B.; Nilsson, U. Production of genetically improved silver birch plantations in southern and central Sweden. *Silva Fenn.* **2022**, *56*, 10512. [CrossRef]
31. Lojewski, N.R.; Fischer, D.G.; Bailey, J.K.; Schweitzer, J.A.; Whitham, T.G.; Hart, S.C. Genetic basis of aboveground productivity in two native *Populus* species and their hybrids. *Tree Physiol.* **2009**, *29*, 1133–1142. [CrossRef]
32. Zanewich, K.P.; Pearce, D.W.; Rood, S.B. Heterosis in poplar involves phenotypic stability: Cottonwood hybrids outperform their parental species at suboptimal temperatures. *Tree Physiol.* **2018**, *38*, 789–800. [CrossRef] [PubMed]
33. Zeltiņš, P.; Matisons, R.; Gailis, A.; Jansons, J.; Katrevičs, J.; Jansons, Ā. Genetic parameters of growth traits and stem quality of silver birch in a low-density clonal plantation. *Forests* **2018**, *9*, 52. [CrossRef]
34. Ciadamidaro, L.; Girardclos, O.; Bert, V.; Zappelini, C.; Yung, L.; Foulon, J.; Papin, A.; Sébastien Roy, S.; Blaudez, D.; Chalot, M. Poplar biomass production at phytomanagement sites is significantly enhanced by mycorrhizal inoculation. *Environ. Exp. Bot.* **2017**, *139*, 48–56. [CrossRef]
35. Dučic, T.; Berthold, D.; Langenfeld-Heyser, R.; Beese, F.; Polle, A. Mycorrhizal communities in relation to biomass production and nutrient use efficiency in two varieties of Douglas fir (*Pseudotsuga menziesii* var. menziesii and var. glauca) in different forest soils. *Soil Biol. Biochem.* **2009**, *41*, 742–753. [CrossRef]
36. Luo, S.; Phillips, R.P.; Jo, I.; Fei, S.; Liang, J.; Schmid, B.; Eisenhauer, N. Higher productivity in forests with mixed mycorrhizal strategies. *Nat. Commun.* **2023**, *14*, 1377. [CrossRef] [PubMed]
37. Rähn, E.; Lutter, R.; Riit, T.; Tullus, T.; Tullus, A.; Tedersoo, L.; Drenkhan, R.; Tullus, H. Soil mycobiomes in native European aspen forests and hybrid aspen plantations have a similar fungal richness but different compositions, mainly driven by edaphic and floristic factors. *Front. Microbiol.* **2024**, *15*, 1372938. [CrossRef] [PubMed]
38. Davison, J.; Öpik, M.; Zobel, M.; Vasar, M.; Metsis, M.; Moora, M. Communities of arbuscular mycorrhizal fungi detected in forest soil are spatially heterogeneous but do not vary throughout the growing season. *PLoS ONE* **2012**, *7*, e41938. [CrossRef]

39. Dietrich, P.; Ferlian, O.; Huang, Y.; Luo, S.; Quosh, J.; Eisenhauer, N. Tree diversity effects on productivity depend on mycorrhizae and life strategies in a temperate forest experiment. *Ecology* **2023**, *104*, e3896. [CrossRef]
40. Öpik, M.; Moora, M.; Liira, J.; Zobel, M. Composition of root-colonizing arbuscular mycorrhizal fungal communities in different ecosystems around the globe. *J. Ecol.* **2006**, *94*, 778–790. [CrossRef]
41. Rozek, K.; Rola, K.; Błaszkowski, J.; Leski, T.; Zubek, S. How do monocultures of fourteen forest tree species affect arbuscular mycorrhizal fungi abundance and species richness and composition in soil? *For. Ecol. Manag.* **2020**, *465*, 118091. [CrossRef]
42. Suz, L.M.; Barsoum, N.; Benham, S.; Dietrich, H.; Fetzer, K.D.; Fischer, R.; García, P.; Gehrman, J.; Kristöfel, F.; Manninger, M.; et al. Environmental drivers of ectomycorrhizal communities in Europe's temperate oak forests. *Mol. Ecol.* **2014**, *23*, 5628–5644. [CrossRef] [PubMed]
43. Bierza, W.; Bierza, K.; Trzebny, A.; Greń, I.; Dabert, M.; Ciepał, R.; Trocha, L.K. The communities of ectomycorrhizal fungal species associated with *Betula pendula* Roth and *Pinus sylvestris* L. growing in heavy-metal contaminated soils. *Plant Soil* **2020**, *457*, 321–338. [CrossRef]
44. Hrynkiewicz, K.; Szymańska, S.; Piernik, A.; Thiem, D. Ectomycorrhizal community structure of *Salix* and *Betula* spp. at a saline site in central Poland in relation to the seasons and soil parameters. *Water Air Soil Pollut.* **2015**, *226*, 99. [CrossRef] [PubMed]
45. Lutter, R.; Riit, T.; Agan, A.; Rähn, E.; Tullus, A.; Sopp, R.; Ots, K.; Kaivapalu, M.; Täll, K.; Tullus, T.; et al. Soil fungal diversity of birch plantations on former agricultural land resembles naturally regenerated birch stands on agricultural and forest land. *For. Ecol. Manag.* **2023**, *542*, 121100. [CrossRef]
46. Rudawska, M.; Kujawska, M.; Leski, T.; Janowski, D.; Karliński, L.; Wilgan, R. Ectomycorrhizal community structure of the admixture tree species *Betula pendula, Carpinus betulus*, and *Tilia cordata* grown in bare-root forest nurseries. *For. Ecol. Manag.* **2019**, *437*, 113–125. [CrossRef]
47. Ruotsalainen, A.L.; Markkola, A.M.; Kozlov, M.V. Mycorrhizal colonisation of mountain birch (*Betula pubescens* ssp. czerepanovii) along three environmental gradients: Does life in harsh environments alter plant-fungal relationships? *Environ. Monit. Assess.* **2009**, *148*, 215–232. [CrossRef]
48. Mangeot-Peter, L.; Tschaplinski, T.J.; Engle, N.L.; Veneault-Fourrey, C.; Martin, F.; Deveau, A. Impacts of soil microbiome variations on root colonization by fungi and bacteria and on the metabolome of *Populus tremula* × *alba*. *Phytobiomes J.* **2020**, *4*, 142–155. [CrossRef]
49. Avis, P.G. The forest typology of Latvia: An overview and comparison. *Proc. Latv. Acad. Sci. Sect. B Nat. Exact Appl. Sci.* **1997**, *51*, 195–199.
50. Kottek, M.; Grieser, J.; Beck, C.; Rudolf, B.; Rubel, F. World map of the Köppen-Geiger climate classification updated. *Meteorol. Z.* **2006**, *15*, 259–263. [CrossRef]
51. Harris, I.; Osborn, T.J.; Jones, P.; Lister, D. Version 4 of the CRU TS monthly high-resolution gridded multivariate climate dataset. *Sci. Data* **2020**, *7*, 109. [CrossRef] [PubMed]
52. Meier, H.E.M.; Kniebusch, M.; Dieterich, C.; Gröger, M.; Zorita, E.; Elmgren, R.; Myrberg, K.; Ahola, M.P.; Bartosova, A.; Bonsdorff, E.; et al. Climate change in the Baltic Sea region: A summary. *Earth Syst. Dyn.* **2022**, *13*, 457–593. [CrossRef]
53. Šēnhofa, S.; Zeps, M.; Gailis, A.; Kāpostiņš, R.; Jansons, Ā. Development of stem cracks in young hybrid aspen plantations. *For. Stud.* **2016**, *65*, 16–23. [CrossRef]
54. Klavina, D.; Tedersoo, L.; Agan, A.; Zaluma, A.; Bitenieks, K.; Polmanis, K.; Daugaviete, M.; Gaitnieks, T.; Drenkhan, R. Effect of stand thinning, former land use and individual tree parameters on wood inhabiting fungal community composition in young living Norway spruce. *Fungal Ecol.* **2023**, *65*, 101281. [CrossRef]
55. Tedersoo, L.; Lindahl, B. Fungal identification biases in microbiome projects. *Environ. Microbiol. Rep.* **2016**, *8*, 774–779. [CrossRef]
56. Vetrovsky, T.; Baldrian, P.; Morais, D. SEED 2: A user-friendly platform for amplicon high-throughput sequencing data analyses. *Bioinformatics* **2018**, *34*, 2292–2294. [CrossRef]
57. Rognes, T.; Flouri, T.; Nichols, B.; Quince, C.; Mahé, F. VSEARCH: A versatile open source tool for metagenomics. *PeerJ* **2016**, *4*, e2584. [CrossRef]
58. Bengtsson-Palme, J.; Ryberg, M.; Hartmann, M.; Branco, S.; Wang, Z.; Godhe, A.; Wit, P.D.; Sánchez-García, M.; Ebersberger, I.; de Sousa, F.; et al. Improved software detection and extraction of ITS1 and ITS 2 from ribosomal ITS sequences of fungi and other eukaryotes for analysis of environmental sequencing data. *Methods Ecol. Evol.* **2013**, *4*, 914–919. [CrossRef]
59. Curren, E.; Yoshida, T.; Kuwahara, V.S.; Leong, S.C.Y. Rapid profiling of tropical marine cyanobacterial communities. *Reg. Stud. Mar. Sci.* **2019**, *25*, 100485. [CrossRef]
60. Nguyen, N.H.; Williams, L.J.; Vincent, J.B.; Stefanski, A.; Cavender-Bares, J.; Messier, C.; Paquette, A.; Gravel, D.; Reich, P.B.; Kennedy, P.G. Ectomycorrhizal fungal diversity and saprotrophic fungal diversity are linked to different tree community attributes in a field-based tree experiment. *Mol. Ecol.* **2016**, *25*, 4032–4046. [CrossRef] [PubMed]
61. le Cao, K.A.; Boitard, S.; Besse, P. Sparse PLS discriminant analysis: Biologically relevant feature selection and graphical displays for multiclass problems. *BMC Bioinform.* **2011**, *12*, 253. [CrossRef] [PubMed]
62. Rohart, F.; Gautier, B.; Singh, A.; Le Cao, K.-A. mixOmics: An R package for 'omics feature selection and multiple data integration. *PLoS Comput. Biol.* **2017**, *13*, e1005752. [CrossRef] [PubMed]
63. R Core Team. *R: A Language and Environment for Statistical Computing*; R Foundation for Statistical Computing: Vienna, Austria, 2024; Available online: https://www.R-project.org/ (accessed on 24 November 2024.).

64. Oksanen, J.; Blanchet, F.G.; Kindt, R.; Legendre, P.; O'hara, R.B.; Simpson, G.L.; Solymos, P.; Stevens, M.H.H.; Wagner, H. *Vegan: Community Ecology Package, R package Version 2.6–6.1*; R Foundation for Statistical Computing: Vienna, Austria, 2024; Available online: https://CRAN.R-project.org/package=vegan (accessed on 24 November 2024.).
65. Tedersoo, L.; Bahram, M.; Põlme, S.; Kõljalg, U.; Yorou, N.S.; Wijesundera, R.; Ruiz, L.V.; Vasco-Palacios, A.M.; Thu, P.Q.; Suija, A.; et al. Global diversity and geography of soil fungi. *Science* **2014**, *346*, 1256688. [CrossRef] [PubMed]
66. Lu, X.; Cao, T.; Nguyen TT, T.; Yuan, H.S. Six new species of *Tomentella* (Thelephorales, Basidiomycota) from tropical pine forests in Central Vietnam. *Front. Microbiol.* **2022**, *13*, 864198. [CrossRef] [PubMed]
67. Gross, N.; Bagousse-Pinguet, Y.L.; Liancourt, P.; Berdugo, M.; Gotelli, N.J.; Maestre, F.T. Functional trait diversity maximizes ecosystem multifunctionality. *Nat. Ecol. Evol.* **2017**, *1*, 0132. [CrossRef]
68. Gottel, N.R.; Castro, H.F.; Kerley, M.; Yang, Z.; Pelletier, D.A.; Podar, M.; Karpinets, T.; Uberbacher, E.; Tuskan, G.A.; Vilgaly, R.; et al. Distinct microbial communities within the endosphere and rhizosphere of *Populus deltoides* roots across contrasting soil types. *Appl. Environ. Microbiol.* **2011**, *77*, 5934–5944. [CrossRef]
69. Ferlian, O.; Goldmann, K.; Eisenhauer, N.; Tarkka, M.T.; Buscot, F.; Heintz-Buschart, A. Distinct effects of host and neighbour tree identity on arbuscular and ectomycorrhizal fungi along a tree diversity gradient. *ISME Commun.* **2021**, *1*, N119. [CrossRef]
70. Qiao, X.; Sun, T.; Lei, J.; Xiao, L.; Xue, L.; Zhang, H.; Jia, J.; Bei, S. Arbuscular mycorrhizal fungi contribute to wheat yield in an agroforestry system with different tree ages. *Front. Microbiol.* **2022**, *13*, 1024128. [CrossRef]
71. Faggioli, V.; Menoyo, E.; Geml, J.; Kemppainen, M.; Pardo, A.; Salazar, M.J.; Becerra, A.G. Soil lead pollution modifies the structure of arbuscular mycorrhizal fungal communities. *Mycorrhiza* **2019**, *29*, 363–373. [CrossRef]
72. Baum, C.; Makeschin, F. Effects of nitrogen and phosphorus fertilization on mycorrhizal formation of two poplar clones (*Populus trichocarpa* and *P. tremula* x *tremuloides*). *J. Plant Nutr. Soil Sci.* **2000**, *163*, 491–497. [CrossRef]
73. Bois, G.; Piché, Y.; Fung, M.Y.P.; Khasa, D.P. Mycorrhizal inoculum potentials of pure reclamation materials and revegetated tailing sands from the Canadian oil sand industry. *Mycorrhiza* **2005**, *15*, 149–158. [CrossRef] [PubMed]
74. Guevara, G.; Bonito, G.; Trappe, J.M.; Cázares, E.; Williams, G.; Healy, R.A.; Schadt, C.; Vilgalys, R. New North American truffles (*Tuber* spp.) and their ectomycorrhizal associations. *Mycologia* **2013**, *105*, 194–209. [CrossRef] [PubMed]
75. Podila, G.K.; Sreedasyam, A.; Muratet, M.A. Populus rhizosphere and the ectomycorrhizal interactome. *Crit. Rev. Plant Sci.* **2009**, *28*, 359–367. [CrossRef]
76. Ceulemans, T.; Van Geel, M.; Jacquemyn, H.; Boeraeve, M.; Plue, J.; Saar, L.; Kasari, L.; Peeters, G.; van Acker, K.; Crauwels, S.; et al. Arbuscular mycorrhizal fungi in European grasslands under nutrient pollution. *Glob. Ecol. Biogeogr.* **2019**, *28*, 1796–1805. [CrossRef]
77. Deng, M.; Hu, S.; Guo, L.; Jiang, L.; Huang, Y.; Schmid, B.; Liu, C.; Chang, P.; Li, S.; Liu, X.; et al. Tree mycorrhizal association types control biodiversity-productivity relationship in a subtropical forest. *Sci. Adv.* **2023**, *9*, eadd4468. [CrossRef]
78. Sudheer, S.; Raal, A.; Jürgenson, A.; Bhat, R. Wild Edible Mushroom Trade in Europe. In *Wild Mushrooms and Health: Diversity, Phytochemistry, Medicinal Benefits, and Cultivation*; CRC Press: Boca Raton, FL, USA, 2023; 194p.
79. Nilson, S.; Persson, O. *Fungi of Northern Europe 1: Larger Fungi (Excluding Gill-Fungi)*; Penguin Book: New York, NY, USA, 1977; p. 104.
80. Kajihara, K.T.; Egan, C.P.; Swift, S.O.; Wall, C.B.; Muir, C.D.; Hynson, N.A. Core arbuscular mycorrhizal fungi are predicted by their high abundance–occupancy relationship while host-specific taxa are rare and geographically structured. *New Phytol.* **2022**, *234*, 1464–1476. [CrossRef]
81. Lee, J.; Lee, S.; Young, J.P.W. Improved PCR primers for the detection and identification of arbuscular mycorrhizal fungi. *FEMS Microbiol. Ecol.* **2008**, *65*, 339–349. [CrossRef]

forests

Article

From Lab to Nursery: Novel Approaches of Seed Disinfection for Managing Pine Pitch Canker Propagation

Luís Fernandes [1,†], Diana S. Paiva [1,2,*,†], Ana C. Silva [3,4,5], Cláudia Fernandes [6], Ana Rita Fernandes [7], Dina Ribeiro [7], Luís Martins [6], Helena Bragança [3,5] and António Portugal [1,2,8]

1 Centre for Functional Ecology (CFE)—Science for People & the Planet, Department of Life Sciences, University of Coimbra (UC), Calçada Martim de Freitas, 3000-456 Coimbra, Portugal; luis.dsfernandes@hotmail.com (L.F.); aportuga@bot.uc.pt (A.P.)
2 FitoLab—Laboratory for Phytopathology, Instituto Pedro Nunes (IPN), Rua Pedro Nunes, 3030-199 Coimbra, Portugal
3 National Institute of Agricultural and Veterinary Research, I.P. (INIAV), Quinta do Marquês, 2780-159 Oeiras, Portugal; anacristina.silva@iniav.pt (A.C.S.); helena.braganca@iniav.pt (H.B.)
4 CESAM and Department of Biology, University of Aveiro, Campus Universitário de Santiago, 3810-193 Aveiro, Portugal
5 GREEN-IT Bioresources for Sustainability, ITQB NOVA Instituto de Tecnologia Química e Biológica António Xavier, Av. da República, 2780-157 Oeiras, Portugal
6 Department of Forestry Sciences and Landscape Architecture, University of Trás-os-Montes and Alto Douro (UTAD), Quinta dos Prados, 5000-801 Vila Real, Portugal; klaus2013@gmail.com (C.F.)
7 Institute for the Conservation of Nature and Forests (ICNF), Avenida da República 16, 1050-191 Lisbon, Portugal
8 TERRA—Associate Laboratory for Sustainable Land Use and Ecosystem Services, Department of Life Sciences, University of Coimbra (UC), Calçada Martim de Freitas, 3000-456 Coimbra, Portugal
* Correspondence: dpaivaphd@gmail.com
† These authors contributed equally to this work.

Abstract: *Fusarium circinatum*, the causative agent of pine pitch canker disease, is a pathogenic fungus that poses a significant threat to pine forests globally. It infects various *Pinus* species, causing resinous cankers, needle discoloration, and tree death. The disease severely impacts forest ecosystems, necessitating cost-effective and environmentally friendly management strategies. Contaminated pine seeds and seedlings are the main pathways for introducing this fungus to disease-free areas. To mitigate this disease and prevent its spread, it is crucial to implement new processes in forest plant production systems that align with the existing conditions of forest nurseries, ensuring effective and sustainable management. With this in mind, a national collaborative study involving 14 Portuguese partners was initiated to develop new prevention and mitigation strategies. In this work, four different treatments—MennoFlorades, Captan, ethanol, and hot water—were tested for their ability to eliminate *F. circinatum* from contaminated *Pinus* seeds in vitro. The most effective treatments were selected for further in vitro assays and real-context nursery germination trials to assess their impacts on seed germination, plant production, and certification. MennoFlorades, Captan, and hot water were tested in the nursery, with hot water showing the most promising results due to its negligible impact on seedlings, eco-friendly nature, ease of implementation, and cost-effectiveness. These findings offer promising prospects for preventing pine pitch canker outbreaks in nurseries and, consequently, in forests.

Keywords: forests; *Fusarium circinatum*; *Pathogenic fungus*; *Pinus* trees; preventive strategies; quarantine pest; seeds

Citation: Fernandes, L.; Paiva, D.S.; Silva, A.C.; Fernandes, C.; Fernandes, A.R.; Ribeiro, D.; Martins, L.; Bragança, H.; Portugal, A. From Lab to Nursery: Novel Approaches of Seed Disinfection for Managing Pine Pitch Canker Propagation. *Forests* **2024**, *15*, 1154. https://doi.org/10.3390/f15071154

Academic Editors: Katarína Pastirčáková and Rostislav Zemek

Received: 5 June 2024
Revised: 27 June 2024
Accepted: 2 July 2024
Published: 3 July 2024

1. Introduction

Fusarium circinatum Nirenberg & O'Donnell, the causal agent of pitch canker disease, is recognized as one of the most severe diseases affecting pine trees, posing a significant threat to forest nurseries and plantations worldwide [1–3]. This pathogenic fungus targets *Pinus* species and *Pseudotsuga menziesii* (Douglas fir), displaying varying degrees of virulence, depending on the host species and prevailing biotic and abiotic conditions [3–7]. The pathogen's impact is pervasive, affecting all stages of pine growth. Common symptoms in seedlings include blight and damping-off, while on young and mature trees, dieback of branches and stems is frequent, often accompanied by conspicuous cankers and resin exudation. Additionally, dieback symptoms in the crown can also occur due to water flow obstruction caused by cankers and the saturation of xylem by excessive resin, potentially strangling the treetop and leading to tree death [2,3,8]. This pathogen was first recorded in the US in 1946, and since then, sporadic outbreaks and epidemics have been reported in numerous countries [4,9,10]. In Europe, occurrences have been reported in Portugal, Spain, France, and Italy among various susceptible *Pinus* species [11–15].

In Portugal, the coniferous forest holds undeniable social, economic, and environmental importance, with native species like maritime pine (*Pinus pinaster*) and stone pine (*Pinus pinea*) dominating the territorial landscape. The maritime pine stands as the third most significant forest species in the country, covering 22.1% of the mainland area. However, it has experienced a declining trend, primarily attributed to rural wildfires, surpassing the expansion of new plantations and further compounded by forest pests like pine pitch canker and pine wood nematode (*Bursaphelenchus xylophilus*), impacting productivity and investment confidence in this species. In recent years, this decrease in area seems to have been slowing down, showcasing the resilience of these ecosystems to disturbances [16]. The maritime pine industry, a major carbon reservoir in the Portuguese forest (90.3 Gg CO_2), accounts for 80% of employment and 88% of industrial companies in the forestry sector, contributing to 52% of the Gross Value Added (GVA) and 3.2% of national exports of goods [16,17]. Besides the maritime pine, the stone pine also holds a significant representation in the national forest, ranking as the fifth forest species in Portugal and covering 6% of the mainland area, which corresponds to over 20% of its global distribution [16]. While other species are also naturally present, they have less prominence. Noteworthy about the stone pine is its potential for use in multifunctional agroforestry systems, ensuring a variety of products and the provision of ecosystem services related to soil and water protection, carbon sequestration, and biodiversity maintenance. Similarly to fruit species, it offers the possibility of orchard cultivation in grafted plantations dedicated to pine nut production. With an estimated pine nut production ranging between 70,000 and 120,000 tons annually, equivalent to a value between EUR 75 million and EUR 128.5 million, exports of this product reach approximately EUR 15 million per year [16,17]. The sustainability of these vital economic sectors relies on forest resources and their effective management, including afforestation and reforestation actions, which require access to high-quality plants free from phytosanitary issues.

Efforts to manage the spread of tree pathogens, including *F. circinatum*, face significant challenges due to the lack of economically viable treatments for large-scale control. Strategies primarily revolve around preventing the pathogen's introduction, early detection, and eradication of outbreaks in previously disease-free areas. Since 2007, *F. circinatum* has been categorized as a quarantine fungus within the European Union (EU), which has imposed strict measures to prevent its transmission through infected materials and to curb disease expansion [18,19]. Consequently, a zero-tolerance policy is enforced. According to the pest risk assessment conducted by the European Food and Environment Safety Agency (EFSA), contaminated pine seeds and seedlings represent the primary pathways for potential fungus spread to disease-free areas [20]. Implementing efficient management measures is crucial to minimize the likelihood of pathogen introduction and reduce the associated costs of eradication and control efforts.

In Portugal, since its initial detection in 2008 in nursery plants [15], *F. circinatum* has been identified in 27 locations across the country, including both nurseries and pine forests, with a higher incidence in the northern and central regions. This has resulted in the destruction of 1.8 million plants and the quarantine of 2500 kg of seeds from host species. The presence of this fungus continues to pose both ecological and economic threats, emphasizing the urgent need to develop mechanisms to prevent or minimize its risk of dispersal. Additionally, as previously mentioned, while current efforts to control the spread of this disease focus heavily on the early reaction to the presence of the fungus to mitigate outbreaks, they should instead be redirected towards preventing infection altogether. In this context, the operational group GO +PrevCRP was established as part of the Rural Development Program PDR2020, bringing together 14 national partners. The action plan, spanning 4 years (2017–2021), aimed to develop new, effective, and environmentally friendly prevention and mitigation strategies across various crucial factors contributing to the spread of the disease within nurseries, such as contaminated irrigation water and substrates, the results of which were recently published [21–24]. These preventive measures must be grounded in experimental findings, ensuring their feasibility and seamless integration into nursery operations without compromising seed germination rates or the development and quality of the resulting plants. Given that the primary means of pitch canker disease dispersal are contaminated pine seeds and seedlings, the main goals of this study were (i) to assess the effectiveness of certain treatments in eliminating *F. circinatum* inoculum from seeds; (ii) to evaluate their impacts on seed germination; and (iii) to apply these experimental findings in practical field settings. This included conducting large-scale experiments in nurseries to gather a wider sample size and assess the feasibility of implementing treatment measures in real-world scenarios. For this purpose, several treatments were selected based on their previously reported fungicide activity. To the best of our knowledge, the efficacy of some of these treatments as seed disinfectants against *F. circinatum* is being tested for the first time, particularly on *Pinus* species commonly found in Portuguese production and conservation forests.

2. Materials and Methods

2.1. Artificial Seed Inoculation

Seeds from different *Pinus* species, specifically *Pinus pinaster*, *P. pinea* and *P. radiata*, potential hosts of *Fusarium circinatum*, were provided by the National Centre for Forest Seeds (CENASEF). *Pinus pinaster* seeds were collected from 80-year-old parental trees, *P. radiata* seeds from 27-year-old parental trees, and *P. pinea* seeds from 53-year-old parental trees. All parental trees were healthy and showed no signs of disease. In a controlled laboratory environment, these seeds underwent artificial inoculation (excluding a portion kept for use as a control in later analyses) and subsequent assessment for successful inoculation following the methodology outlined in the OEPP/EPPO Bulletin (2019) 49 (2), 228–247 [25] for fungal detection in seeds. Given the quarantine status of this fungus in the European Union (EU), this procedure was conducted by the National Institute of Agricultural and Veterinary Research, I.P. (INIAV), a state laboratory designated as the National Reference Laboratory (LNR). Upon confirmation of successful seed inoculation, preliminary seed disinfection treatments were initiated.

2.2. Treatment Selection

The tested treatments included MennoFlorades (Menno Chemie–Vertrieb GMBH, Norderstedt, Germany), Captan 800 WDG (Adama, Lisbon, Portugal), ethanol (Biochem Iberica, Montijo, Portugal) at various concentrations, and hot water treatment at different temperatures (Table 1). This selection was made based on their overall fungicide efficacy, as well as their accessibility and user-friendliness. All products were supplied by Biochem Iberica—Agricultural and Industrial Chemicals, Lda.

Table 1. Details of the treatments used for disinfecting *Pinus* seeds.

Treatment	Concentration/Temperature	Time (min)	No. of Treated Seeds (per *Pinus* Species)
MennoFlorades	4% (v/v)	60	100
Captan	1.9 g/L	5	100
Ethanol	50% (v/v)	5	100
	60% (v/v)	5	100
	70% (v/v)	5	100
Hot Water	52 °C	30	100
	55 °C	30	100
	60 °C	15	100

Action mechanisms of the tested treatments are as follows:

MennoFlorades is a surface disinfectant with proven fungicidal efficacy, including against various *Fusarium* species. Its active ingredient, benzoic acid (C_6H_5COOH), acts on glycolysis, specifically on phosphofructokinase, which is inhibited by the acidification of the intracellular content caused by the extracellular accumulation of benzoate. This inhibition triggers a cascade effect inhibiting metabolic pathways and a consequent reduction in the ATP concentration and production [26].

Captan (N-trichloromethylthio]-4-cyclohexene-1,2-dicarboximide) is a widely used agricultural fungicide, acting at the mitochondrial level as an uncoupler of oxidative phosphorylation. Its fungicidal capabilities also extend to its affinity for sulfhydryl groups common in mitochondrial enzymes such as NADH dehydrogenase and β–hydroxybutyrate dehydrogenase, leading to their inhibition and an ability to inhibit other crucial processes in respiration [27].

Ethanol, like other alcohol molecules, serves as a base for many fungicides and exhibits antifungal activity on its own. Its primary target is the fungal cell membrane, inducing stress by interacting at the polar–apolar interface, weakening the hydrophobic barrier, and allowing free exchange of polar molecules, thereby disrupting the membrane structure and function. Additionally, it has other effects, such as protein denaturation and inhibition of nutrient uptake [28].

Hot water treatment is a technique already used to combat various phytopathogenic fungi [29] using high temperatures to inhibit fungal development (e.g., through protein denaturation and cell membrane damage, among other mechanisms), and has shown promising efficacy in eliminating *Fusarium circinatum* from *Pinus radiata* seeds [30]. According to Liao et al. [31], the lethal temperatures for conidia and mycelium are reported as 52 °C and 55 °C, respectively.

2.3. Treatment Application

The treatments were prepared and applied as specified in Table 1 provided below.

Sterile distilled water (dH_2O) was employed to prepare the treatment solutions, and was used for the hot water treatments. The treatments were conducted in sterile glass containers, with seeds being agitated during the application time, following the conditions as indicated in Table 1. In temperature-dependent treatments, the temperature was adjusted and maintained using magnetic stirring hot plates. After the total treatment time, the seeds were collected, allowed to dry at room temperature for 10 min, and then plated onto Dichloran Chloramphenicol Peptone Agar (DCPA) medium plates supplemented with 0.5 mg/mL of streptomycin following the methodology outlined in the OEPP/EPPO Bulletin (2019) 49 (2), 228–247 for *F. circinatum* detection in seeds [25]. Additionally, for each *Pinus* species considered in the study, two controls were included: a positive control (C+), where 100 artificially contaminated seeds (see Section 2.1) were left untreated and were plated under the same conditions; and a negative control (C−), where 100 uncontaminated and untreated seeds were plated under the same conditions (Figure 1).

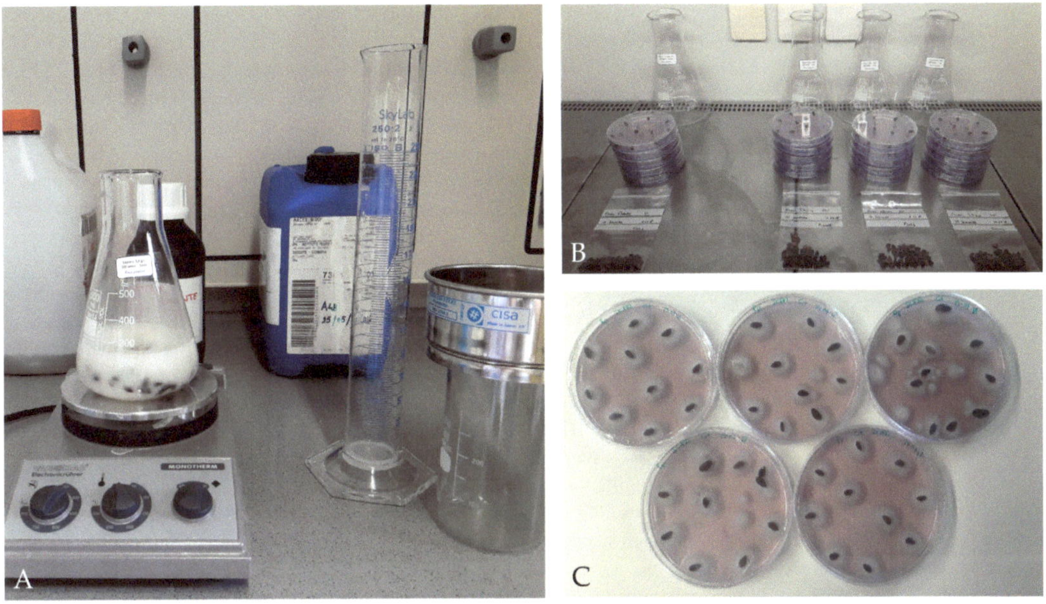

Figure 1. Stages of *Pinus pinaster* seed disinfection: (**A**) treatment with 1.9 g/L Captan; (**B**) plating on DCPA media following treatment application; (**C**) observation of *Fusarium circinatum* growth on the positive control after 5 days of incubation.

The disinfection percentage was calculated by considering zero contaminated seeds (zero *F. circinatum* isolates) as achieving 100% successful disinfection and applying the following formula when the count was greater than zero: % of disinfection = (no. of contaminated seeds/total plated seeds) × 100. The results of the disinfection treatment were analyzed via the non-parametric Kruskal–Wallis test and Dunn's post hoc analysis with $p < 0.01$ using PAST software, version 4.03 [32].

2.4. Fungal Isolation and Identification

For all treatments, the inoculated media plates were aerobically incubated in the dark at 25 °C for 30 days. All emerging colonies were isolated into axenic cultures in duplicate onto Potato Dextrose Agar (PDA, Difco™, Sparks, MD, USA) plates suitable for isolation, and Synthetic Nutrient Agar (SNA) plates suitable for identifying *F. circinatum* based on morphological features. These cultures were aerobically incubated in the dark at 25 °C for a minimum of 10 days. After this incubation period, the number of *F. circinatum* isolates and the number of isolates of other fungi were documented, and photographic records were also captured. The identification of *F. circinatum* was carried out according to the protocol following the EPPO guidelines [OEPP/EPPO Bulletin (2019) 49 (2), 228–247, protocol available at https://gd.eppo.int/taxon/GIBBCI/documents (accessed on 23 April 2024)] [25]. Colonies displaying morphologies significantly different from *Fusarium* spp. were also identified using a combination of molecular and morphological analyses following the standard procedures described below.

Morphological and Molecular Identification of Non-Fusarium Species

Genomic DNA was extracted from PDA pure cultures with the REDextract–N.Amp™ Plant PCR Kit (Sigma Aldrich, St. Louis, MO, USA), with several modifications. A small portion (~1 mm^2) of the colonies was scraped from the agar surface into a PCR–style microtube, submerged in 20 µL of extraction solution, and incubated in the thermocycler using the following protocol: 94 °C for 10 min, followed by 60 °C for 13 min and 10 °C for

15 min. After the incubation, 20 μL of dilution solution was added, and the resulting mixture was vortexed for 2 min [33,34]. The obtained DNA was used for the amplification of the ITS–rDNA region by PCR using the fungal universal primer pair ITS1–F and ITS4 [35,36]. PCR reactions, consisting of a final amplification volume of 25 μL, with 12.5 μL of NZYTaq Green Master Mix (NZYTech™, Lisbon, Portugal), 1 μL of each primer (10 mM), 9.5 μL of ultra-pure water, and 1 μL of template DNA, were performed using an ABI GeneAmp™ 9700 PCR System (Applied Biosystems, Waltham, MA, USA) with the following conditions: initial denaturation at 95 °C for 5 min, followed by 35 cycles of denaturation at 94 °C for 30 s, annealing at 52 °C for 30 s, and extension at 72 °C for 1 min, with a final extension at 72 °C for 8 min. Visual confirmation of the overall amplification of the ITS region was performed using agarose gel electrophoresis (1.2%) stained with GreenSafe Premium (NZYTech™, Lisbon, Portugal) and visualized in an Molecular Imager Bio–Rad Gel Doc XR™ (Bio–Rad, Hercules, CA, USA). The obtained amplicons were purified with the EXO/SAP Go PCR Purification Kit (GRISP, Porto, Portugal), following the manufacturer's recommendations, and sequenced using an ABI 3730xl DNA Analyzer system (96 capillary instruments) at STABVIDA, Portugal.

Sequence reads were quality checked and trimmed at both ends using BioEdit Sequence Alignment Editor© v.7.2.5 (https://bioedit.software.informer.com/download/ (accessed on 7 January 2019)). The newly generated sequences were deposited in the GenBank database, and their accession numbers are included in Supplementary Table S1. Similarity searches were performed, and sequences were queried against the National Center of Biotechnology Information (NCBI) nucleotide database using a BLASTn search algorithm [37]. For genera like *Penicillium* and *Aspergillus*, where the ITS region does not provide sufficient differentiation at the species level, isolates were identified only to the genus level. To ensure accurate species identification, molecular results were verified through comprehensive macroscopic and microscopic analysis of taxonomic traits, resorting to Index Fungorum (www.indexfungorum.org (accessed on 20 January 2019)) and Mycobank (https://www.mycobank.org/ (accessed on 20 January 2019)) [38,39].

2.5. Nursery Germination Assays

The best-performing treatments identified in the previous disinfection phase were chosen for nursery germination trials. Collaborative field trials were conducted in partnership with nurseries involved in the project to accommodate larger-scale assessments of the selected treatments. Field trials exclusively employed *P. pinaster* seeds. Although both *P. pinea* and *P. radiata* were included in the preliminary laboratory testing, their seeds were unavailable during the nursery experimental period due to low seed stocks of these species and the necessity to ensure seed availability for nurseries and private stakeholders outside of this project. Nonetheless, as previously noted, *P. pinaster* is one of the most prevalent species in Portuguese forests, thus representing the highest interest in field trials. Certified seeds intended for commercial distribution (without inoculation) were distributed by CENACEF to the nurseries designated for the trials, and the procedures were carried out in accordance with the specifications outlined in Table 2. After applying treatments to replicate laboratory procedures, the treated seeds were planted in containers filled with substrate, with one seed per container cell. Regular irrigation was provided, and seed germination was monitored at one and two months post-sowing (Figure 2).

Forest reproductive materials (FRM) must possess a genetic quality that enables them to originate stable, adapted, resistant, and resilient forest ecosystems capable of addressing both biotic and abiotic challenges that forests face. General standards applicable to the production and commercialization of FRM are defined in accordance with the applicable Portuguese legislation, Decree–Law No. 205/2003, of 12 September, amended and republished by Decree–Law No. 13/2019, of 21 January. To be deemed marketable, a batch of plants must ensure that 95% of its plants exhibit intact and marketable quality and are devoid of any of the following defects: injuries unrelated to pruning or caused by damage during uprooting; absence of buds with potential for producing a main shoot;

presence of multiple stems; malformed root systems; indications of desiccation, overheating, mold, rot, or other harmful organisms; or unbalanced growth. Additional criteria, such as the maximum age and size, must also be considered for plant marketing, with specific limits defined for height (ranging from 7 to 45 cm, depending on the species) and for the minimum root collar diameter (ranging from 2 to 3 mm, also species-dependent). To ensure compliance with the established general standards for FRM production and commercialization of plants produced from seeds subjected to the disinfection treatments, the procedures used by ICNF, I.P., for plant certification, considering the aforementioned parameters, were applied. For that, plants from each trial were observed 7 months after sowing to assess their average height, average collar diameter, and the presence or absence of any deficiencies. Statistical analysis of the data obtained in nursery A for the germination results was undertaken via a *t* test and via one-way analysis of variance (ANOVA) for nursery B. For the latter, whenever significant differences were found ($p < 0.05$), post hoc Tukey's pairwise test was performed to further elucidate differences between treatments at a significance level of $\alpha = 0.05$ using PAST software, version 4.03 [32].

Table 2. Number of containers per treatment, number of cells per container, and total number of cells per treatment used in the nursery germination assays for *Pinus pinaster* seeds.

Nursery	Treatment		No. of Containers per Treatment	No. of Cells per Container	Total No. of Cells per Treatment
A	MennoFlorades	4%—1 h	25	54	1350
	Control	N/A	25	54	1350
B	Captan	1.9 g/L—5 min	24	60	1440
	Hot Water	60 °C—15 min	24	60	1440
	Control	N/A	24	60	1440

N/A—not applicable.

Figure 2. Records of the different stages of the field trials: (**A**) preparation of containers for sowing; (**B**) preparation of treatments; and (**C**) plant certification process.

2.6. *In Vitro Germination Assay*

Concurrently with the field trials, the top-performing treatments identified in the previous disinfection stage were also subjected to a preliminary germination assay to evaluate their impacts on *P. pinaster* seed germination. This assessment was conducted by CENASEF in accordance with the standards of the International Seed Testing Association (ISTA, https://www.seedtest.org/en/home.html (accessed on 4 June 2020)). The general conditions of the germination tests were as follows: 200 seeds in sets of 100 seeds each; filter paper substrate soaked in water; temperature set at 20 °C; humidity maintained at 85% inside the germination chamber; photoperiod of 8 h of light followed by 16 h of darkness; and a test duration of 35 days with counts performed every 7 days. Finally, the germination rate was assessed for all treatments and statistically analyzed using one-way analysis of variance (ANOVA).

An important note regarding the field trials and their respective germination tests must be highlighted. Both trials were conducted during the period declared as a pandemic due to the ongoing COVID–19 crisis. Consequently, experiments involving ethanol at various concentrations, previously conducted in the laboratory, could not be reproduced in the field because of limited access to ethanol. Therefore, although promising results were obtained in the laboratory, it was decided to retain this information, despite the inability to conduct further tests. Consequently, no additional conclusions could be drawn regarding its impact on germination or its performance in the field.

3. Results and Discussion

3.1. *Effect of Treatments on Fusarium Circinatum-Contaminated Pinus Seeds*

The isolation of *F. circinatum* from seeds was assessed for each *Pinus* species and treatment. The resulting data, showing the average number of contaminated seeds for each combination, are presented in Figure 3 and in Supplementary Table S2.

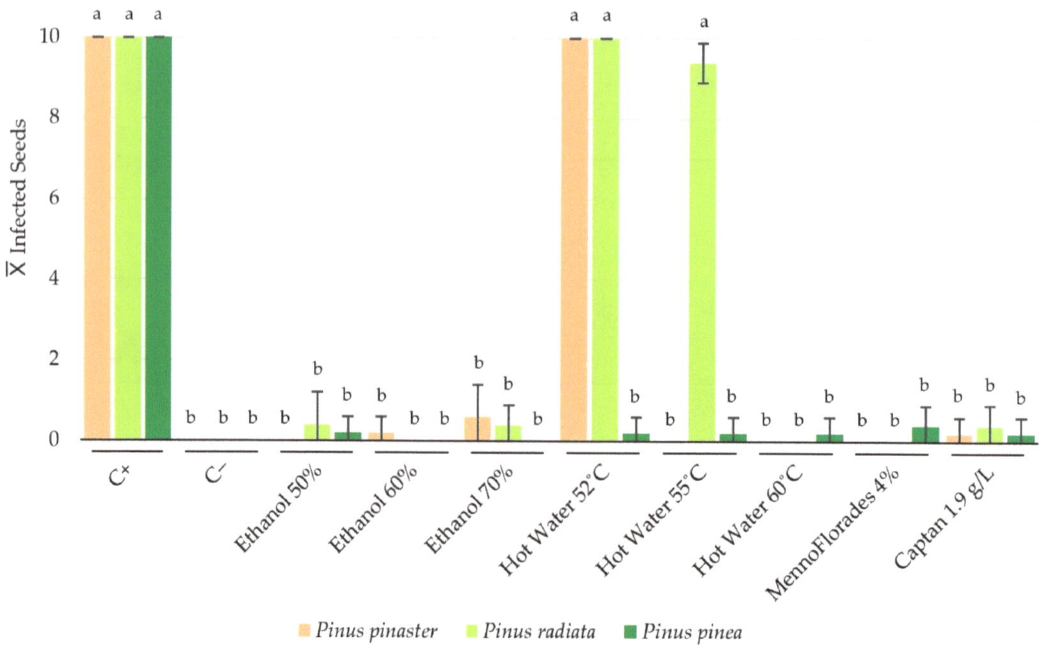

Figure 3. Average numbers of *Pinus pinaster, P. radiata* and *P. pinea* seeds contaminated with *F. circinatum* after treatment application. Error bars shown in the graph correspond to the standard deviations. For each species, data (n = 10) associated with different letters represent significantly different results in accordance with Dunn's post hoc test, $p < 0.01$.

As observed, multiple treatments per species achieved a 100% seed disinfection rate, significantly reducing *F. circinatum* contamination. For *P. pinaster*, treatments with 50% ethanol (v/v), hot water at 55 °C and 60 °C, and MennoFlorades at a 4% (v/v) concentration achieved complete disinfection in all replicates. The statistical analysis indicated that there were no significant differences between the following treatments: ethanol at 60% and 70% concentrations, 1.9 g/L Captan, and the negative control. Additionally, significant differences were found between the previous group and the group composed of the positive control and hot water at 52 °C treatment ($p < 0.01$). A similar trend was seen in *P. radiata* seeds, where hot water at 60 °C, MennoFlorades at 4% (v/v), and 60% ethanol (v/v) all showed no *F. circinatum* growth. The data analysis confirmed that there were no significant differences between the following treatments: ethanol at 50% and 70%, 1.9 g/L Captan, and the negative control. Furthermore, significant differences were observed between the previous group and a second data group comprising hot water at 52 °C and 55 °C, and the positive control ($p < 0.01$). For *P. pinea* seeds, the statistical analysis identified two significantly different groups: one solely comprising the positive control and another including all evaluated treatments and the negative control ($p < 0.01$). In the latter group, only treatments with 60% and 70% ethanol (v/v) showed no *F. circinatum* colonies. An interesting aspect of these results is the apparent importance of seed morphology. *Pinus pinea* seeds, which are larger and possess smoother surfaces, responded better to all treatments, with each treatment significantly reducing the number of contaminated seeds. In contrast, *P. pinaster* and *P. radiata* seeds, which are smaller and have more rugged surfaces, showed varied responses to two of the three tested temperatures for the hot water treatments. For ethanol, Captan, and MennoFlorades treatments, similar results were observed across all concentrations and *Pinus* species.

Previous studies on disinfection of *Pinus* seeds contaminated with *F. circinatum* have explored various methods, including hydrogen peroxide [40], non-thermal plasma [41], commercial fungicides, antagonistic organisms, thermotherapy, and essential oils [42]. Non-thermal plasma seed disinfection, despite showing promising results, was found to negatively affect *P. radiata* seed germination when exposed for 60 s or more [41]. However, this negative impact was not observed in *P. sylvestris* seeds contaminated with *F. oxysporum* [43]. Nevertheless, implementing non-thermal plasma in a nursery setting would be impractical. The use of antagonistic organisms like *Trichoderma* spp. has proven to be ineffective in seed disinfection, despite their effectiveness in in vitro plate assays [42]. Essential oils have also shown generalized seed germination inhibition, limiting their potential as viable treatments [42].

Among the treatments tested in this study, ethanol, particularly at a 60% concentration, effectively eliminated the *F. circinatum* inoculum from all three tested *Pinus* species. This efficacy aligns with previous reports and is attributed to protein denaturation and the disruption of cell membranes [28,41]. The results of hot water treatments were similar to those of previous studies [30,31], but we found that higher temperatures (above 60 °C) yielded the best outcomes. This indicates the potential of hot water treatments for use in *Pinus* plant production to prevent phytopathogenic fungi infection, similar to their successful application in vineyard production where they effectively eliminate fungal pathogens from grapevine cuttings [29]. The efficacy of Captan, known for its fungicidal activity against *F. circinatum* [42], was confirmed in our assays. MennoFlorades also demonstrated significant results across all tested seeds, suggesting that its metabolic inhibition, primarily due to benzoic acid [26], effectively suppresses fungal growth, as previously reported [44], including *F. circinatum* growth.

Across all *Pinus* species, treatments including 60% ethanol, hot water at 60 °C, MennoFlorades at 4% (v/v), and Captan at 1.9 g/L showed promising results and were selected for further assays to assess their impacts on germination both in vitro and in a nursery context. However, due to an ethanol shortage mentioned in Section 2.6, it was not feasible to continue using it in subsequent germination tests. Therefore, only hot water at 60 °C, MennoFlorades at 4% (v/v), and Captan at 1.9 g/L were utilized for the subsequent steps of the study.

3.2. Isolated and Identified Non-Fusarium Species

On *P. pinaster* seeds, the identified isolates included *Aspergillus* sp., *Botrytis cinerea*, and *Leptobacillium chinense*. *Botrytis cinerea* was only detected in the negative control, while *L. chinense* was found only in the ethanol 50% treatment. *Aspergillus* sp. was found in all treatments except for the positive control, hot water at 52 °C, and 1.9 g/L Captan. The absence of *B. cinerea* on treated seeds, despite its presence in the negative control, is noteworthy, as this species can pose a threat to nurseries due to its ability to infect container grown seedlings, potentially leading to their death [45].

Pinus radiata seeds showed higher fungal diversity, with 4% MennoFlorades being the only treatment with no fungal growth. The following species were identified on *P. radiata* seeds: *Echria macrotheca*, *Irpex laceratus*, *Chromelosporiopsis carneum*, *Cladosporium cladosporioides*, *Neurospora crassa*, *Peziza varia*, *Pezizaceae* sp., *Aequabiliella effusa*, *Sistotrema* sp., *Coniochaeta decumbens*, *Coniochaeta hoffmannii*, *Coniochaeta acaciae*, and *Trichodelitschia bisporula*. None of these species have been reported as potential *Pinus* pathogens, with most being endophytes or saprophytes [46–53].

No fungal growth was detected on *P. pinea* seeds, aside from the *F. circinatum* colonies and some *Trichoderma* sp. isolates on the negative control.

3.3. Nursery Germination Assays

To evaluate the impacts of the selected treatments (4% MennoFlorades, hot water at 60 °C, and 1.9 g/L Captan) on *Pinus pinaster* seed germination and plant quality, large-scale germination assays were conducted in partnering nurseries. The germination rate for each treatment was determined at 1 and 2 months post-sowing, with a control group of non-treated seeds included in each nursery, and these results are presented in Table 3 and in Supplementary Materials Table S3.

Table 3. Average germination rate (%) after *Pinus pinaster* seed disinfection with the different treatments in nurseries at 1 and 2 months post-sowing. In each column, data followed by different letters were significantly different according to Tukey's pairwise test ($p < 0.05$).

Germination Rate Post-Sowing (%)				
Nursery	**A**		**B**	
Treatment	**1 Month**	**2 Months**	**1 Month**	**2 Months**
MennoFlorades at 4% for 60 m	54.30 a	63.19 a	-	-
Captan at 1.9 g/L	-	-	15.63 a	48.54 a
Hot Water at 60 °C for 15 m	-	-	16.32 a	59.03 b
Control	59.41 a	66.96 a	17.5 a	55.21 b

In nursery A, the germination rate for seeds treated with MennoFlorades were not significantly different from the control group in both months. In contrast, in nursery B, seeds treated with 1.9 g/L Captan had a significantly lower germination rate compared to the control group. The hot water treatment, while not significantly different from the control group, showed an average germination rate of 59.03% at two months post-sowing, compared to 55.21% observed in the control group. The overall low germination rate observed across both nurseries, for both treated and non-treated seeds, is likely due to the late sowing in June instead of the usual sowing in March as a result of the COVID-19 pandemic and associated restrictions. This delay, combined with the extremely high temperatures during the summer period, negatively impacted germination rates, a phenomenon also observed in our irrigation water nursery assays [21].

The plant certification process evaluated the *P. pinaster* plants that resulted from the germination of treated and control seeds in both nurseries. This evaluation took place 7 months post-sowing and assessed compliance with several standards defined by the applicable Portuguese legislation, as mentioned previously in Section 2.5. For *P. pinaster* plants, the species-specific parameters include a maximum age of 1 year, a minimum height of 7 cm, a maximum height of 30 cm, and a minimum root collar diameter of 2 mm. The results of the plant certification evaluation are presented in Table 4.

Table 4. Number of certified *Pinus pinaster* plants and respective parameters per nursery and treatment.

	Certification				
Nursery	**A**		**B**		
Treatment	**4% MennoFlorades**	**Control**	**Hot Water at 60 °C**	**1.9 g/L Captan**	**Control**
Alveoli (total)	1350	1350	1440	1440	1440
No. of plants/lot	853	904	1236	1163	1156
No. of certified plants	0	10	1150	1000	1075
% of certified plants	0.00	1.11	93.04	85.98	92.99
Mean height (cm)	4.88	4.63	17.11	16.78	18.33
Mean diameter (mm)	1.04	0.96	2.03	2.00	2.10

In nursery B, a high percentage of plants met the certification standards across all treatments and the control group. The lowest certification rate was observed in plants from 1.9 g/L Captan-treated seeds, where approximately 86% of the plants fulfilled the certification requirements. The mean height of the plants was significantly above the legal minimum of 7 cm, and the mean diameter was equal to or greater than the legal minimum of 2 mm for all tests. Conversely, in nursery A, the number of certified plants was significantly lower. Only ten plants from the control group met the certification parameters, and no plants from the MennoFlorades treatment did so. Both the mean height and diameter were below the legal minimums in both the treatment and control groups. The late sowing and intense heat during the period are likely reasons for these poor results obtained from this nursery. However, it remains inconclusive whether MennoFlorades would negatively affect the certification of *Pinus* seedlings under regular conditions.

Unlike a different study that had previously reported on *P. radiata* seeds treated with hot water, where the germination percentage was as low as 45%–50%, even at lower temperatures (54 °C) [30], our study showed promising results for *P. pinaster* seeds. These differences could be attributed to the shorter exposure time of seeds to high temperatures, which was one-half to one-third of that tested by Agustí-Brisach et al. [30]. Among all the treatments, hot water at 60 °C for 15 min demonstrated no negative impacts on seed germination or the percentage of certified plants and had a negligible effect on the mean height and diameter of the resulting plants. These findings, combined with its low economic impact for forest reproductive material (FRM) producers and its eco-friendly nature, make this treatment an ideal candidate for application in nurseries as a preventive measure against *F. circinatum*.

3.4. *In Vitro Germination Assays*

Alongside the nursery trials, in vitro germination assays were conducted by CENASEF in accordance with ISTA standards. The treatments evaluated in these assays were the same as those tested in the nurseries: MennoFlorades at 4% (v/v), Captan at 1.9 g/L, and hot water at 60 °C. The results, shown in Figure 4 and in Supplementary Table S4, indicate that the germination rate was close to 90% for most treatments and the control group, with no significant differences between them (F = 3.87; df = 2.09; p = 0.20), with 1.9 g/L Captan showing an average germination rate of 84% compared to the 88.5% verified for the control group. These findings are consistent with the ones of the nursery assays, where Captan was the only treatment resulting in a pointedly lower germination rate compared to the control. Moreover, hot water treatment did not negatively impact seed germination, corroborating our nursery results. This is further supported by the study of Jones et al. [54], which reported similar outcomes for seeds of another *Pinus* species, namely, *P. palustris*. However, extended periods of hot water treatment (30 to 45 min), even at lower temperatures (50–54 °C), have been reported to significantly impact *P. radiata* seed germination [30].

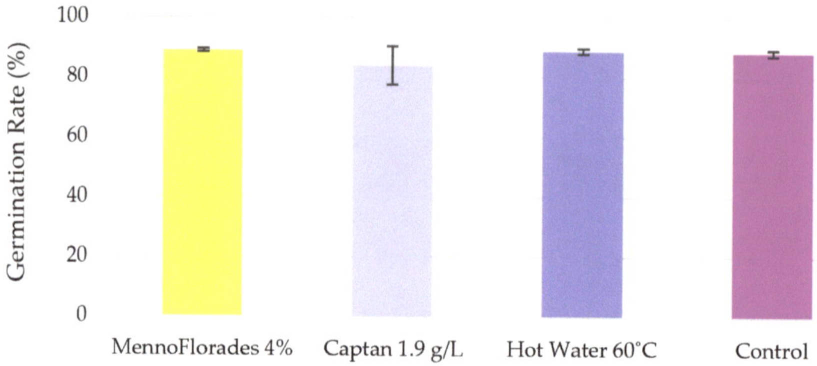

Figure 4. Average germination rate of *Pinus pinaster* per treatment from the in vitro CENASEF assays. Error bars shown in the graph correspond to the standard deviations.

Higher germination rates were observed in the in vitro tests compared to the nursery assays, which was expected given that ISTA standards ensure ideal conditions for seed germination, conditions that are not always replicable in a real nursery context.

4. Conclusions

Given the significance of *Pinus* forests to the Portuguese environment and economy, any strategy that can mitigate threats like potential contamination by *Fusarium circinatum* is highly valuable. This is particularly true if the strategy is effective, environmentally friendly, and economically viable on a large scale, such as in nurseries where most forest reproductive material (FRM) for reforestation is produced.

Our study took an innovative lab to nursery approach, thus ensuring that the strategies developed were based on scientific data and results, but always working in close collaboration with FRM producers to ensure the viability of said strategies in practical and economical contexts. This approach allowed us to identify several promising treatments for seed disinfection that can be implemented in nurseries to prevent the spread of pine pitch canker. When combined with other measures, such as irrigation water disinfection, alternative substrates, as presented in our previously published work [21–24], and good nursery practices, these treatments can significantly reduce the spread of *F. circinatum* in both nurseries and forests. A technical manual based on our findings was composed [55] and distributed to FRM producers' nationwide, who are already implementing some of the suggested strategies to prevent *F. circinatum* outbreaks in their nurseries, thus ensuring healthier and more resilient pine populations and contributing to the long-term sustainability of forest ecosystems.

Supplementary Materials: The following supporting information can be downloaded at https://www.mdpi.com/article/10.3390/f15071154/s1, Table S1: Origin and GenBank accession numbers of all non-*Fusarium* isolates obtained in this study; Table S2: Numbers of *Pinus pinaster*, *P. radiata*, and *P. pinea* seeds contaminated with *Fusarim circinatum* after disinfection treatment application; Table S3: Germination rate (%) after *Pinus pinaster* seed disinfection with the different treatments in nurseries at 1 and 2 months post-sowing; Table S4: Germination rate of *Pinus pinaster* per treatment from the in vitro CENASEF assays.

Author Contributions: Conceptualization, L.F., D.S.P., D.R., L.M., H.B. and A.P.; methodology, L.F., D.S.P., A.C.S., C.F. and A.R.F.; software, L.F. and D.S.P.; validation, L.F., D.S.P., A.C.S., D.R., L.M., H.B. and A.P.; formal analysis, L.F. and D.S.P.; investigation, L.F., D.S.P., A.C.S., C.F. and A.R.F.; resources, D.R., L.M., H.B. and A.P.; data curation, L.F. and D.S.P.; writing—original draft preparation, L.F. and D.S.P.; writing—review and editing, L.F., D.S.P. and A.P.; visualization, L.F. and D.S.P.; supervision, D.R., L.M., H.B. and A.P.; project administration, D.R.; funding acquisition, D.R., L.M., H.B. and A.P. All authors have read and agreed to the published version of the manuscript.

Funding: This research was co-funded by PDR2020, Portugal2020 and FEDER under the framework GO +PrevCRP (PDR2020–101–031058, partnership No. 112, initiative No. 237) and PRR—Recovery and Resilience Plan and by the NextGeneration EU European Funds. Diana Paiva was supported by PhD research grant UI/BD/150843/2021 (https://doi.org/10.54499/UI/BD/150843/2021) awarded by the Centre for Functional Ecology—Science for People and the Planet (CFE) and co–funded by Fundação para a Ciência e Tecnologia, I.P. (FCT) through national funding by the Ministério da Ciência, Tecnologia e Ensino Superior (MCTES) from Fundo Social Europeu (FSE). Ana C. Silva was financially supported by PhD research grant 2022.11826.BD. Helena Bragança acknowledges support from UIDB/04551/2020 and UIDP/04551/2020 center grants to GREEN–IT. The authors thank FCT for funding the R&D Unit Centre for Functional Ecology—Science for People and the Planet (CFE) with reference UIDB/04004/2020 and DOI identifier 10.54499/UIDB/04004/2020 (https://doi.org/10.54499/UIDB/04004/2020) and the Associate Laboratory TERRA with reference LA/P/0092/2020 and DOI identifier 10.54499/LA/P/0092/2020 (https://doi.org/10.54499/LA/P/0092/2020), financed by FCT/MCTES through national funds (PIDDAC).

Data Availability Statement: The original contributions presented in the study are included in the article and Supplementary Materials, further inquiries can be directed to the corresponding author.

Acknowledgments: The authors would like to thank the nurseries APFC-Equipar, Germiplanta, ICNF-Alcácer do Sal, and ICNF-Amarante for their collaboration in this study and for providing the project with the necessary infrastructure and manpower to conduct these assays. The authors also thank Eugénio Diogo from INIAV for his collaboration. Photographs included in Figures 1 and 2 were captured by the article's authors, Luís Fernandes and Diana Paiva, and by Dina Ribeiro from ICNF, respectively.

Conflicts of Interest: The authors declare no conflicts of interest.

References

1. Gordon, T. Pitch Canker Disease of Pines. *Phytopathology* **2006**, *96*, 657–659. [CrossRef] [PubMed]
2. Bezos, D.; Martinez-Alvarez, P.; Fernandez, M.; Diez, J. Epidemiology and Management of Pine Pitch Canker Disease in Europe—A Review. *Balt. For.* **2017**, *23*, 279–293.
3. Wingfield, M.J.; Hammerbacher, A.; Ganley, R.J.; Steenkamp, E.T.; Gordon, T.R.; Wingfield, B.D.; Coutinho, T.A. Pitch canker caused by *Fusarium circinatum*—A growing threat to pine plantations and forests worldwide. *Australas. Plant Pathol.* **2008**, *37*, 319–334. [CrossRef]
4. Drenkhan, R.; Ganley, B.; Martín-García, J.; Vahalík, P.; Adamson, K.; Adamčíková, K.; Ahumada, R.; Blank, L.; Bragança, H.; Capretti, P.; et al. Global Geographic Distribution and Host Range of *Fusarium circinatum*, the Causal Agent of Pine Pitch Canker. *Forests* **2020**, *11*, 724. [CrossRef]
5. Schmale, D.G., III; Gordon, T.R. Variation in susceptibility to pitch canker disease, caused by *Fusarium circinatum*, in native stands of *Pinus muricata*. *Plant Pathol.* **2003**, *52*, 720–725. [CrossRef]
6. Gordon, T.R.; Kirkpatrick, S.C.; Aegerter, B.J.; Wood, D.L.; Storer, A.J. Susceptibility of Douglas fir (*Pseudotsuga menziesii*) to pitch canker, caused by *Gibberella circinata* (anamorph = *Fusarium circinatum*). *Plant Pathol.* **2006**, *55*, 231–237. [CrossRef]
7. Martínez-Álvarez, P.; Pando, V.; Diez, J.J. Alternative species to replace Monterey pine plantations affected by pitch canker caused by *Fusarium circinatum* in northern Spain. *Plant Pathol.* **2014**, *63*, 1086–1094. [CrossRef]
8. Solel, Z.; Bruck, R.I. Relation between wilt rate and obstruction of water flow in stems of two families of loblolly pine affected by pitch canker. *Eur. J. For. Pathol.* **1990**, *20*, 317–320. [CrossRef]
9. Hepting, G.H.; Roth, E.R. Pitch canker, a new disease of some southern pines. *J. For.* **1946**, *44*, 724–744.
10. Berbegal, M.; Pérez-Sierra, A.; Armengol, J.; Grünwald, N.J. Evidence for Multiple Introductions and Clonality in Spanish Populations of *Fusarium circinatum*. *Phytopathology* **2013**, *103*, 851–861. [CrossRef]
11. Landeras, E.; Garcia, P.; Fernández, Y.; Braña, M.; Fernández-Alonso, O.; Méndez-Lodos, S.; Pérez-Sierra, A.; Leon, M.; Abad-Campos, P.; Berbegal, M.; et al. Outbreak of Pitch Canker Caused by *Fusarium circinatum* on *Pinus* spp. in Northern Spain. *Plant Dis.* **2005**, *89*, 1015. [CrossRef]
12. Dwinell, D. Global Distribution of the Pitch Canker Fungus. In *Current and Potential Impacts of Pitch Canker in Radiata Pine*; IMPACT Monterey Workshop: Monterey, CA, USA; Citeseer: Princeton, NJ, USA, 1999; Volume 30, pp. 54–57.
13. European and Mediterranean Plant Protection Organization (EPPO) Global Database. *First Report of Gibberella circinata in France*; EPPO: Paris, France, 2006; p. 104.
14. Carlucci, A.; Colatruglio, L.; Frisullo, S. First report of Pitch Canker caused by *Fusarium circinatum* on *Pinus halepensis* and *P. pinea* in Apulia (southern Italy). *Plant Dis.* **2007**, *91*, 1683. [CrossRef]
15. Bragança, H.; Diogo, E.; Moniz, F.; Amparo, P. First report of Pitch canker on Pines caused by *Fusarium circinatum* in Portugal. *Plant Dis.* **2009**, *93*, 1079. [CrossRef]

16. 6° Inventário Florestal Nacional—IFN6. Available online: https://inforcna.pt/Media/Files/201979_Ifn6PrincipaisResultadosJun2019.pdf (accessed on 13 May 2024).
17. Instituto Nacional de Estatística—Statistics Portugal, Estatísticas do Comércio Internacional. Available online: https://www.ine.pt/xportal/xmain?xpid=INE&xpgid=ine_publicacoes&PUBLICACOESpub_boui=280862956&PUBLICACOESmodo=2 (accessed on 13 May 2024).
18. Vettraino, A.; Potting, R.; Raposo, R. EU legislation on forest plant health: An overview with a focus on *Fusarium circinatum*. *Forests* **2018**, *9*, 568. [CrossRef]
19. European and Mediterranean Plant Protection Organization (EPPO) Global Database. EPPO Datasheet: Fusarium circinatum. Available online: https://gd.eppo.int/taxon/GIBBCI/datasheet (accessed on 13 May 2024).
20. European Food and Environment Safety Agency (EFSA) Panel on Plant Health (PLH). Risk assessment of *Gibberella circinata* for the EU territory and identification and evaluation of risk management options. *EFSA J.* **2010**, *8*, 1620. [CrossRef]
21. Fernandes, L.; Paiva, D.; Roxo, I.; Fernandes, A.R.; Ribeiro, D.; Ribeiro, H.; Portugal, A. Development of New Preventive Strategies for Pine Pitch Canker Caused by *Fusarium circinatum* in Irrigation Water and Evaluation in a Real Nursery Context. *Forests* **2023**, *14*, 443. [CrossRef]
22. Silva, A.C.; Diogo, E.; Bragança, H. Effect of Substrate Solarization for the Control of Fungi: The Case Study of *Fusarium circinatum*, the Quarantine Agent of Pine Pitch Canker. *Silva Lusit.* **2021**, *29*, 161–175. [CrossRef]
23. Chemetova, C.; Quilhó, T.; Braga, S.; Fabião, A.; Gominho, J.; Ribeiro, H. Aged Acacia melanoxylon bark as an organic peat replacement in container media. *J. Clean. Prod.* **2019**, *232*, 1103–1111. [CrossRef]
24. Chemetova, C.; Mota, D.; Fabião, A.; Gominho, J.; Ribeiro, H. Low-temperature hydrothermally treated Eucalyptus globulus bark: From by-product to horticultural fiber-based growing media viability. *J. Clean. Prod.* **2021**, *319*, 128805. [CrossRef]
25. European and Mediterranean Plant Protection Organization (EPPO). 91 (2) *Fusarium circinatum* (formerly *Giberella circinata*). *EPPO Bull.* **2019**, *49*, 228–247. [CrossRef]
26. Krebs, H.A.; Wigginst, D.; Stubbs, M. Studies on the mechanism of the antifungal action of benzoate. *Biochem. J.* **1983**, *7*, 657–663. [CrossRef] [PubMed]
27. Nelson, B.D. Action of the Fungicides Captan and Folpet on Rat Liver Mitochondria. *Biochem. Pharmacol.* **1971**, *20*, 737–748. [CrossRef] [PubMed]
28. Dao, T.; Dantigny, P. Control of food spoilage fungi by ethanol. *Food Control* **2011**, *22*, 360–368. [CrossRef]
29. Crous, P.; Swart, L.; Coertze, S. The effect of hot-water treatment on fungi occurring in apparently healthy grapevine cuttings. *Phytopathol. Mediterr.* **2001**, *40*, S464–S466.
30. Agustí-Brisach, C.; Pérez-Sierra, A.; Armengol, J.; García-Jiménez, J.; Berbegal, M. Efficacy of hot water treatment to reduce the incidence of *Fusarium circinatum* on *Pinus radiata* seeds. *Forestry* **2012**, *85*, 629–635. [CrossRef]
31. Liao, T.; Ye, J.; Chen, J.; Han, Y.; Wu, C.; An, Y. Biological characteristics of *Fusarium circinatum*. *J. Nanjing For. Univ.* **2008**, *32*, 83–86.
32. Hammer, Ø.; Harper, D.A.T.; Ryan, P.D. PAST: Paleontological statistics software package for education and data analysis. *Palaeontol. Electron.* **2001**, *4*, 9.
33. Paiva, D.S.; Fernandes, L.; Trovão, J.; Mesquita, N.; Tiago, I.; Portugal, A. Uncovering the Fungal Diversity Colonizing Limestone Walls of a Forgotten Monument in the Central Region of Portugal by High-Throughput Sequencing and Culture-Based Methods. *Appl. Sci.* **2022**, *12*, 10650. [CrossRef]
34. Paiva, D.S.; Fernandes, L.; Pereira, E.; Trovão, J.; Mesquita, N.; Tiago, I.; Portugal, A. Exploring Differences in Culturable Fungal Diversity Using Standard Freezing Incubation—A Case Study in the Limestones of Lemos Pantheon (Portugal). *J. Fungi* **2023**, *9*, 501. [CrossRef]
35. White, T.J.; Bruns, T.; Lee, S.; Taylor, J. Amplification and direct sequencing of fungal ribosomal RNA genes for phylogenetics. *PCR Protoc. Guide Methods Appl.* **1990**, *18*, 315–322.
36. Gardes, M.; Bruns, T.D. ITS primers with enhanced specificity for basidiomycetes application to the identification of mycorrhizae and rusts. *Mol. Ecol.* **1993**, *2*, 113–118. [CrossRef]
37. Altschul, S.F.; Madden, T.L.; Schäffer, A.A.; Zhang, J.; Zhang, Z.; Miller, W.; Lipman, D.J. Gapped BLAST and PSI-BLAST: A new generation of protein database search programs. *Nucleic Acids Res.* **1997**, *25*, 3389–3402. [CrossRef] [PubMed]
38. Crous, P.W.; Gams, W.; Stalpers, J.A.; Robert, V.; Stegehuis, G. MycoBank: An online initiative to launch mycology into the 21st century. *Stud. Mycol.* **2004**, *50*, 19–22.
39. Robert, V.; Stegehuis, G.; Stalpers, J. The MycoBank Engine and Related Databases. 2005. Available online: https://www.MycoBank.org/ (accessed on 13 May 2024).
40. Dwinell, L.D. Contamination of *Pinus radiata* seeds in California by *Fusarium circinatum*. In Proceedings of the Annual International Research Conference on Methyl Bromide Alternatives and Emissions Reductions, States Department of Agriculture (USDA) Forest Service, San Diego, CA, USA; 1999; pp. 1–4.
41. Šerá, B.; Zahoranová, A.; Bujdáková, H.; Šerý, M. Disinfection from pine seeds contaminated with *Fusarium circinatum* Nirenberg & O'Donnell using non-thermal plasma treatment. *Rom. Rep. Phys* **2019**, *71*, 701.
42. Iturritxa, E.; Desprez-Loustau, M.L.; García-Serna, I.; Quintana, E.; Mesanza, N.; Aitken, J. Effect of alternative disinfection treatments against fungal canker in seeds of *Pinus radiata*. *Seed Technol.* **2011**, 33, 88–110.

43. Świecimska, M.; Tulik, M.; Šerá, B.; Golińska, P.; Tomeková, J.; Medvecká, V.; Bujdáková, H.; Oszako, T.; Zahoranová, A.; Šerý, M. Non-Thermal Plasma Can Be Used in Disinfection of Scots Pine (*Pinus sylvestris* L.) Seeds Infected with *Fusarium oxysporum*. *Forests* **2020**, *11*, 837. [CrossRef]
44. Gioia, T.; Galinski, A.; Lenz, H.; Müller, C.; Lentz, J.; Heinz, K.; Nagel, K.A. GrowScreen-PaGe, a non-invasive, high-throughput phenotyping system based on germination paper to quantify crop phenotypic diversity and plasticity of root traits under varying nutrient supply. *Funct. Plant Biol.* **2016**, *44*, 76–93. [CrossRef]
45. Capieau, K.; Stenlid, J.; Stenström, E. Potential for biological control of *Botrytis cinerea* in *Pinus sylvestris* seedlings. *Scand. J. For. Res.* **2004**, *19*, 312–319. [CrossRef]
46. Jiao, Y.; Huang, J. Allelopathic effects of aqueous extracts from uncomposted and composted Mexican devil (*Ageratina adenophora*) plants on forest fungal growth and soil nitrogen and phosphorus mobilization. *Weed Sci.* **2024**, *72*, 76–85. [CrossRef]
47. Lygis, V.; Vasiliauskaite, I.; Matelis, A.; Pliūra, A.; Vasaitis, R. Fungi in living and dead stems and stumps of *Pinus mugo* on coastal dunes of the Baltic Sea. *Plant Protect. Sci.* **2014**, *50*, 221–226. [CrossRef]
48. Lygis, V.; Vasiliauskaite, I.; Stenlid, J.; Vasaitis, R. Impact of forest fire on occurrence of *Heterobasidion annosum* ss root rot and other wood–inhabiting fungi in roots of *Pinus mugo*. *Forestry* **2010**, *83*, 83–92. [CrossRef]
49. Stępniewska, S.; Mańka, M. Fungi inhabiting Scots pine (*Pinus sylvestris*) seeds in Wielkopolska region. *Phytopathol. Pol.* **2004**, *31*, 73–76.
50. Sanz–Ros, A.V.; Müller, M.M.; San Martín, R.; Diez, J.J. Fungal endophytic communities on twigs of fast and slow growing Scots pine (*Pinus sylvestris* L.) in northern Spain. *Fungal Biol.* **2015**, *119*, 870–883. [CrossRef] [PubMed]
51. Kuo, H.C.; Hui, S.; Choi, J.; Asiegbu, F.O.; Valkonen, J.P.; Lee, Y.H. Secret lifestyles of *Neurospora crassa*. *Sci. Rep.* **2014**, *4*, 5135. [CrossRef] [PubMed]
52. Botella, L.; Diez, J.J. Phylogenic diversity of fungal endophytes in Spanish stands of *Pinus halepensis*. *Fungal Divers.* **2011**, *47*, 9–18. [CrossRef]
53. Garay–Serrano, E.; del Pilar Ortega–Larrocea, M.; Reverchon, F.; Suárez–Quijada, I. Persistence of ecto– and ectendomycorrhizal fungi associated with *Pinus montezumae* in experimental microcosms. *Symbiosis* **2018**, *74*, 67–78. [CrossRef]
54. Jones, J.P.; Sun, X.; Eckhardt, L.; Webber, A.; Hess, N.; Barnett, J.; McGilvary, J. Longleaf seedling production: Some problems and their solutions. *LA Agric.* **2002**, *45*, 4–11.
55. Fernandes, A.R.; Silva, A.C.; Portugal, A.; Fernandes, C.; Ribeiro, D.; Bragança, H.; Ribeiro, H.; Barbosa, J.N.; Fernandes, L.; Martins, L.; et al. *Prevenção do Cancro-Resinoso-do-Pinheiro (*Fusarium circinatum*)-Manual Técnico para Fornecedores de Materiais Florestais de Reprodução*, 1st ed.; Centro PINUS: Viana do Castelo, Portugal, 2021.

forests

MDPI

Article

Diversity Analysis of Macrofungi and Lichenised Fungi in Pyrenean Oak (*Quercus pyrenaica* Willd.) and Chestnut (*Castanea sativa* L.) Forests: Implications for the Conservation of Forest Habitats in Castilla y León (Central-Northwest Spain)

David Rodríguez-de la Cruz [1,2,*], Sonia Perfecto-Arribas [2] and Luis Delgado-Sánchez [1,2]

[1] Centre for Environmental Studies and Rural Dynamization (CEADIR), Universidad de Salamanca, Filiberto Villalobos 119, 37007 Salamanca, Spain; ldelsan@usal.es

[2] Department of Botany and Plant Physiology, Universidad de Salamanca, Licenciado Méndez Nieto s/n, 37007 Salamanca, Spain; soniaperfecto@usal.es

* Correspondence: droc@usal.es; Tel.: +0034-677-584-172

Abstract: Fungi can be used as bioindicators to assess the biodiversity and conservation status of different habitats because of their high adaptability and sensitivity to changes in water, air, and soil quality. In this study, records of macrofungi and lichenised fungi were extracted from GBIF, surveyed using GIS software, and used to compare the fungal diversity of the Pyrenean oak and chestnut groves of Castilla y León, analysing the possible implications of their presence for the conservation of these forest habitats. In *Quercus pyrenaica* forests, a greater number of lichen and macrofungi species and records were recorded than in *Castanea sativa* forests, although the greater area occupied by the former could have influenced this diversity. The higher presence of ectomycorrhizal macrofungal species in chestnut groves, as well as the higher sensitivity to pollution of lichens in Pyrenean oak-dominated environments, showed the relevance of the analysis of these data for a better understanding of the conservation status of forest habitats. However, in order to obtain more accurate results, it would be necessary to carry out specific studies on a smaller scale.

Keywords: chestnut groves; fungal diversity; Iberian Peninsula; lichens; macrofungi; lichens; fungal diversity; Pyrenean oak

Academic Editors: Katarína Pastirčáková and Rostislav Zemek

Received: 30 October 2024
Revised: 17 December 2024
Accepted: 20 December 2024
Published: 24 December 2024

Citation: Rodríguez-de la Cruz, D.; Perfecto-Arribas, S.; Delgado-Sánchez, L. Diversity Analysis of Macrofungi and Lichenised Fungi in Pyrenean Oak (*Quercus pyrenaica* Willd.) and Chestnut (*Castanea sativa* L.) Forests: Implications for the Conservation of Forest Habitats in Castilla y León (Central-Northwest Spain). *Forests* **2025**, *16*, 9. https://doi.org/10.3390/f16010009

1. Introduction

The fungi kingdom is highly diverse, both taxonomically and in terms of life forms [1]. Macrofungi, which are distinguished by having spore-bearing structures visible to the naked eye, belong mostly to the division Basidiomycota (although some may be Ascomycota or even zygomycetes fungi). These organisms are usually saprobes or mycorrhizal symbionts, but some are also pathogens of plants or other fungi [2]. Lichenised fungi are an ecologically diverse and successful group living in symbiosis with photoautotrophic organisms (algae or cyanobacteria) [3]. It is one of the most important life forms among the Ascomycota, although some Basidiomycota also live as lichens [4]. The study of fungal diversity is very valuable for assessing ecosystem quality because fungi participate in the recycling of organic matter and improve the living standards of plant species through symbiotic associations [5].

The detection of anthropogenic pollution impact is usually conducted through chemical tests, and physical parameters are measured directly from the environment [6]. However, in recent decades, bio-indicators have gained a lot of relevance and have been widely applied to assess alterations in the environment because they are cheaper than physical and chemical indicators of environmental quality and can provide additional information on the health status of ecosystems [7]. Biological indicators can be defined as species, groups of species, or communities of a wide range of organisms whose presence, quantity, and nature allow us to make inferences about the quality of the environment [8]. Usually, a group of organisms or even a species is selected as the most appropriate bio-indicator in each environment, depending on the studied habitats, the local conditions, and disturbances [9]. A good bioindicator must have several characteristics, such as being easily identifiable, being sensitive to small changes in the environment, or playing a relevant role in the dynamics of ecosystems, among others [10]. However, it must be taken into account that no taxon fits all these premises [11]. The main reflection of these biological indicators is manifested through the biodiversity observed in an established area [12], and they can also act as early warning signals as they provide information on measurable qualitative environmental aspects according to their presence, absence, abundance, activity, morphology, physiology, or behaviour [8]. This information could be used to identify negative or positive effects on the environment so that conservation strategies may be considered [12].

The most commonly used biological indicators are plants, algae, and animals [13], with fewer references to the use of fungi as bioindicators. Despite this, it is considered that fungi possess the potential to act as efficient bioindicators due to their life cycle characteristics, such as ubiquitous distribution, sensitivity, survivability, and tolerance to a changing environment [14]. Their adaptability and sensitivity enable them to show real-time responses to environmental changes and stressors in their habitats, making them versatile and reliable tools for assessing the quality of the environment [5].

Fungi, given their diversity and wide range of ecological adaptations, are suitable as bio-indicators for monitoring the following: water, air, and soil quality; the state of agricultural and forest habitats; and the impact of climate change [15]. For example, lichens are used as bioindicators of air quality because they acquire all their nutrients from direct exposure to the atmosphere, which promotes the accumulation of air pollutants [16]. On the other hand, the presence or absence of certain species of macrofungi can be used as an indicator of the degree of disturbance of forest habitats [17], with poorly growing trees producing fewer sporocarps from mycorrhizal species than healthy ones. In this context, both lichens [18] and macrofungi [19] have been used in forest governance to identify forest habitats in need of management measures.

The main objective of the present work is to analyse the diversity of macrofungi and lichenised fungi in the chestnut (*Castanea sativa* Mill.) and Pyrenean oak (*Quercus pyrenaica* Willd.) forests of the Castilla y León region (Central Spain) through the combined use of records in databases and Geographic Information Systems (GIS). This diversity analysis also makes it possible to assess the application of these organisms as bioindicators and to understand the implications for the conservation of forest habitats.

2. Materials and Methods

2.1. Study Area

The study area was limited to chestnut and oak forest formations in the Castilla y León region. For this purpose, the digital cartography of the Spanish Forest Map at a scale of 1:25,000 [20] was obtained. The forest areas located within the boundaries of Castilla y León were selected using the ArcGIS10.8 software package [21]. The main information related to the forests analysed is summarised in Table 1. Then, the two types of habitats studied were chosen from this area (Figure 1) and assigned to codes 9230 and 9260 within the codification of the Habitats of Community Interest (HCI) of the European Union [22].

Table 1. Main characteristics of the analysed forests in Castilla y León (C-NW Iberian Peninsula). BioR: Biogeographic region; Nplot: number of plots; Has: number of hectares; Age: percentage of trees with the indicated age; Other trees: second most reported tree species. Source: MFE [20].

Forest	BioR	Nplot	Has	Age (in %) <30 Years	30–70 Years	>70 Years	Other Trees
Pyrenean oak	Med	18,116	449,384	7.55	56.90	35.66	*Quercus ilex* L., *Pinus pinaster* Ait., *Castanea sativa*, *Pinus sylvestris* L.
	At	4460	96,006	13.69	57.72	28.68	*Quercus petraea* (Matt.) Liebl., *Betula pendula* Roth., *Castanea sativa*, *Fagus sylvatica* L.
Chestnut	Med	1094	15,499	3.84	14.01	82.17	*Quercus pyrenaica*, *Quercus ilex*, *Pinus pinaster*
	At	305	4013	0.08	3.42	98.50	*Quercus pyrenaica*, *Betula pendula*, *Quercus robur* L.

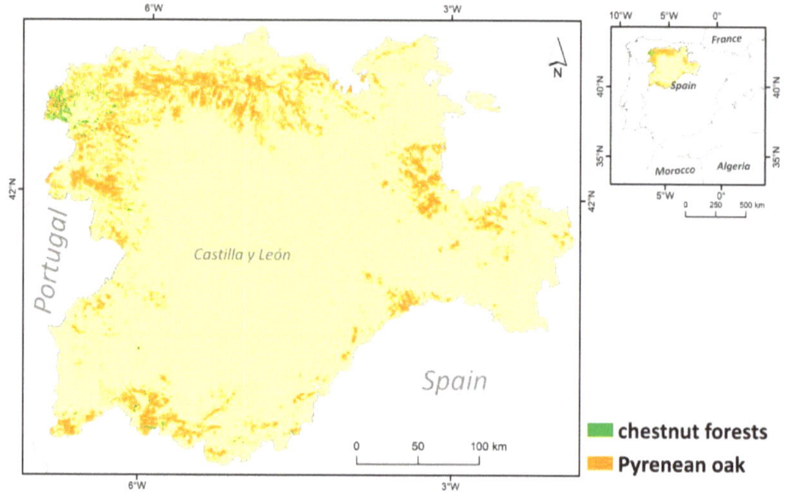

Figure 1. Distribution of Pyrenean oak (*Quercus pyrenaica*) and chestnut (*Castanea sativa*) forests in Castilla y León region (C-NW Spain).

The territory of Castilla y León is spread over two different biogeographical regions, the Atlantic region and the Mediterranean region, although the former is restricted to a small area in the north of the region [23]. Precisely, the presence of these two biogeographical regions underlines the relevance of the selected territory, as these two biogeographical regions cover a large part of the European territory, with consequent climatic and ecological differences. In order to compare the data from the two regions, the hectares occupied by chestnut and Pyrenean oak in both biogeographical regions were calculated by GIS analysis.

2.2. Data Collection

The GBIF (Global Biodiversity Information Facility) database was used to obtain the records of macrofungi and lichens [24,25]. First, all records of the fungi kingdom and the phyla Ascomycota and Basidiomycota were downloaded. From this dataset, the citations involved in the analysed area were selected (16,684 records), checking that they were correctly georeferenced (non-georeferenced records and those with erroneous georeferencing were excluded: 2020 records, mainly due to confusion between zone 29 and the correct zone 30) as well as their correct taxonomic categorisation [26] (information was supplemented for some citations in which their Phylum, Class, Order, or Family was not specified). Once the dataset was reviewed, occurrence citations were separated into lichens (lichenised fungi) and macrofungi. In addition, the lichens and macrofungi present in the Pyrenean oak forests or chestnut groves of the studied area were selected separately, with a buffer of 200 metres to avoid data loss due to inaccurate georeferencing because of the orography. Thus, we worked with 5706 records (1484 records of macrofungi and 4222 records of lichens). Of these records, 93% were preserved records, and 7% were human observations. Regarding the contribution of collections or institutions in this study, 90% of the data came from the herbaria of the University of León and the Royal Botanical Garden of Madrid. A few fungal species were also eliminated from the records obtained (80 records), which could fall under the definition of macrofungi [2] but are pathogens of crop plants, e.g., *Taphrina populina* Fr.

2.3. Data Analysis

Different aspects such as species richness (alpha diversity), similarity between the two habitats (beta diversity), macrofungi lifestyles, as well as lichen habit and sensitivity to pollution were evaluated. The alpha diversity was calculated separately for both forest habitats through the elaboration of lists of lichen and macrofungi species, both at the general level for the whole study area and in the 10×10 km grids. In the latter case, differences in alpha diversity in the study territory can be assessed within each of the habitats and groups analysed. Beta diversity was assessed using the Sørensen Similarity Index to compare the likeness in species of each fungal group in both forest types [27,28].

The assessment of the conservation status of the forest habitats was carried out by analysing the life history of the macrofungi species and the sensitivity to pollution of the different species of lichenised fungi present in the two forest types. To determine the mode of life (mycorrhizal, saprophytic, and parasitic), basic literature was reviewed [29,30], while data from various sources were used to assess the habit (corticolous, epiphytic, lignicolous, muscicolous, saxicolous, and terricolous) and sensitivity to lichen contamination (low, medium and high), according to scientific literature [31,32].

3. Results

3.1. Species Richness

The number of macrofungal species present in *Quercus pyrenaica* forests was 560 and 127 in *Castanea sativa* forests, while the number of lichen species was 526 in the former and 85 in the latter (Table S1). Therefore, fungal composition showed a higher alpha diversity in Pyrenean oak forest formations for the two fungal groups, about 4.5 times higher in the case of macrofungi and up to 6 times higher for lichenised fungi. In any case, these data must be contextualised in relation to the surface area of each forest formation. The spatial analysis showed that the hectares occupied by this species of oak in Castilla and León region were about 545,390 ha, which is 5.78% of the total area of this region. Chestnut groves covered about 19,512 ha, i.e., 0.21% of the study area. In relation to the number of species per UTM 10×10 grid squares (Figure 2), in *Castanea sativa* forests, the highest number of species for lichens and macrofungi corresponded to the distribution of these forest ecosystems (NW and SW of the region analysed). In the case of environments dominated by *Quercus pyrenaica*, differences were detected for both groups of fungal organisms. There were more species of macrofungi in the north and south-west of the study area, while lichenised fungi were more numerous in the north-west and some south-eastern squares.

In both forest environments, species of macrofungi belonging to the phylum Basidiomycota predominated, with percentages above 90%, although the presence of species belonging to Ascomycota was higher in the ecosystems dominated by Pyrenean oak, where they accounted for 7.5% of the species recorded. As for the diversity of macrofungal orders, 10 orders were present in chestnut groves (one of Ascomycota), while 21 were present in *Quercus pyrenaica* communities (three belonging to Ascomycota). The most represented orders were the same in both plant formations: Agaricales, Boletales, Polyporales, and Russulales, except for the order Pezizales, which was not present in *Castanea sativa* forests. The first four orders mentioned accounted for 71.5% and 78.7% of the macrofungal species recorded in Pyrenean oak and chestnut groves, respectively, with the order Agaricales being the most diverse, hosting just over 43% of the species in both ecosystems (Table S2).

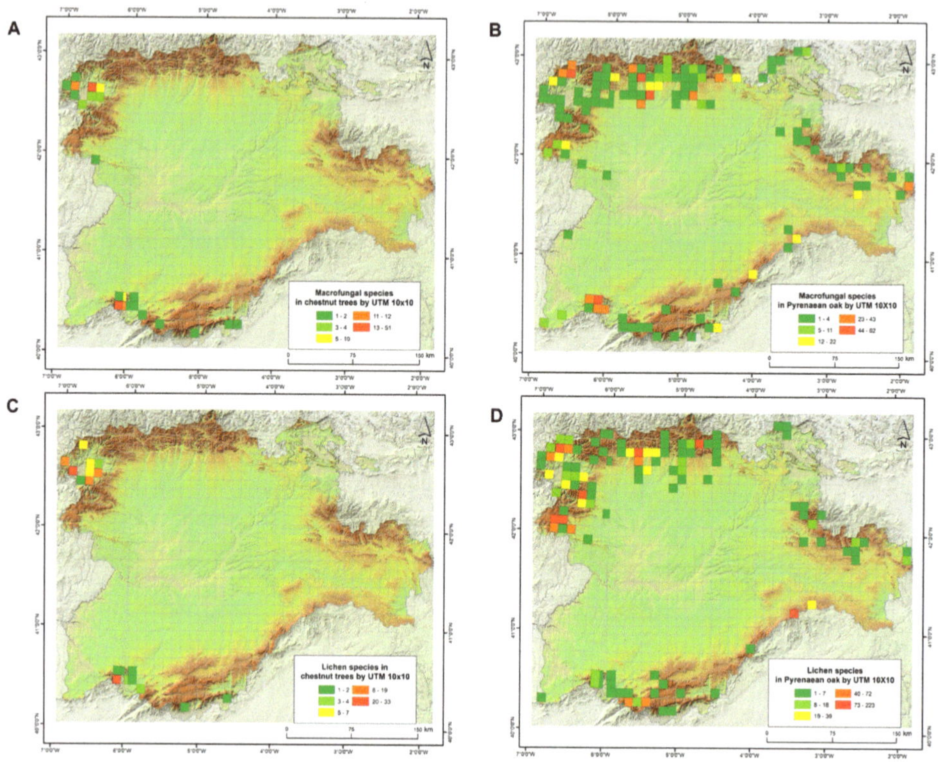

Figure 2. Number of species of macrofungi (upper part) and lichenised fungi (lower part) in forest formations of *Castanea sativa* (**A**,**C**) and *Quercus pyrenaica* (**B**,**D**) in Castilla y León (C–NW Spain).

In lichenised fungi, all belonging to the phylum Ascomycota, this pattern of diversity was repeated in relation to the orders, with 19 in Pyrenean oak and 11 in chestnut woods. The four most represented orders were the same in both environments (Caliciales, Lecanorales, Peltigerales, and Pertusariales), accounting for 69.1% and 83.6% of the lichens recorded. Among these orders, Lecanorales was notable, as it included 38.6% of lichenicolous species in oak communities and 48.2% in chestnut communities.

The fifteen most abundant species of macrofungi and lichens were identified according to the records assigned to Pyrenean oak and chestnut forests, and it was observed that only five species coincide in both environments. These species are shown in Table 2 with grey background. For lichens, among the fifteen most cited species in the two forest habitats, twelve species coincided, as also shown in Table 2. Therefore, when comparing the two environments, there was greater overlap between the dominant species of lichens than of macrofungi.

Table 2. Most frequent species of lichenised fungal macrofungi on Pyrenean oak and chestnut forest formations in the study area. In brackets, number of records. In grey background, species identified in both ecosystems.

Macrofungi		Lichenised Fungi	
Pyrenean Oak	**Chestnut**	**Pyrenean Oak**	**Chestnut**
Peniophora quercina (Pers.) Cooke (26)	*Hymenochaete rubiginosa* (Dicks.) Lév. (5)	*Parmelia sulcata* Taylor (137)	*Evernia prunastri* (14)
Athelia epiphylla Pers. (22)	*Amanita muscaria* (4)	*Evernia prunastri* (L.) Ach. (108)	*Parmelia sulcata* (12)
Peniophorella praetermissa (P. Karst.) K.H. Larss. (17)	*Hygrophoropsis aurantiaca* (Wulfen) Maire ex Martin-Sans (4)	*Physconia distorta* (With.) J.R. Laundon (95)	*Lepra albescens* (10)
Botryobasidium subcoronatum (Höhn. & Litsch.) Donk (16)	*Hyphoderma anthracophilum* (Bourdot) Jülich (4)	*Parmelina tiliacea* (Hoffm.) Hale (88)	*Hypogymnia physodes* (L.) Nyl. (8)
Ramaria formosa (Pers.) Quél. (16)	*Phanerochaete sordida* (4)	*Ramalina fraxinea* (L.) Ach. (86)	*Melanelixia glabra* (7)
Phanerochaete sordida (P. Karst.) J. Erikss. & Ryvarden (15)	*Amanita pantherina* (DC.) Krombh. (3)	*Lecanora chlarotera* Nyl. (78)	*Melanohalea exasperata* (6)
Auriscalpium vulgare Gray(14)	*Byssomerulius corium* (Pers.) Parmasto (3)	*Lepra albescens* (Huds.) Hafellner (75)	*Pertusaria flavida* (DC.) J.R. Laundon (6)
Rhizopogon roseolus (Corda) Th. Fr. (13)	*Fistulina hepatica* (Schaeff.) With. (3)	*Pseudevernia furfuracea* (L.) Zopf (75)	*Hypocenomyce scalaris* (Ach.) M. Choisy (5)
Stereum hirsutum (Willd.) Pers. (13)	*Lyomyces crustosus* (Pers.) P. Karst. (3)	*Physcia aipolia* (Ehrh. ex Humb.) Fürnr. (70)	*Lecanora chlarotera* (5)
Efibula tuberculata (P. Karst.) Zmitr. & Spirin (12)	*Peniophora quercina* (3)	*Ramalina farinacea* (L.) Ach. (65)	*Lobarina scrobiculata* (Scop.) Nyl. (5)
Mycena polygramma (Bull.) Gray (11)	*Radulomyces confluens* (Fr.) M.P. Christ. (3)	*Xanthoria parietina* (L.) Th. Fr. (60)	*Nephroma laevigatum* (5)
Phanerochaete velutina (DC.) P. Karst. (11)	*Stereum hirsutum* (13)	*Melanelixia glabra* (Schaer.) O. Blanco, A. Crespo, Divakar, Essl., D. Hawksw. & Lumbsch (59)	*Pertusaria pertusa* (L.) Tuck. (4)
Amanita muscaria (L.) Lam. (10)	*Trametes versicolor* (3)	*Melanohalea exasperata* (De Not.) O. Blanco, A. Crespo, Divakar, Essl., D. Hawksw. & Lumbsch (55)	*Physconia distorta* (4)
Amphinema byssoides (Pers.) J. Erikss. (10)	*Amanita gemmata* (Fr.) Bertill. (2)	*Peltigera collina* (Ach.) Schrad. (51)	*Ramalina farinacea* (4)
Trametes versicolor (L.) Lloyd (9)	*Amanita rubescens* Pers. (2)	*Nephroma laevigatum* Ach. (50)	*Lecanora intumescens* (Rebent.) Rabenh. (3)

When comparing species richness between biogeographical regions, both in *Quercus pyrenaica* and *Castanea sativa* communities, it was observed that the species richness of macrofungi and lichens is much higher in the Mediterranean region than in the Atlantic region (Table 3).

Table 3. Beta diversity between the two biogeographical regions of Castilla y León, both of lichenised fungi and macrofungi in *Castanea sativa* and *Quercus pyrenaica* forests. SpAt: number of species in the Atlantic region; SpMed: number of species in the Mediterranean region; SC: Species in common; Is: Sørensen similarity index.

Forest Type	Fungal Group	SpAt	SpMed	SC	Is
Pyrenean oak	Macrofungi	240	408	88	0.27
	Lichenised fungi	284	433	191	0.53
Chestnut	Macrofungi	2	126	1	0.01
	Lichenised fungi	24	76	15	0.30

3.2. Beta Diversity

Macrofungi and lichenised fungi shared 87 and 79 species, respectively, between the two forests. In the case of chestnut groves, they accounted for 68.5% and 92.9% of the total number of species referenced for both groups of organisms. The results obtained for Sørensen's index in these two fungal groups were almost identical, 0.25 and 0.26, respectively. The results obtained for Sørensen's index were higher for lichens in Pyrenean oak forests and showed very low values for macrofungi in chestnut forests (Table 3).

3.3. Macrofungi Lifestyle, Habit and Environmental Sensitivity of Lichenised Fungi

Most of the macrofungal species were found to be saprophytic in both studied forest formations. However, the proportion of saprophytic species was higher in Pyrenean oak forests, with 71.8% compared to 55.1% in chestnut forests (Figure 3A). Similarly, chestnut groves showed a higher proportion of mycorrhizal species (44.1% vs. 27.7%), while the representation of parasitic species was less than 1%.

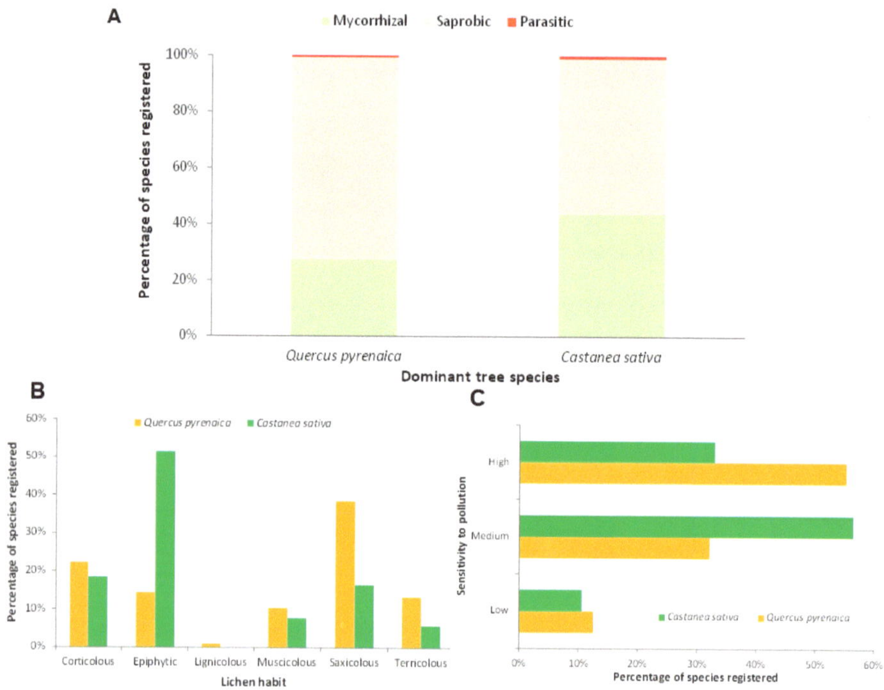

Figure 3. Lifestyle percentage of macrofungi registered species (**A**), percentage of lichen reported species habit (**B**) and sensitivity to contamination of lichen registered species (in %) (**C**) in *Castanea sativa* and *Quercus pyrenaica* forest ecosystems of Castilla y León region (C-NW Spain).

When comparing the habits of the lichenised fungi present in the two forest formations, it was observed that the three most common habits were epiphytic, saxicolous, and corticolous (Figure 3B). However, in Pyrenean oak, saxicolous species dominated, whereas in *Castanea sativa* trees, most were epiphytic. In addition, differences were also observed in terricolous species, which were twice as common in *Quercus pyrenaica*. When analysing the sensitivity to pollution of the lichen species present in both forest habitats, it was observed that most of the Pyrenean oak species had a high sensitivity, in contrast to the chestnut species, which mostly had a medium sensitivity (Figure 3C).

4. Discussion

The higher number of species of macrofungi and lichenised fungi in *Quercus pyrenaica* forests compared to those recorded in *Castanea sativa* forests seems to be related to the extent of these ecosystems in the studied region [2]. However, when assessing the relationship between the species number and hectares of each forest formation, the chestnut groves revealed higher species richness per hectare for both fungal groups. Similar results were obtained in field works carried out in the NW Spain for epiphytic lichens [33] and in the CW of the Iberian Peninsula for macrofungi [34]. Other studies carried out in smaller plots occupied by the analysed forest formations revealed higher alpha diversity associated with *Quercus pyrenaica* than with *Castanea sativa* [35]. These results showed the importance of local studies for a more accurate understanding of fungal diversity [36], but larger-scale studies may have greater applicability for analysing phenomena such as global change [37]. In any case, both forest types had a higher specific fungal richness than other forest environments, such as pine forests of different species [33,38], although in this case, greater human intervention in the management of these formations must be taken into account [39]. Differences in the most frequent species might have been caused by specific habitat characteristics such as resource availability, soil composition, or tree species characteristics such as pH and bark texture [40].

The analysis of the species with the highest number of records in the consulted databases led to the presence of typical species in other forest environments than that of these two trees of the family Fagaceae. In *Quercus pyrenaica* forests, *Auriscalpium vulgare* or *Rhizopogon roseolus* were frequently cited, while in chestnut forests, *Hygrophoropsis aurantiaca* was registered (Table 2). All of these species are typical of coniferous forests [39]. The presence of macrofungi associated with various pine species can be explained by the composition of Pyrenean oak forests in many areas of the Iberian Peninsula, where some species, such as *Pinus sylvestris* L. may occur, sometimes to a remarkable extent [41]. The choice of a 200 m buffer in the delimitation of mapped areas for *Quercus pyrenaica* habitats may also have influenced a higher number of records of pine-associated macrofungal species, although this type of buffer is commonly used in geographic vegetation analysis [42]. It should be noted that other fungal species were also recorded in association with other deciduous forest species, e.g., *Xylaria carpophila* (Pers.) Fr. on beechnuts of *Fagus sylvatica* L. The growth of certain macrofungi associated with dominant tree species, such as *Peniophora quercina* or *Lanzia echinophila* (Bull.) Korf indicates their ecosystemic role as recyclers of plant material and their influence on the proper functioning of both environments [17].

The assessment of the most diverse macrofungal genera on chestnut groves revealed that *Amanita* Pers., *Cortinarius* (Pers.) Gray., and *Lactarius* Pers., all essentially mycorrhizal [30], had the highest number of species. Other studies carried out in the same environments in the northwest of the Iberian Peninsula [43] also showed a great diversity of mycorrhizal genera. The higher species richness of the genus *Cortinarius* has been highlighted as an indicator of mature forests [44], which for *Castanea sativa* forests in the studied area could indicate a better conservation status compared to those in northeastern Portugal [43]. The presence of ectomycorrhizal macrofungi also seems to have a relevant role in the prevention of forest pathologies in chestnut groves, such as those caused by various species of the genus *Phytophthora* de Bary [45]. In Pyrenean oak forest formations, mycorrhizal genera such as *Cortinarius*, *Russula* Pers., and *Lactarius* also stood out for their specific richness, although other saprophytic genera appeared, such as *Mycena* (Pers.) Roussel and *Marasmius* Fr. The presence of saprophytic genera among the most diverse has already been reported in other macrofungal analyses carried out in *Quercus pyrenaica* forests in the Iberian Peninsula [46], which may indicate a greater use of these formations for livestock [47] in relation to the smaller areas occupied by chestnut forests. A high stocking rate could lead to a lower degree of conservation of Pyrenean oak ecosystems [48] and a higher presence of saprophytic species [49].

The dominance of Basidiomycota taxa has already been reported in other studies [49,50], although a higher diversity of Ascomycota associated with Pyrenean oak was observed. This could be influenced by habitat-specific factors or particular ecological interactions, as Ascomycota decrease with stand age and are displaced by Basidiomycota at those developmental stages where mycorrhizal fungi are more abundant [50,51]. The lower richness of Ascomycota in chestnut groves coincides with the findings of other studies [52] and could, according to these records, be related to a higher degree of maturity of chestnut groves compared to Pyrenean oak forests in the Iberian Peninsula.

Some studies revealed a greater interest in fungal diversity in peninsular forest formations in the Mediterranean region [53], which may explain the results obtained in the analysis of records for the studied area, although the large surface area represented by the Mediterranean region compared to the Atlantic region must also be taken into account. The species number per hectare, however, was higher in the Atlantic region, except for the case of the macrofungi that developed their fruiting bodies in *Castanea sativa* forests. These data could be biased by the low number of citations in the Atlantic region, probably due to lower sampling in that area and perhaps conditioned by the locations of chestnut groves in these areas [54]. The higher species richness per unit area in the Atlantic biogeographic region is probably due to the existence of more favourable climatic conditions for the development of favourable habitats, as has been shown by comparing fungal diversity between the two biogeographic regions [55] in other geographical areas where they coexist.

Beta diversity between the two forest environments showed a low similarity when using any of the two fungal groups considered. The values obtained are similar to those presented in other peninsular studies comparing macrofungal diversity in tree habitats of various species of the genus *Quercus* L. [56] or even other wooded ecosystems in several areas of the Mediterranean Basin [28]. In the case of lichenised fungi, some studies suggest a greater species similarity between native trees in relation to other exotic species [57], although the presence of many taxa is conditioned by factors other than the species acting as a phorophyte, such as the age of the trees [58]. The use of other biological indicators assessing the influence of bioclimatic regions on species similarity or divergence also revealed that species distribution is influenced by other factors [59], such as the dominant tree species [38]. Forest structure and dynamics, with circumstances such as forest age or tree age, promote forests with more diverse structures, different substrate types and microhabitats, which positively influences species diversity, especially rare, threatened, or late successional species [40].

The analysis of macrofungal life forms in these two forest formations was carried out in order to contribute to the assessment of the conservation status of the Pyrenean oak and chestnut forests of the studied area. The predominance of saprophytic taxa in Pyrenean oak ecosystems, as well as that of mycorrhizal species in chestnut forests, coincided with analyses carried out in NE Portugal [35], although the percentage of saprophytic species present in *Quercus pyrenaica* formations was higher in the studied area. An optimal conservation status in Mediterranean forest environments estimates an approximate presence of 51% of saprophytic species, 47% of symbiotic species, and 2% of parasitic species [60], which would indicate that the Pyrenean oak formations in the administrative region of Castilla y León did not have an optimal conservation status [19]. The high percentage of saprobes in the Pyrenean oak habitats was very similar to that obtained in studies carried out in Mediterranean mixed *Quercus* forests, in which livestock use is considered to be the main cause that may have conditioned the increase in macrofungal species with this lifestyle [48,56]. The poor state of conservation of Pyrenean oak forest formations could be due to the fact that many of them have been configured as pastures and a high stocking rate, which would increase the level of organic matter and thus the amount of saprophytic fungi, as pointed out in the analysis of the diversity of macrofungal genera. In contrast, similar levels of saprophytic and mycorrhizal species were observed in the chestnut groves of the study area, as was the case in a study carried out in a mature Mediterranean forest dominated by holm oak in which the percentage of saprophytes decreased because the forests were more closed with a diverse shrub substrate [49]. These factors, together with the orography and other issues [54], could have influenced *Castanea sativa* forests to show a better degree of conservation than *Quercus pyrenaica* forests [19].

The most common habit of the lichenicolous species differed according to the forest environment studied. This difference in the dominant habit might be conditioned by the nature of each of these habitats since the distribution pattern of epiphytes and saxicolous is very different, with saxicolous being more abundant in exposed rocky substrates of mountainous habitats and epiphytes in mixed-use landscapes with a variety of substrates [61,62]. Thus, the dominance of saxicolous lichens in Pyrenean oak could be related to more rocky substrates that favour these preferences [33]. Moreover, epiphytes need a certain degree of shade and humidity to live, so in Mediterranean areas characterised by low rainfall, high temperatures, and seasonal droughts, they would be more dependent on the nature of the forest canopy and shrubs that provide shade and higher humidity to tree trunks [61,63]. Forests of both trees are able to create distinct microclimates under their canopy [64], so the abundance of epiphytic lichens in chestnut groves could be due to the more suitable environmental conditions for this habit, with sufficient light and less water deficit [65].

The known sensitivity of lichenised fungi to air quality, mainly sulphur and nitrogen dioxide deposition, influences the richness and composition of forest lichens [66]. The occurrence of species more vulnerable to pollution in Pyrenean oak might suggest that these areas have better air quality in the study area. Air quality limits the possibility of recolonisation of highly sensitive species characteristic of unpolluted environments [67], which could have an impact on lower air quality in environments linked to chestnut forests. In a study carried out in chestnut groves in southern Italy [68], it was observed that lichen species less tolerant to environmental pollution were found further away from agroforestry areas. Part of these tree formations in the study area were related to various human uses of this tree (e.g., chestnut harvesting, basketry, etc.), as shown in a study carried out in the northwest of the Iberian Peninsula [69].

Studies combining the use of databases and GIS are very rare, although this work aims to show that databases with records of presence could be very useful for diversity studies and for assessing the degree of conservation of forest formations, at least for initial or preliminary evaluations. It should be noted that further studies and fieldwork on this group of organisms are needed to better understand their ecology and their efficient use in the management of the plant communities in which they develop their life cycles.

5. Conclusions

The use of biodiversity databases or repositories could be very useful for preliminary analyses of the diversity of a given group of organisms over a large area. This type of analysis with data taken in the field would be extremely complex to obtain for large territories, so the use of these databases combined with the management of geographic information systems would be a valuable tool.

Similarly, the use of these tools to use a group of organisms as assessors of the conservation status of forest ecosystems could be interesting as a preliminary step in more comprehensive studies in more localised areas. More focused work planning in specific areas could then be carried out in a more systematised and effective way.

Supplementary Materials: The following supporting information can be downloaded at: https://www.mdpi.com/article/10.3390/f16010009/s1, Table S1. Species of macrofungi and lichens occurring in *Quercus pyrenaica* and *Castanea sativa* forests of Castilla y León (C-NW Spain; in alphabetical order). Source: GBIF (2023). Table S2. Main fungal Orders (in number of species) registered in *Quercus pyrenaica* and *Castanea sativa* forests of the Castilla y León region (Spain). Source: GBIF (2023).

Author Contributions: Conceptualization, D.R.-d.l.C. and L.D.-S.; methodology, D.R.-d.l.C. and L.D.-S.; software, L.D.-S. and S.P.-A.; validation, D.R.-d.l.C., S.P.-A. and L.D.-S.; formal analysis, D.R.-d.l.C., S.P.-A. and L.D.-S.; investigation, D.R.-d.l.C., S.P.-A. and L.D.-S.; resources, D.R.-d.l.C. and L.D.-S.; data curation, D.R.-d.l.C., S.P.-A. and L.D.-S.; writing—original draft preparation, D.R.-d.l.C., S.P.-A. and L.D.-S.; writing—review and editing, D.R.-d.l.C., S.P.-A. and L.D.-S.; visualization, D.R.-d.l.C. and L.D.-S.; supervision, D.R.-d.l.C. and L.D.-S.; project administration, D.R.-d.l.C. and L.D.-S.; funding acquisition, D.R.-d.l.C. and L.D.-S. All authors have read and agreed to the published version of the manuscript.

Funding: This research received no external funding.

Data Availability Statement: The data used for this article are available at https://www.gbif.org/occurrence/download/0088666-230530130749713.

Conflicts of Interest: The authors declare no conflict of interest.

References

1. Naranjo-Ortiz, M.A.; Gabaldón, T. Fungal evolution: Diversity, taxonomy and phylogeny of the Fungi. *Biol. Rev.* **2019**, *94*, 2101–2137. [CrossRef]
2. Mueller, G.M.; Schmit, J.P.; Leacock, P.R.; Buyck, B.; Cifuentes, J.; Desjardin, D.E.; Halling, R.E.; Hjortstam, K.; Iturriaga, T.; Larsson, K.H.; et al. Global diversity and distribution of macrofungi. *Biodivers. Conserv.* **2007**, *16*, 37–48. [CrossRef]
3. Lumbsch, H.T.; Leavitt, S.D. Goodbye morphology? A paradigm shift in the delimitation of species in lichenized fungi. *Fungal Divers.* **2011**, *50*, 59–72. [CrossRef]
4. Sipman, H.J.; Aptroot, A. Where are the missing lichens? *Mycol. Res.* **2002**, *105*, 1433–1439. [CrossRef]
5. Warnasuriya, S.D.; Udayanga, D.; Manamgoda, D.S.; Biles, C. Fungi as environmental bioindicators. *Sci. Total Environ.* **2023**, *892*, 164583. [CrossRef] [PubMed]
6. Joimel, S.; Cortet, J.; Jolivet, C.C.; Saby, N.P.A.; Chenot, E.D.; Branchu, P.; Consalès, J.N.; Lefort, C.; Morel, J.L.; Schwartz, C. Physico-chemical characteristics of topsoil for contrasted forest, agricultural, urban and industrial land uses in France. *Sci. Total Environ.* **2016**, *545*, 40–47. [CrossRef]
7. Zaghloul, A.; Saber, M.; Gadow, S. Biological indicators for pollution detection in terrestrial and aquatic ecosystems. *Bull. Natl. Res. Cent.* **2020**, *44*, 127. [CrossRef]
8. Gerhardt, A. Bioindicator species and their use in biomonitoring. *Environ. Monit. Assess.* **2002**, *1*, 77–123.
9. Holt, E.A.; Miller, S.W. Bioindicators: Using organisms to measure. *Nature* **2011**, *3*, 8–13.
10. McGeoch, M.A. The selection, testing and application of terrestrial insects as bioindicators. *Biol. Rev.* **1998**, *73*, 181–201. [CrossRef]
11. Puig-Gironès, R.; Real, J. A comprehensive but practical methodology for selecting biological indicators for long-term monitoring. *PLoS ONE* **2022**, *17*, e0265246. [CrossRef] [PubMed]
12. Parmar, T.K.; Rawtani, D.; Agrawal, Y.K. Bioindicators: The natural indicator of environmental pollution. *Front. Life Sci.* **2016**, *9*, 110–118. [CrossRef]
13. Stankovic, S.; Kalaba, P.; Stankovic, A.R. Biota as toxic metal indicators. *Environ. Chem. Lett.* **2014**, *12*, 63–84. [CrossRef]
14. Soares, D.M.; Procópio, D.P.; Zamuner, C.K.; Nóbrega, B.B.; Bettim, M.R.; de Rezende, G.; Lopes, P.M.; Pereira, A.B.D.; Bechara, E.J.H.; Oliveira, A.G.; et al. Fungal bioassays for environmental monitoring. *Front. Bioeng. Biotechnol.* **2022**, *10*, 954579. [CrossRef]
15. Lelli, C.; Bruun, H.H.; Chiarucci, A.; Donati, D.; Frascaroli, F.; Fritz, Ö.; Goldberg, I.; Nascimbene, J.; Tøttrup, A.P.; Rahbek, C.; et al. Biodiversity response to forest structure and management: Comparing species richness, conservation relevant species and functional diversity as metrics in forest conservation. *For. Ecol. Manag.* **2019**, *432*, 707–717. [CrossRef]
16. Ristić, S.; Šajn, R.; Stamenković, S. Lichens as the Main Indicator in Biological Monitoring of Air Quality. In *Contaminant Levels and Ecological Effects. Emerging Contaminants and Associated Treatment Technologies*; Balabanova, B., Stafilov, T., Eds.; Springer: Cham, Switzerland, 2021; pp. 121–129. ISSN 2524-6410.
17. Egli, S. Mycorrhizal mushroom diversity and productivity—An indicator of forest health? *Ann. For. Sci.* **2011**, *68*, 81–88. [CrossRef]
18. Will-Wolf, S. *Analyzing Lichen Indicator Data in the Forest Inventory and Analysis Program*; US Department of Agriculture, Forest Service, Pacific Northwest Research Station: Portland, OR, USA, 2011; 61p, ISBN 978-1-5059-0726-1.
19. Sánchez-Martínez, C.; Benito-Peñil, D.; García de Enterría, S.; Barajas-Castro, I.; Martín-Herrero, N.; Pérez-Ruiz, C.; Sánchez-Sánchez, J.; Sánchez-Agudo, J.A.; Rodríguez-de la Cruz, D.; Galante-Patiño, E. *Manual de Gestión Sostenible de Bosques Abiertos Mediterráneos*; Castilla Tradicional: Valladolid, Spain, 2012; pp. 68–79. ISBN 978-84-940714-0-9.
20. Forestry Map of Spain (MFE), Spanish Ministry for the Ecological Transition and the Demographic Challenge. Available online: https://www.miteco.gob.es/es/cartografia-y-sig/ide/descargas/biodiversidad/mfe.aspx (accessed on 27 January 2023).

21. ESRI. *ArcGIS Desktop: Release 10.8*; Environmental Systems Research Institute: Redlands, CA, USA, 2023.
22. European Commission. Council Directive 92/43/EEC of 21 May 1992 on the conservation of natural habitats and of wild fauna and flora. *Off. J. Eur. Union* **1992**, *L206*, 50.
23. European Commission. Council Directive 2007/2/EC of the European Parliament and of the Council of 14 March 2007 establishing an Infrastructure for Spatial Information in the European Community (INSPIRE). *Off. J. Eur. Union* **2007**, *L108*, 1.
24. GBIF.org. GBIF Occurrence Download. Available online: https://www.gbif.org/occurrence/download/0088666-230530130749713 (accessed on 5 January 2023).
25. GBIF Data Standards. Available online: https://www.gbif.org/standards (accessed on 5 January 2023).
26. CABI Index Fungorum. Available online: http://www.indexfungorum.org/names/names.asp (accessed on 8 May 2023).
27. Belguidoum, A.; Lograda, T.; Ramdani, M. Diversity and distribution of epiphytic lichens on *Cedrus atlantica* and *Quercus faginea* in Mount Babor Forest, Algeria. *Biodiversitas* **2021**, *22*, 887–899. [CrossRef]
28. Fernández Ruiz, A.; Rodríguez de la Cruz, D.; Vicente Villardón, J.L.; Sánchez Durán, S.; García Jiménez, P.; Sánchez Sánchez, J. Considerations on Field Methodology for Macrofungi Studies in Fragmented Forests of Mediterranean Agricultural Landscapes. *Agronomy* **2022**, *12*, 528. [CrossRef]
29. Eyssartier, G.; Roux, P. *Le Guide des Champignons France et Europe*; Éditions Belin: Paris, France, 2017; 1152p, ISBN 978-2-410-01042-8.
30. Salcedo, I. Diversidad taxonómica y funcional de los hongos. *Artikutza* **2019**, *100*, 151–175.
31. Barreno Rodríguez, E.; Pérez-Ortega, S. *Líquenes de la Reserva Natural Integral de Muniellos, Asturias*; KRK: Oviedo, Spain, 2003; 595p, ISBN 84-96119-36-X.
32. Martellos, S.; Conti, M.; Nimis, P.L. Agregación de datos de líquenes italianos en ITALIC 7.0. *Rev. De Hongos* **2023**, *9*, 556.
33. Maceda-Veiga, A.; Gómez-Bolea, A. Small, fragmented native oak forests have better preserved epiphytic lichen communities than tree plantations in a temperate sub-oceanic Mediterranean climate region. *Bryologist* **2017**, *120*, 191–201. [CrossRef]
34. Santos-Silva, C.; Natário, B.; Andrade, J.; Louro, R. Serra de São Mamede Natural Park, a macrofungal diversity hotspot in the Mediterranean region. *Check List* **2022**, *18*, 109–137. [CrossRef]
35. Baptista, P.; Sousa, M.J.; Dias, R.; Matos, M.; Rodrigues, P.; Martins, A. Macrofungi from *Castanea sativa* Mill. and Quercus pyrenaica Wild.: Evaluation of mycorrhizal vs nonmycorrhizal fungi biodiversity—Project AGRO 689. In Proceedings of the 5th International Conference on Mycorrhiza (ICOM5), Granada, Spain, 23–27 July 2006.
36. Toljander, J.F.; Eberhardt, U.; Toljander, Y.K.; Paul, L.R.; Taylor, A.F. Species composition of an ectomycorrhizal fungal community along a local nutrient gradient in a boreal forest. *New Phytol.* **2006**, *170*, 873–884. [CrossRef]
37. Yu, H.; Wang, T.; Skidmore, A.; Heurich, M.; Bässler, C. How future climate and tree distribution changes shape the biodiversity of macrofungi across Europe. *Divers. Distrib.* **2023**, *29*, 666–682. [CrossRef]
38. González-Montelongo, C.; Pérez-Vargas, I. Is an invasive alien tree able to sustain a similar lichen diversity as the native forest? The case of the sweet chestnut (*Castanea sativa* Mill.) and the laurel forest in Macaronesia. *For. Ecol. Manag.* **2021**, *488*, 119009. [CrossRef]
39. Collado, E.; Bonet, J.A.; Alday, J.G.; de Aragón, J.M.; De-Miguel, S. Impact of forest thinning on aboveground macrofungal community composition and diversity in Mediterranean pine stands. *Ecol. Indic.* **2021**, *133*, 108340. [CrossRef]
40. Nascimbene, J.; Thor, G.; Nimis, P.L. Effects of forest management on epiphytic lichens in temperate deciduous forests of Europe–A review. *For. Ecol. Manag.* **2013**, *298*, 27–38. [CrossRef]
41. Vilches de la Serna, B.; Sánchez-Mata, D.; Gavilán, R.G. Marcescent *Quercus pyrenaica* forest on the Iberian Peninsula. In *Vegetation Structure and Function at Multiple Spatial, Temporal and Conceptual Scales*; Box, E.O., Ed.; Springer International Publishing: Berlin/Heidelberg, Germany, 2016; pp. 257–283. ISSN 2198-2562.
42. Fernandes, M.R.; Aguiar, F.C.; Ferreira, M.T. Assessing riparian vegetation structure and the influence of land use using landscape metrics and geostatistical tools. *Landsc. Urban Plan.* **2011**, *99*, 166–177. [CrossRef]
43. Baptista, P.; Martins, A.; Tavares, R.M.; Lino-Neto, T. Diversity and fruiting pattern of macrofungi associated with chestnut (*Castanea sativa*) in the Trás-os-Montes region (Northeast Portugal). *Fungal Ecol.* **2010**, *3*, 9–19. [CrossRef]
44. Richard, F.; Moreau, P.A.; Selosse, M.A.; Gardes, M. Diversity and fruiting patterns of ectomycorrhizal and saprobic fungi in an old-growth Mediterranean forest dominated by *Quercus ilex* L. *Can. J. Bot.* **2004**, *82*, 1711–1729. [CrossRef]
45. Blom, J.M.; Vannini, A.; Vettraino, A.M.; Hale, M.D.; Godbold, D.L. Ectomycorrhizal community structure in a healthy and a *Phytophthora*-infected chestnut (*Castanea sativa* Mill.) stand in central Italy. *Mycorrhiza* **2009**, *20*, 25–38. [CrossRef] [PubMed]
46. Branco, S.; Baptista, P.; Martins, A.; Rodrigues, A.P. Estudo da comunidade macrofúngica associada a *Quercus pyrenaica* Willd. no Parque Natural de Montesinho. In Proceedings of the VII Congresso Luso-Galaico de Macromicologia. Macrofungos: Diversidade e Biotecnología, Vila Real, Portugal, 13–15 October 2005.
47. Núñez, V.; Hernando, A.; Velázquez, J.; Tejera, R. Livestock management in Natura 2000: A case study in a *Quercus pyrenaica* neglected coppice forest. *J. Nat. Conserv.* **2012**, *20*, 1–9. [CrossRef]

48. García Jiménez, P.; Fernández Ruiz, A.; Sánchez Sánchez, J.; Rodríguez de la Cruz, D. Mycological Indicators in Evaluating Conservation Status: The Case of *Quercus* spp. Dehesas in the Middle-West of the Iberian Peninsula (Spain). *Sustainability* **2020**, *12*, 10442. [CrossRef]
49. Fernández, A.; Sánchez, S.; García, P.; Sánchez, J. Macrofungal diversity in an isolated and fragmented Mediterranean Forest ecosystem. *Plant Biosyst.* **2020**, *154*, 139–148. [CrossRef]
50. Martin-Pinto, P.; Sanz-Benito, I.; Santos, M.; Oria-de-Rueda, J.A.; Geml, J. Anthropological impacts determine the soil fungal distribution of Mediterranean oak stands. *Ecol. Indic.* **2021**, *132*, 108343. [CrossRef]
51. Zhang, K.; Cheng, X.; Shu, X.; Liu, Y.; Zhang, Q. Linking soil bacterial and fungal communities to vegetation succession following agricultural abandonment. *Plant Soil* **2018**, *431*, 19–36. [CrossRef]
52. Anthony, M.A.; Crowther, T.W.; Van Der Linde, S.; Suz, L.M.; Bidartondo, M.I.; Cox, F.; Schaub, M.; Rautio, P.; Ferretti, M.; Vesterdal, L.; et al. Forest tree growth is linked to mycorrhizal fungal composition and function across Europe. *ISME J.* **2022**, *16*, 1327–1336. [CrossRef] [PubMed]
53. Domínguez-Núñez, J.A.; Oliet, J.A. Management of mushroom resources in Spanish forests: A review. *Forestry* **2023**, *96*, 135–154. [CrossRef]
54. Rubio, A. 9260 Bosques de *Castanea sativa*. In *Bases Ecológicas Preliminares Para la Conservación de los Tipos de Hábitat de Interés Comunitario en España*; Dirección General de Medio Natural y Política Forestal, Ministerio de Medio Ambiente, y Medio Rural y Marino, Ed.; Ministerio de Medio Ambiente, y Medio Rural y Marino: Madrid, Spain.
55. Djemiel, C.; Dequiedt, S.; Horrigue, W.; Bailly, A.; Lelièvre, M.; Tripied, J.; Guilland, C.; Perrin, S.; Comment, G.; Saby, N.P.A.; et al. Unraveling biogeographical patterns and environmental drivers of soil fungal diversity at the French national scale. *SOIL* **2023**, *10*, 251–273. [CrossRef]
56. Sanz, A.; Fernández Ruiz, A.; García Jiménez, P.; Sánchez Sánchez, J.; Rodríguez de la Cruz, D. Estudio preliminar de la diversidad de macrohongos en El Monte de Villoria (Salamanca, España). *Bot. Compl.* **2022**, *46*, e80421. [CrossRef]
57. Nascimbene, J.; Lazzaro, L.; Benesperi, R. Patterns of β-diversity and similarity reveal biotic homogenization of epiphytic lichen communities associated with the spread of black locust forests. *Fungal Ecol.* **2015**, *14*, 1–7. [CrossRef]
58. Brunialti, G.; Giordani, P.; Ravera, S.; Frati, L. The reproductive strategy as an important trait for the distribution of lower-trunk epiphytic lichens in old-growth vs. non-old growth forests. *Forests* **2020**, *12*, 27. [CrossRef]
59. Baselga, A.; Jiménez-Valverde, A. Environmental and geographical determinants of beta diversity of leaf beetles (Coleoptera: Chrysomelidae) in the Iberian Peninsula. *Ecol. Entomol.* **2007**, *32*, 312–318. [CrossRef]
60. Moreno, G. *Setas Micorrrizógenas, Parásitas y Saprófitas; Una Forma de Valorar el Impacto Ambiental en Nuestros Bosques*; Congress Communication: Laredo, Spain, 1996.
61. Aragón, G.; Martínez, I.; Izquierdo, P.; Belinchón, R.; Escudero, A. Effects of forest management on epiphytic lichen diversity in Mediterranean forests. *Appl. Veg. Sci.* **2010**, *13*, 183–194. [CrossRef]
62. Wolseley, P.A.; Stofer, S.; Mitchell, R.; Truscott, A.M.; Vanbergen, A.; Chimonides, J.; Scheidegger, C. Variation of lichen communities with landuse in Aberdeenshire, UK. *Lichenologist* **2006**, *38*, 307–322. [CrossRef]
63. Király, I.; Nascimbene, J.; Tinya, F.; Ódor, P. Factors influencing epiphytic bryophyte and lichen species richness at different spatial scales in managed temperate forests. *Biodivers. Conserv.* **2013**, *22*, 209–223. [CrossRef]
64. Belinchón, R.; Martínez, I.; Escudero, A.; Aragón, G.; Valladares, F. Edge effects on epiphytic communities in a Mediterranean *Quercus pyrenaica* forest. *J. Veg. Sci.* **2007**, *18*, 81–90. [CrossRef]
65. Tzenkova, A.; Ivancheva, J. Climate and microclimate in the region of the chestnut forests along the northern slopes of Belasitza Mountain. In Proceedings of the Sustainable Management of Sweet Chestnut Ecosystems-CAST Bul, Blagoevgrad, Bulgaria, 2–5 November 2005; pp. 37–44.
66. Nascimbene, J.; Ylisirniö, A.L.; Pykälä, J.; Giordani, P. Lichens: Sensitive indicators of changes in the forest environment. In *Integrative approaches as an opportunity for the conservation of forest biodiversity*; Kraus, D., Krumm, F., Eds.; European Forest Institute: Joensuu, Finland, 2013; pp. 180–185.
67. Paoli, L.; Fačkovcová, Z.; Lackovičová, A.; Guttová, A. Air pollution in Slovakia (Central Europe): A story told by lichens (1960–2020). *Biologia* **2021**, *76*, 3235–3255. [CrossRef]
68. Aprile, G.G.; Catalano, I.; Migliozzi, A.; Mingo, A. Monitoring epiphytic lichen biodiversity to detect environmental quality and air pollution: The case study of Roccamonfina Park (Campania Region–Italy). *Air Pollut. New Dev.* **2011**, *10*, 179007.
69. Roces-Diaz, J.V.; Díaz-Varela, E.R.; Barrio-Anta, M.; Álvarez-Álvarez, P. Sweet chestnut agroforestry systems in North-western Spain: Classification, spatial distribution and an ecosystem services assessment. *For. Syst.* **2018**, *27*, 10. [CrossRef]

forests

MDPI

Article

Bursaphelenchus xylophilus in *Pinus sylvestris*—The First Report in Europe

Luís Fonseca [1,2,*], Hugo Silva [1], Joana M. S. Cardoso [1], Ivânia Esteves [1], Carla Maleita [1,3], Sónia Lopes [4] and Isabel Abrantes [1]

1 Centre for Functional Ecology-CFE, Associate Laboratory for Sustainable Land Use and Ecosystem Services-TERRA, Department of Life Sciences, University of Coimbra, 3000-456 Coimbra, Portugal; hugoffs.95@gmail.com (H.S.); joana.cardoso@uc.pt (J.M.S.C.); iesteves@uc.pt (I.E.); carla.maleira@uc.pt (C.M.); isabel.abrantes@uc.pt (I.A.)

2 FitoLab—Laboratory for Phytopathology, Instituto Pedro Nunes (IPN), 3030-199 Coimbra, Portugal

3 Chemical Engineering and Renewable Resources for Sustainability (CERES), Department of Chemical Engineering, University of Coimbra, 3030-790 Coimbra, Portugal

4 Institute for Nature Conservation and Forests (ICNF, IP), 3200-219 Lousã, Portugal; sonia.lopes@icnf.pt

* Correspondence: luis.fonseca@uc.pt

Abstract: The pinewood nematode (PWN), *Bursaphelenchus xylophilus*, is the causal agent of pine wilt disease and is considered an A2 quarantine organism by the European Plant Protection Organisation. In Europe, this nematode has been reported in *Pinus pinaster*, *P. radiata*, and *P. nigra*. In May 2024, severe wilting symptoms were observed in *P. sylvestris* trees at Serra da Lousã (Coimbra, the central area of continental Portugal). Wood samples were collected from six wilted trees, and the presence of PWN was investigated. From these, *B. xylophilus* specimens were detected in five out of the six trees. Species identification was performed based on species-specific morphological diagnostic characters, and this was confirmed by real-time PCR using species-specific primers targeting the *B. xylophilus* satellite DNA region. This study presents the first detection of *B. xylophilus* in *P. sylvestris* in Portugal and in Europe.

Keywords: detection; morphology; pinewood nematode; *Pinus sylvestris*; real-time PCR

Citation: Fonseca, L.; Silva, H.; Cardoso, J.M.S.; Esteves, I.; Maleita, C.; Lopes, S.; Abrantes, I. *Bursaphelenchus xylophilus* in *Pinus sylvestris*—The First Report in Europe. *Forests* **2024**, *15*, 1556. https://doi.org/10.3390/f15091556

Academic Editor: Young-Seuk Park

Received: 12 July 2024
Revised: 29 August 2024
Accepted: 2 September 2024
Published: 4 September 2024

1. Introduction

The pinewood nematode (PWN), *Bursaphelenchus xylophilus* (Steiner & Buhrer, 1934) Nickle, 1970, an A2 quarantine organism according to the European Plant Protection Organisation (EPPO) is the causal agent of pine wilt disease (PWD) and is a major pathogen of conifers, which has negative impacts on forest health, natural ecosystem stability and on international wood trade. Species belonging to the genus *Pinus* are the main PWN hosts, and PWN transmission is carried out by insects, mostly belonging to the genus *Monochamus* (Coleoptera-Cerambycidae) [1]. *Pinus mugo* Turra, 1764 (dwarf mountain pine), *P. nigra* J.F. Arnold, 1785 (black pine), *P. pinaster* Aiton, 1789 (maritime pine) and *P. sylvestris* Linnaeus, 1753 (Scots pine) are among the most susceptible European *Pinus* species [2]. The *P. sylvestris* species are considered at risk in the northern and central areas of the EPPO region, whereas *P. mugo*, *P. nigra*, and *P. pinaster* are threatened in central and southern areas [1].

Bursaphelenchus xylophilus is a native species from North America, and it spread to Japan in the early XXth century and then into China, Korea, and Taiwan [1]. In Europe, the first detection was in 1999, in Continental Portugal, associated with *P. pinaster* [3]; in 2008, it was detected for the first time in Spain [4,5]; and in 2009, in Madeira Island in *P. pinaster* [6].

Following the detection of PWN in Portugal and in Europe, strict measures and control actions were immediately designed and implemented to prevent further PWN dispersal into other European countries. According to national and European Union rules, thousands of *Pinus* spp. wood samples have been collected and analyzed. Since then, the PWN was also reported in *P. radiata* D. Don, 1836 in Spain [7] and in *P. nigra*, in Portugal [8].

The Scots pine (*P. sylvestris*) is the most widespread species of the *Pinus* genus in the world and is native to Eurasia. It occupies naturally a territorial range from the Iberian Peninsula in the west to the far east of Russia [9,10]. It grows at a wide range of altitudes, from sea level to high mountains, in different soils, and in areas with different mean annual precipitation levels and extreme temperatures. Portugal represents the outermost western limit of Scots pine's natural range, and this species appears naturally in forest areas of high-altitude mountains from the north and central regions of Portugal (Marão, Peneda-Gerês, Lousã, Estrela) [11].

In the United States of America (USA), Canada, and Japan, *B. xylophilus* was already detected in *P. sylvestris* declining trees [12–17]. In Europe, to date, several field surveys have been conducted to determine the incidence and distribution of *Bursaphelenchus* species in *P. sylvestris* stands. Although there were no records of PWN in *P. sylvestris*, several other species, including *B. mucronatus* Mamiya & Enda, 1979 of the Xylophilus group, have been found associated with trees showing different degrees of decline [17–20]. This study has a main objective to assess the presence of *B. xylophilus* in *P. sylvestris* trees displaying wilting symptoms at Serra da Lousã, Portugal, using species-specific morphological diagnostic characters and real-time PCR.

2. Materials and Methods

2.1. Wood Sampling and Nematode Extraction

Six *P. sylvestris* trees displaying severe wilting symptoms, without needles, were sampled for the presence of PWN at Serra da Lousã, a mountain located at the south-western extremity of the European Central Cordillera with 15,158 ha and with 1204 m at the highest point [21]. Wood samples (one sample per tree) were collected at different points of the main trunk, at breast height, using a low-speed drill, and kept in plastic bags until nematode extraction. Nematodes were extracted by the tray method [22]. After 48 h, nematodes were collected, handpicked for morphological and molecular studies, and transferred to the fungus *Botrytis cinerea* Persoon, 1794 grown at 25 °C on malt extract agar medium (BD Difco, Franklin Lakes, NJ, USA) to obtain a nematode isolate [23].

2.2. Morphological Characterization

Females and males extracted from the wood samples were killed by heat in a drop of water on a cavity glass slide, mounted in water, immediately photographed, and morphologically characterized based on the species-specific diagnostic characters [17,24,25]. Photographs were taken with a Leica DM 2500 bright field light microscope using a Leica DFC 450 digital camera (Wetzlar, Germany).

2.3. Real-Time PCR

DNA from five nematodes collected from each wood sample was extracted by a simplified procedure described by [26] and included in EPPO standard diagnostic PM 7/4 (4) *Bursaphelenchus xylophilus* [25]. Real-time PCR reactions were performed using the CFX96 Touch™ Real-Time PCR Detection System (Bio-Rad, Hercules, CA, USA) in 20 µL reactions containing 5 µL of extracted genomic DNA, 10 µL of 2× NZYSpeedy qPCR Probe Master Mix (NZYtech, Lisbon, Portugal) and 200 nM of each primer and probe. Primers BSatF (5′-TGACGGAGTGAATTGACAAGACA-3′) and BSatRV (5′-AAGCTGAAACTTGCCATGCTAAA-3′) and the fluorogenic TaqMan probe BSatS (5′-ACACCATTCGAAAGCTAATCGCCTGAGA-3′) were designed by [27] for *B. xylophilus* species-specific detection based on the determined sequence of *B. xylophilus* satellite DNA and were used in this assay for the specific detection of PWN by real-time PCR, as recommended by EPPO [25]. Thermal cycling conditions consisted of 3 min at 95 °C, followed by 30 cycles of 5 s at 95 °C and 30 s at 60 °C. Data were analyzed using Bio-Rad CFX Manager 3.1 software, according to the manufacturer's instructions. Ct values were calculated by the software by determining the PCR cycle number at which the reporter fluorescence surpassed the background. The following controls included (i) negative isolation control (NIC) to monitor contaminations during DNA extraction—DNA extracted from five nematodes isolated from wood samples belonging to other families; (ii) positive isolation control (PIC) to ensure that DNA of sufficient quantity and quality is isolated—DNA extracted from five cultured *B. xylophilus* nematodes (isolate BxPt17AS); (iii) negative amplification control (NAC) to discard false positives due to contamination during reaction mix preparation—with molecular grade water used to prepare reaction mixs; and (iv) positive amplification control (PAC) to monitor the efficiency of the amplification—DNA previously extracted from cultured *B. xylophilus* (isolate BxPt17AS). Triplicate reactions for each control and sample were performed.

3. Results

Morphologically, the presence of *B. xylophilus* was detected in five out of the six wood samples collected from *P. sylvestris* trees. In trees 3 and 6, different developmental stages were detected (adults, propagative, and JIII dispersive juvenile stages) (Table 1). The males (Figure 1A–D) and females (Figure 1E–G) exhibited the main characteristics of *B. xylophilus* [17,24,25]: all specimens displayed a high cephalic region offset by a constriction, with six lips; a stylet with weakly developed basal knobs; a large medium bulb (Figure 1B,C); and an excretory pore either at or behind the median bulb (Figure 1C). Males (Figure 1A) had a strongly curved ventral tail, conoid, with a small terminal bursa, which could be seen in the dorso-ventral position, long spicules, a capitulum flattened with small condylus and a distinct rostrum, with a disc-like projection (cucullus) at the distal extremity (Figure 1D). Females presented a long post-uterine sac and vulva, usually at 70%–80% of the body length; a distinct vulval flap (Figure 1F); and a rounded tail with the absence of a mucro was observed (Figure 1G).

Table 1. Number of *Bursaphelenchus xylophilus* and other nematodes/100 g of wood collected from six *Pinus sylvestris* trees in Serra da Lousã (Coimbra, Portugal).

P. sylvestris	Wood Sample (g)	*B. xylophilus*/100 g			Other Nematodes/100 g
		Juveniles	Females	Males	
Tree 1	140	3 *	----	----	185
Tree 2	145	----	----	----	131
Tree 3	134	212 **	345	370	376
Tree 4	136	4 *	----	----	52
Tree 5	133	5 *	----	----	79
Tree 6	139	91 **	29	28	286

* JIII dispersive juvenile stages; ** propagative and JIII dispersive juvenile stages.

In trees 1, 4, and 5, only JIII dispersive juvenile stages were detected. In all samples, other nematodes (bacterivores and fungivores) belonging to the orders of Rhabditida and Panagrolaimida [28] were detected (Table 1).

Real-time PCR amplifications (Figure 2) confirmed the morphological identification. Genomic DNA from the five nematodes extracted from trees 1, 3, 4, 5, and 6 was specifically amplified as in the positive controls (PIC and PAC), while fluorescence remained below the threshold values for the negative controls (NIC and NAC) (Figure 2).

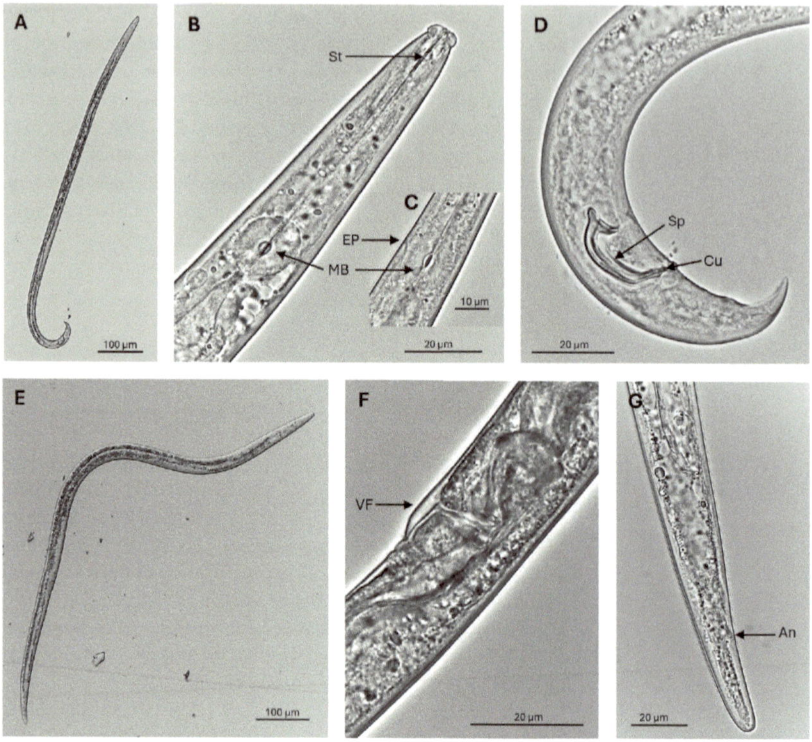

Figure 1. Light microscopic photographs of *Bursaphelenchus xylophilus* from *Pinus sylvestris*, collected at Serra da Lousã (Coimbra, Portugal), killed with heat and mounted in water. (**A**) Male (entire body); (**B**) anterior region; (**C**) medium bulb region; and (**D**) male tail. (**E**) Female (entire body); (**F**) vulvar region; and (**G**) female rounded tail. St: stylet; EP: excretory pore; MB: median bulb; Sp: spicules; Cu: cucullus; VF: vulvar flap; and An: anus.

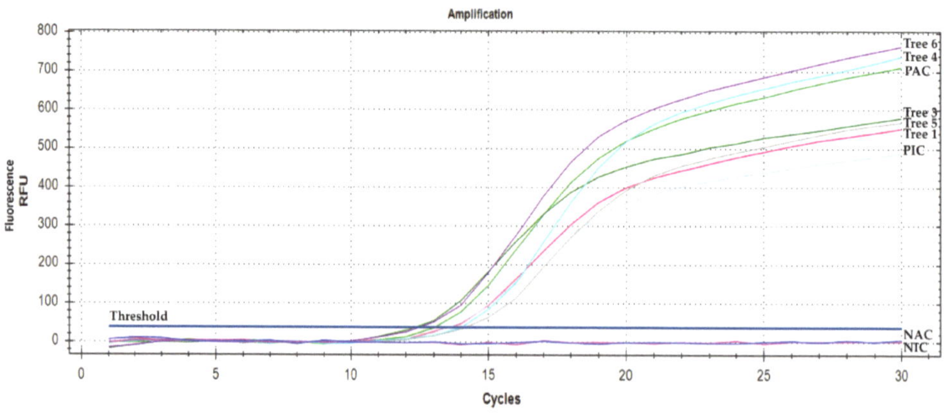

Figure 2. Real-time polymerase chain reaction assay for *Bursaphelenchus xylophilus* for species-specific detection based on a sequence of *B. xylophilus* satellite DNA using the Bio-Rad CFX96 Touch™ Real-Time PCR Detection System. The assays were conducted with five nematodes collected from suspensions obtained from six *Pinus sylvestris* trees (trees 1, 3, 4, 5, and 6—see Table 1). NIC: negative isolation control; NAC: negative amplification control; PIC: positive isolation control; and PAC: positive amplification control. Data from 1 of 3 technical replicates.

4. Discussion

This study represents the first report of PWN, an A2 EPPO quarantine organism, in *P. sylvestris* in Europe's natural environment. The presence of the PWN in Europe exemplifies the severe environmental and economic impacts that an invasive forest organism can have on pine forests and forest-based industries. The detection of PWN in Portugal has prompted severe control measures on the movement of susceptible wood products and coniferous wood within the EU. To prevent the introduction of PWN and its spread to new areas, Portugal has faced strict restrictions on the movement of plants, woody material, and forest products and implemented control and management strategies in forested areas, including (i) prospection and field surveys; (ii) the eradication of infected trees (iii) the management of tree decline and monitoring of the insect vector and other agents of decline; (iv) insect trap installation; and (v) the treatment and processing of host coniferous products [29]. To date, in Europe (Portugal and Spain), the PWN has been reported in *P. pinaster*, *P. radiata*, and *P. nigra* [3–8].

Scots pine is one of the most commercially important species, not only in Central Europe but also across Eurasia. The wood is workable and is one of the strongest softwoods. It is widely used in building and construction, furniture, pulp, and paper. The wood is durable in wet conditions, making it historically valuable for mining props, waterwheels, and piles. Additionally, Scots pine is often employed in land reclamation and for stabilizing loose sand due to its tolerance for poor soils [10,30].

The PWN detection in *P. sylvestris* within the context of climate change presents an ecological and economic threat to European forests due to the high susceptibility of this pine species to the PWN and due to its great distribution in European territories. Warmer temperatures and altered precipitation create more favorable conditions for PWN dissemination. In Europe, under current climate conditions, only 4.7% of *P. sylvestris* habitats are predicted to be at risk of being infected by the PWN, but by 2070, this percentage will rise to 61.9% on average, according to a predicted climate change scenario [31].

The high susceptibility of this pine species to the PWN has been confirmed in several research studies under artificial conditions [32–37]. Moreover, injured roots and stems of *P. sylvestris* seedlings could be infected with the PWN through nematode-infected sawdust [38], and PWN can persist for at least six years in 20-year-old inoculated *P. sylvestris* trees [39]. Concerning the insect vector, in Europe, *Monochamus galloprovincialis* (Olivier, 1795) is the main vector, and the most important hosts of *M. galloprovincialis* are pines, mainly *P. sylvestris*, *P. nigra*, *P. halepensis* Mill., 1768, and *P. pinaster* [1,40]. In Portugal, *P. pinaster* seems to be the favorite host for *M. galloprovincialis*; however, studies on feeding, oviposition, and attractive volatiles revealed that *P. sylvestris* can also be a good host for the insect vector [40–43].

5. Conclusions

This study presents the first detection of PWN in *P. sylvestris* in Portugal and in Europe. Species identification was based on species-specific morphological diagnostic characters and on real-time PCR, using species-specific primers targeting the *B. xylophilus* satellite DNA region. This study also adds valuable information to the current situation of the PWN in the EPPO zone, underlines the need for comprehensive PWN surveys across various pine species in Europe, and highlights the need for the implementation of integrated preventive measures by European countries to prevent the spread of PWN to uninfected areas where *P. sylvestris* is predominant, especially considering the challenges posed by climate change scenarios.

Author Contributions: Conceptualization, L.F. and I.A.; methodology, L.F., H.S., J.M.S.C., C.M., I.E. and S.L.; writing—original draft preparation, L.F.; writing—review and editing, all authors; supervision, L.F. and I.A.; project administration, L.F.; funding acquisition, L.F. and I.A. All authors have read and agreed to the published version of the manuscript.

Funding: This research supported by FEDER funds through the Portugal 2020 (PT 2020) COMPETE 2020 and by Fundação para a Ciência e Tecnologia (FCT), under contracts UIDB/04004/2020 (https://doi.org/10.54499/UIDB/04004/2020); UIDP/04004/2020 (https://doi.org/10.54499/UIDP/04004/2020); LA/P/0092/2020 (https://doi.org/10.54499/LA/P/0092/2020); UIDB/00102/2020 (https://doi.org/10.54499/UIDB/00102/2020), UIDP/00102/2020 (https://doi.org/10.54499/UIDP/00102/2020), CEECIND/02082/2017/CP1460/CT0004 (https://doi.org/10.54499/CEECIND/02082/2017/CP1460/CT0004, Ivânia Esteves), and Instituto do Ambiente, Tecnologia e Vida. Hugo Silva (Grant—2023.03527.BD) is funded by FCT, the European Social Fund (ESF), under the Programa Demografia, Qualificações e Inclusão (PDQI)—Portugal2030.

Data Availability Statement: The original contributions presented in this study are included in the article; further inquiries can be directed to the corresponding author.

Acknowledgments: The authors would like to thank Raúl Brito (ICNF, I.P.) for his assistance in collecting *P. sylvestris* wood samples.

Conflicts of Interest: The authors declare no conflicts of interest. The funders had no role in the design of the study; in the collection, analyses, or interpretation of data; in the writing of the manuscript; or in the decision to publish the results.

References

1. EPPO. *Bursaphelenchus xylophilus*. EPPO Datasheets on Pests Recommended for Regulation. 2024. Available online: https://gd.eppo.int (accessed on 27 May 2024).
2. Evans, H.F.; McNamara, D.G.; Braasch, H.; Chadoeuf, J.; Magnusson, C. Pest Risk Analysis (PRA) for the territories of the European Union (as PRA area) on *Bursaphelenchus xylophilus* and its vectors in the genus *Monochamus*. *EPPO Bull.* **1996**, *26*, 199–249. [CrossRef]
3. Mota, M.M.; Braasch, H.; Bravo, M.A.; Penas, A.C.; Burgemeister, W.; Metge, K.; Sousa, E. First report of *Bursaphelenchus xylophilus* in Portugal and in Europe. *Nematology* **1999**, *1*, 727–734. [CrossRef]
4. Abelleira, A.; Picoaga, A.; Mansilla, J.P.; Aguin, O. Detection of *Bursaphelenchus xylophilus*, causal agent of pine wilt disease on *Pinus pinaster* in Northwestern Spain. *Plant Dis.* **2011**, *95*, 776. [CrossRef] [PubMed]
5. Robertson, L.; Arcos, S.C.; Escuer, M.; Merino, R.S.; Esparrago, G.; Abelleira, A.; Navas, A. Incidence of the pinewood nematode *Bursaphelenchus xylophlius* Steiner & Buhrer, 1934 (Nickle, 1970) in Spain. *Nematology* **2011**, *13*, 755–757.
6. Fonseca, L.; Cardoso, J.M.S.; Lopes, A.; Pestana, M.; Abreu, F.; Nunes, N.; Mota, M.; Abrantes, I. The pinewood nematode, *Bursaphelenchus xylophilus*, in Madeira Island. *Helminthologia* **2012**, *49*, 96–103. [CrossRef]
7. Zamora, P.; Rodríguez, V.; Renedo, F.; Sanz, A.V.; Domínguez, J.C.; Pérez-Escolar, G.; Miranda, J.; Álvarez, B.; González-Casas, A.; Mayor, E.; et al. First report of *Bursaphelenchus xylophilus* causing pine wilt disease on *Pinus radiata* in Spain. *Plant Dis.* **2015**, *99*, 1449. [CrossRef]
8. Inácio, M.L.; Nóbrega, F.; Vieira, P.; Bonifácio, L.; Naves, P.; Sousa, E.; Mota, M. First detection of *Bursaphelenchus xylophilus* associated with *Pinus nigra* in Portugal and in Europe. *For. Pathol.* **2015**, *45*, 235–238. [CrossRef]
9. Skilling, D.D. *Pinus sylvestris* (Scotch pine). In *Silvics of North America*, 1st ed.; Burns, R.M., Honkala, B.H., Eds.; U.S. Department of Agriculture, Forest Service: Washington, DC, USA, 1990; pp. 489–496.
10. Houston, D.T.; de Rigo, D.; Caudullo, G. *Pinus sylvestris* in Europe: Distribution, habitat, usage, and threats. In *European Atlas of Forest Tree Species*; San-Miguel-Ayanz, J., de Rigo, D., Caudullo, G., Houston Durrant, T., Mauri, A., Eds.; European Union Official Publication: Luxembourg, 2016; pp. 132–133.
11. Fernandes, C.; Gaspar, M.J.; Pires, J.; Alves, A.; Simões, R.; Rodrigues, J.C.; Silva, M.E.; Carvalho, A.; Brito, J.E.; Lousada, J.L. Physical, chemical and mechanical properties of *Pinus sylvestris* wood at five sites in Portugal. *IForest* **2017**, *10*, 669–679. [CrossRef]
12. Dropkin, V.H.; Foudin, A.S. Report of the occurrence of *Bursaphelenchus lignicolus* induced pine wilt disease in Missouri. *Plant Dis. Rep.* **1979**, *63*, 904–905.
13. Dropkin, V.H.; Foudin, A.; Kondo, E.; Linit, M.; Smith, M.; Robbins, K. Pinewood nematode: A threat to U.S. forests? *Plant Dis.* **1981**, *65*, 1022–1027. [CrossRef]
14. Adams, J.C.; Morehart, A.L. Decline, and death of *Pinus* spp. in Delaware caused by *Bursaphelenchus xylophilus*. *J. Nematol.* **1982**, *14*, 382–385. [PubMed]
15. Carling, D.E. Some insects associate of the pinewood nematode in eastern Virgínia. *Can. J. Forest Res.* **1984**, *14*, 826–829. [CrossRef]
16. Malek, R.B.; Appleby, J.E. Epidemiology of pine wilt in Illinois. *Plant Dis.* **1984**, *68*, 180–186. [CrossRef]
17. Ryss, A.; Vieira, P.; Mota, M.; Kulinich, O. A synopsis of the genus *Bursaphelenchus* Fuchs, 1937 (Aphelenchida: Parasitaphelenchidae) with key to species. *Nematology* **2005**, *7*, 393–458.
18. Braasch, H. *Bursaphelenchus* species in conifers in Europe: Distribution and morphological relationships. *EPPO Bull.* **2001**, *31*, 127–142. [CrossRef]
19. Polomski, J.; Schönfeld, U.; Braasch, H.; Dobbertin, M.; Burgermeister, W.; Rigling, D. Occurrence of *Bursaphelenchus* species in declining *Pinus sylvestris* in a dry Alpine valley in Switzerland. *For. Pathol.* **2006**, *36*, 110–118. [CrossRef]

20. Errico, G.; Carletti, B.; Schröder, T.; Mota, M.; Vieira, P.; Roversi, P.F. An update on the occurrence of nematodes belonging to the genus *Bursaphelenchus* in the Mediterranean area. *Forestry* **2008**, *88*, 509–520. [CrossRef]
21. Aldeias do Xisto. Available online: https://www.aldeiasdoxisto.pt/pt/aldeias/lousa/ (accessed on 6 June 2024).
22. Whitehead, A.G.; Hemming, J.R. A comparison of some quantitative methods of extracting small vermiform nematodes from soil. *Ann. Appl. Biol.* **1965**, *55*, 25–38. [CrossRef]
23. Fonseca, L.; Santos, M.C.V.; Santos, M.S.; Curtis, R.H.C.; Abrantes, I.M.O. Morpho-biometrical characterisation of Portuguese *Bursaphelenchus xylophilus* isolates with mucronate, digitate or round tailed females. *Phytopathol. Mediterr.* **2008**, *47*, 223–233.
24. Nickle, W.R.; Golden, A.M.; Mamiya, Y.; Wergin, W.P. On the taxonomy and morphology of pine wood nematode, *Bursaphelenchus xylophilus* (Steiner & Buhrer, 1934) Nickle, 1970. *J. Nematol.* **1981**, *13*, 385–392.
25. EPPO. PM 7/44(4) *Bursaphelenchus xylophilus*. *EPPO Bull.* **2023**, *53*, 156–183. [CrossRef]
26. Castagnone, S.; Abad, P.; Castagnone-Sereno, P. Satellite DNA-based species-specific identification of single individuals of the pinewood nematode *Bursaphelenchus xylophilus* (Nematoda: Aphelenchoididae). *Eur. J. Plant Pathol.* **2005**, *112*, 191–193. [CrossRef]
27. François, C.; Castagnone, C.; Boonham, N.; Tomlinson, J.; Lawson, R.; Hockland, S.; Quill, J.; Vieira, P.; Mota, M.; Castagnone-Sereno, P. Satellite DNA as a target for Taqman real-time PCR detection of the pinewood nematode, *Bursaphelenchus xylophilus*. *Mol. Plant Pathol.* **2007**, *8*, 803–809. [CrossRef]
28. Hodda, M. Phylum Nematoda: A classification, catalogue, and index of valid genera, with a census of valid species. *Zootaxa* **2022**, *5114*, 1–289.
29. Rodrigues, J.M.; Sousa, E.; Abrantes, I. Pine Wilt Disease: Historical Review. In *Pine Wil Disease in Europe: Biological Interactions and Integrated Management*, 1st ed.; Sousa, E., Vale, F., Abrantes, I., Eds.; FNAPF: Oeiras, Portugal, 2015; pp. 13–32.
30. Brichta, J.; Stanislav Vacek, S.; Vacek, Z.; Cukor, J.; Mikeska, M.; Bílek, L.; Šimůnek, V.; Josef Gallo, J.; Brabec, P. Importance and potential of Scots pine (*Pinus sylvestris* L.) in 21st century. *Cent. Eur. For. J.* **2023**, *69*, 3–20. [CrossRef]
31. Ikegamia, M.; Jenkins, T.A.R. Estimate global risks of a forest disease under current and future climates using species distribution model and simple thermal model—Pine Wilt disease as a model case. *For. Ecol. Manag.* **2018**, *409*, 343–352. [CrossRef]
32. Panesar, T.S.; Sutherland, J.R. Pathogenicity of canadian isolates of the *Bursaphelenchus xylophilus* (pinewood nematode) to provenances of *Pinus sylvestris* and *Pinus contorta* as grown in Finland: A greenhouse study. *Scand. J. Forest Res.* **1989**, *4*, 549–557. [CrossRef]
33. Melakeberhan, H.; Webster, J.M. Relationship of *Bursaphelenchus xylophilus* population density to mortality of *Pinus sylvestris*. *J. Nematol.* **1990**, *22*, 297–302. [PubMed]
34. Filipiak, A. Pathogenicity of selected isolates of the quarantine pinewood nematode *Bursaphelenchus xylophilus* to Scots pine (*Pinus sylvestris* L.). *J. Plant Prot. Res.* **2015**, *55*, 378–382. [CrossRef]
35. Nunes da Silva, M.; Solla, A.; Sampedro, L.; Zas, R.; Vasconcelos, M.W. Susceptibility to the pinewood nematode (PWN) of four pine species involved in potential range expansion across Europe. *Tree Physiol.* **2015**, *35*, 987–999. [CrossRef]
36. Rodríguez-García, A.; Martín, J.A.; Gil, L.; Menéndez-Gutiérrez, M.; Díaz, R. Resin canal traits variation in *Pinus* spp. with different susceptibility to the pine wood nematode. *Forests* **2023**, *14*, 925. [CrossRef]
37. Kulinich, O.A.; Elena, N.; Arbuzova, E.N.; Chalkin, A.A.; Kozyreva, N.I. Experimental confirmation of *Bursaphelenchus xylophilus* survival and propagation in birch logs. *Russ. J. Nematol.* **2024**, *32*, 67–73.
38. Arbuzova, E.N.; Kulinich, O.A.; Chalkin, A.A.; Kozyreva, N.I.; Gorbach, V.V.; Ryss, A.Y. Infestation of pine (*Pinus sylvestris* L.) seedlings with the pinewood nematode *Bursaphelenchus xylophilus* Steiner and Buhrer (Nickle) through wood sawdust. *Ann. For. Sci.* **2023**, *80*, 6. [CrossRef]
39. Halik, S.; Bergdahl, D.R. Long-term survival of *Bursaphelenchus xylophilus* in living *Pinus sylvestris* in an established plantation. *Eur. J. For. Pathol.* **1984**, *24*, 357–363. [CrossRef]
40. Naves, P.M.; Sousa, E.M.; Quartau, J.M. Feeding and oviposition preferences of *Monochamus galloprovincialis* for certain conifers under laboratory conditions. *Entomol. Exp. Appl.* **2006**, *120*, 99–104. [CrossRef]
41. Koutroumpa, F.A.; Vincent, B.; Roux-Morabito, G.; Martin, C.; Lieutier, F. Fecundity and larval development of *Monochamus galloprovincialis* (Coleoptera Cerambycidae) in experimental breeding. *Ann. For. Sci.* **2008**, *65*, 707. [CrossRef]
42. Szmigielski, R.; Cieslak, M.; Rudziński, K.J.; Maciejewska, B. Identification of volatiles from *Pinus sylvestris* attractive for *Monochamus galloprovincialis* using a SPME-GC/MS platform. *Environ. Sci. Pollut. Res.* **2011**, *19*, 2860–2869. [CrossRef]
43. Sanchez-Husillos, E.; Álvarez-Baz, G.; Etxebeste, I.; Pajares, J.A. Shoot feeding, oviposition, and development of *Monochamus galloprovincialis* on *Pinus pinea* relative to other pine species. *Entomol. Exp. Appl.* **2013**, *149*, 1–10. [CrossRef]

forests

MDPI

Article

Distribution and Prolonged Diapause of the Rowan Seed Predators *Argyresthia conjugella* (Lepidoptera: Yponomeutidae) and *Megastigmus brevicaudis* (Hymenoptera: Torymidae) and their Parasitoids in Norway

Nina Trandem *, Karin Westrum, Trond Hofsvang and Sverre Kobro *

Division of Biotechnology and Plant Health, Norwegian Institute of Bioeconomy Research (NIBIO), P.O. Box 115, 1431 Ås, Norway
* Correspondence: nina.trandem@nibio.no (N.T.); s-kobro@online.no (S.K.)

Abstract: The seed predator *Argyresthia conjugella* Zeller has rowan as its preferred host plant. In years of poor fruiting in rowan, it oviposits on apples. To improve the knowledge of this apple pest, rowanberries were collected from localities all over Norway from 1971 to 1985, and seed predators and their parasitoids were allowed to emerge for up to five years. Two species of seed predators, *A. conjugella* and *Megastimus brevicaudis* Ratzeburg, and seven species of parasitic Hymenoptera were common. The distribution of these species is shown on EIS (European Invertebrate Survey) maps of Norway. The biology of the parasitoids is summarized based on the published literature and their behavior during emergence. The tendency for delayed emergence, which is an indication of prolonged diapause, was more pronounced in *M. brevicaudis* than in *A. conjugella,* the former appearing in all five years. Five of the parasitoids also delayed their emergence, and three of them to a high degree, up to five years. Prolonged diapause must be taken into account in studies of rowanberry insect guilds.

Keywords: apple fruit moth; masting; *Microgaster polita*; mountain ash seed chalcid; natural enemies; *Sorbus aucuparia*; *Torymus aucupariae*

Citation: Trandem, N.; Westrum, K.; Hofsvang, T.; Kobro, S. Distribution and Prolonged Diapause of the Rowan Seed Predators *Argyresthia conjugella* (Lepidoptera: Yponomeutidae) and *Megastigmus brevicaudis* (Hymenoptera: Torymidae) and their Parasitoids in Norway. *Forests* **2023**, *14*, 847. https://doi.org/10.3390/f14040847

Academic Editors: Katarína Pastirčáková and Rostislav Zemek

Received: 28 February 2023
Revised: 10 April 2023
Accepted: 13 April 2023
Published: 20 April 2023

1. Introduction

Rowan (mountain ash), *Sorbus aucuparia* L., is distributed in the forest belt of the northern temperate zone [1]. Rowan grows along the forest edges, in mixed stands of deciduous trees, or in open places in the forest [2]. The fruiting of rowan is intermittent, which is a phenomenon called masting [3]. The masting of rowan is synchronous over large areas and has probably evolved as a defense against seed predators [4]. In Scandinavia, two univoltine seed predators are common in rowan, the apple fruit moth, *Argyresthia conjugella* Zeller, and the mountain ash seed chalcid, *Megastigmus brevicaudis* Ratzeburg [4–6]. Both species are seed predators in their larval stage but differ in their overwintering habits. When fully grown in the autumn, *A. conjugella* larvae vacate the berries and hibernate as cocooned pupae in the litter layer [5,7], whereas *M. brevicaudis* larvae stay in the seeds through the winter and pupate there in the spring [8–10].

Prolonged diapause is a counter-adaptation to masting that allows seed predators to escape the poor crop years [11,12]. This trait is common in seed predators [11,13–15] and has also been found in *M. brevicaudis* [4,5] and *A. conjugella* [4,16,17]. Both masting and prolonged diapause will obviously affect parasitoids and other natural enemies of *A. conjugella* and *M. brevicaudis* [18–20], and prolonged diapause has been reported for some of the species at this trophic level as well [4,5].

Another way for the seed predators to escape poor crop years is to search for other host plants. In Fennoscandia, *A. conjugella* has long been known as a serious pest of apple in the intermasting years of rowan [21]. However, *A. conjugella* is usually not able to complete its life cycle in apple ([17], N.T. and S.K. pers. obs.). Thus, the risk of *A. conjugella* attacking apple at a certain location depends on the relationship between the local population size of viable moth larvae emerging from rowan one year and that of the rowanberry crop next year [22,23]. Based on this principle, a warning system for Norwegian apple growers, now incorporated in the VIPS platform [24,25], was developed during the 1970s [23]. This warning system has also generated a valuable long-term data series on rowanberries [26,27], apple fruit moth larvae, and their parasitization rate [4,19].

The research made back in the 1970s and 80s to increase the knowledge on *A. conjugella* also produced some data that are still largely unpublished. In this paper, we summarize the data from two such studies, one on the geographical distribution of rowanberry seed predators and their parasitoids, and one on the occurrence of prolonged diapause in these insects. Only some of the data on the four most common species have been briefly presented elsewhere as support for a co-evolutionary relationship between rowan and associated insects [4]. Our objectives are to (1) increase the knowledge of seed predators and their parasitoids in Norwegian rowanberries and compare it to a Swedish rowanberry study conducted by O. Ahlberg 50 years earlier [5], and (2) explore the propensity for prolonged diapause in these species, thereby indicating which ones may be underestimated in (1).

2. Materials and Methods

2.1. Sampling of Rowanberries

The first Norwegian large-scale sampling of rowanberries to map the apple fruit moth and other insects in the berries was carried out in August 1971, when the research entomologist T. Edland collected berries from more than a hundred sites in South Norway. The mapping continued in 1975 and lasted till 1984, aiming to cover the whole European Invertebrate Survey (EIS) grid of Norway. This grid consisted of 189 modified 50 × 50 km UTM squares [28]. Most samples contained 1–3 kg of berries, preferably from more than one tree per site, and were collected by agricultural advisors through the Norwegian warning system for apple fruit moth [23]. In the study period, this system encompassed 26–105 sites with rowan trees in regions with commercial apple production, and all of them were in South Norway. From each site, samples of 100 berry clusters were collected in early August and transported to NIBIO at Ås, which is 30 km south of Oslo. The surplus berries from these samples not spent in the warning system were used to rear adult insects for the mapping. In addition, hundreds of berry samples were collected from North Norway and parts of South Norway not covered by the warning system, mostly in the period 1975–1982.

2.2. Rearing of Insects for the Mapping

Before 1978, each berry sample was put in a clay pot with *Sphagnum* in the bottom and overwintered at 3 °C in the dark. From 1978, the overwintering took place under outdoor conditions in an insectarium, storing each sample in a 30 × 60 cm paper bag with a roll of corrugated cardboard (diameter 5–8 cm, length 5 cm). The *Spaghnum* and the cardboard served as preferred pupation sites for *A. conjugella* emerging from the berries, enabling easy sampling of this stage for the warning system. In early spring, about 1 kg of berries from each clay pot or paper bag was transferred to flowerpots (diameter 14 cm) with a transparent lid. The cardboard roll was also transferred. The inside of the flowerpots was lined with black cloth at the bottom and white filter paper on the walls for easier inspection. The flowerpots were kept in a greenhouse until the night frost had ceased, and then under outdoor conditions. They were checked for living imagines at least three times a week for as long as *A. conjugella* or any hymenopterans emerged.

2.3. Study of Prolonged Diapause

Some of the berry samples collected through the warning system in the years 1978–1984, i.e., from apple production areas in South Norway, were kept to study delayed emergence after they had been used for the mapping described above. They were pooled into plastic buckets from the flowerpots and kept for 4 more years. There were three such buckets from each year in the period, with each bucket containing berries and cardboard rolls equivalent to about 5 kg of fresh material. The buckets were wrapped in black plastic and stored at 3 °C in the dark. In mid-May, the buckets were exposed to outdoor conditions for as long as *A. conjugella* or any hymenopterans emerged in a glass collector mounted at the top, usually for two to three months. Emerging insects were removed and counted.

2.4. Identification of Emerged Imagines

Usually, more than 50 mounted specimens of each Hymenoptera species were sent to K.-J. Hedqvist, Swedish Museum of Natural History, Stockholm, for identification. Some were also sent to K. Horstmann, Zoologisches Institut, Würzburg. Based on reference specimens and additional information given by Hedqvist, T. Edland and S. Kobro identified and counted the emerging Hymenoptera and *A. conjugella*. Voucher specimens were deposited at the entomological collections at NIBIO, the University of Oslo, and the University of Bergen.

3. Results

3.1. The Species Found and Their Geographical Distribution

In addition to the two seed predators, seven species of hymenopteran parasitoids emerged from the rowanberry samples in significant numbers (Figures 1 and 2). The fruit moth was more numerous than the seed chalcid, even if some fruit moths had been removed with the cardboard rolls. All nine species emerged from berries, and only *A. conjugella* and its specialist parasitoid, *Microgaster polita* Marshall [5], emerged from the corrugated cardboard.

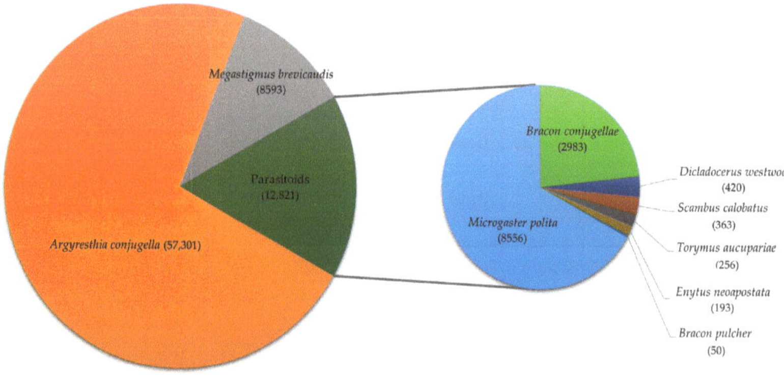

Figure 1. Numbers of the two seed predators and seven parasitoids emerging from rowanberry samples the year after berry collection. Pooled data for berry samples collected 1977–1981 and used for EIS-mapping. The two species of *Bracon* were not separated before 1980.

The distribution of the nine species found during this study is shown in Figure 3. Most EIS grid squares were sampled at least twice, but some were not sampled due to a lack of rowan or rowan with berries. *A. conjugella* was found north of 70° N in all three northernmost squares where it occurred. A few berry samples were collected in two squares north of this (no. 182 and 183), but no insects emerged. *M. brevicaudis* was found north of 69° N in two squares.

The parasitoid *M. polita* closely followed the distribution of its host *A. conjugella*, also in the far north. The host associations of the other parasitoids are probably looser and will be discussed later. Only two of them were confined to South Norway, but one of these

(*Bracon pulcher* Bengtsson) was also the least abundant species (Figure 1); moreover, it was not distinguished from *Bracon conjugellae* Bengtsson before 1980.

Figure 2. Photos of the nine species commonly reared from rowanberries (date of hatching): Two seed predators, *Argyresthia conjugella* (living specimen) and *Megastigmus brevicaudis* (25 April 1980), and seven parasitoids, *Bracon conjugellae* (10 May 1982), *Bracon pulcher* (August 1982), *Dicladocerus westwoodii* (3 August 1980), *Enytus neoapostata* (20 May 1985), *Microgaster polita* (10 May 1976), *Scambus calobatus* (30 March 1981), and *Torymus aucupariae* (24 April 1981). Photos by K. Westrum/NIBIO, except for *A. conjugella* by S. Kobro/NIBIO.

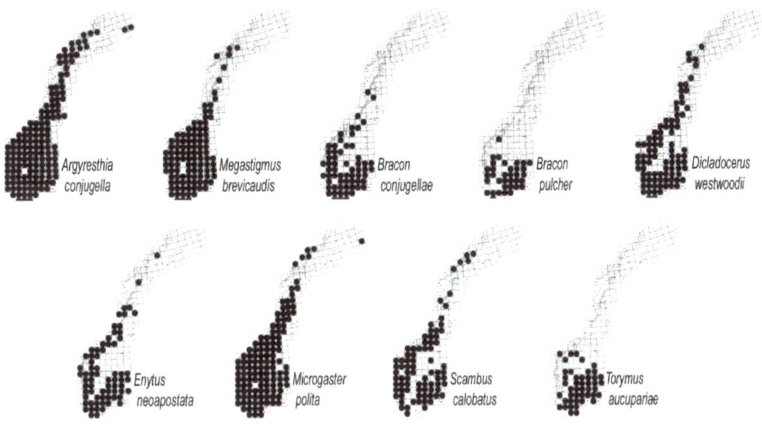

Figure 3. Geographical distribution of the insects commonly reared from rowanberries collected 1971–1984, plotted in The European Invertebrate Survey (EIS) grid of Norway [28]. *Argyresthia conjugella* and *Megastigmus brevicaudis* are seed predators; the others are parasitoids.

3.2. Prolonged Diapause

The ability of prolonged diapause, as indicated by the adult emergence two or more seasons after collecting the berries, was found in both species of the seed predators and in five of the seven parasitoids. A much bigger fraction of *M. brevicaudis* than *A. conjugella* delayed their emergence, and for a longer period (four vs. two years of maximum delay; Figure 4a). The fraction of delayers in the five parasitoid species was either small or large: two species delayed to a small degree (<20%) and three to a large (>80%; Figure 4b). One species, *Torymus aucupariae* Rodzianko, even peaked three years after sampling and also emerged in the fifth year. *Dicladocerus westwoodii* Westwood and *Scambus calobatus* Gravenhorst were not found to delay their emergence at all.

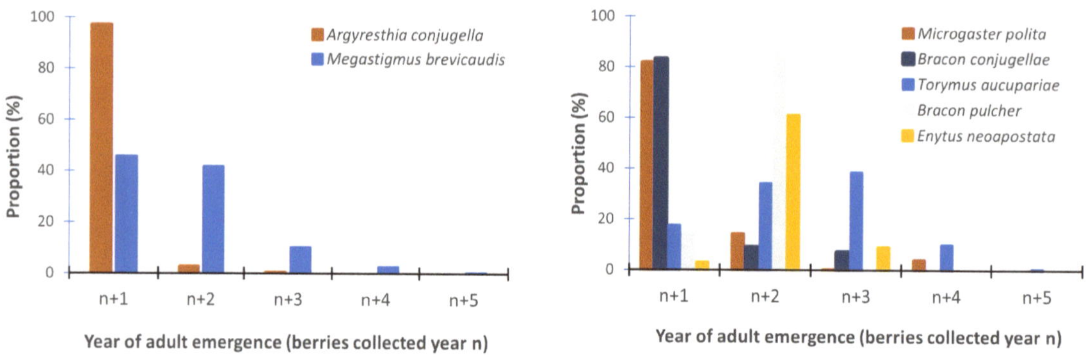

(**a**) (**b**)

Figure 4. Relative annual emergence of adults for each of the insect species exhibiting delayed emergence from the rowanberry samples kept for five years. Pooled data for berries collected 1978–1984; species sorted by proportion emergence in year n + 1: (**a**) The two seed predators. (**b**) The five parasitoids.

4. Discussion

4.1. Geographical Distribution of the Seed Predators

The two seed predators of rowan, *A. conjugella* and *M. brevicaudis*, were found in most of Norway, with the fruit moth occurring somewhat further north than the seed chalcid (70 vs. 69° N). Such a close relationship between the geographical distribution of *A. conjugella* and its host plant, rowan, was hypothesized in 1906 [29] and was also found almost 100 years ago in the Swedish rowanberry study [5].

A third seed predator of rowan, the sawfly *Hoplocampa alpina* (Zetterstedt), is present in Scandinavia [30], but it did not emerge in the study presented here. This was also the case in the older Swedish study [5]; a possible reason for this being that its larvae exit the berries earlier than *A. conjugella* and thus were absent when samples were taken. A gall midge may also be part of the herbivore complex living in rowanberries ([31], S.K. pers. obs.). Rowanberries could thus be a good model system for studying competitive interactions in a confined space.

4.2. Biology and Distribution of the Parasitoids

The seven parasitoid species regularly emerging from the collected berries were from four families of Hymenoptera (Table 1). As the rowanberries were picked from the trees before *A. conjugella* larvae exited the berries and *M. brevicaudis* does not pupate until spring, all seven species must be larval parasitoids that are able to locate the larvae inside the berries. *M. polita* is a koinobiont as it allowed its host *A. conjugella* to enter the corrugated cardboard before killing it. The preferred host in rowanberries for the other six parasitoids is less clear. In the Swedish study [5], where at least five of these occurred, Ahlberg did make an effort to find the host of each (Table 1). However, most of the parasitoids are idiobionts and, therefore, are likely to be able to exploit more than one of the host taxa present in rowanberries [32,33].

Table 1. Overview of the seven larval parasitoids (Hymenoptera) regularly emerging from rowanberries collected in Norway 1971–1984 and what is known about their biology.

Species	Family	Type of Parasitoid (Host Groups Known)	Host in Rowanberries (Ahlberg 1927) [5]	References
Bracon (Glabrobracon) conjugellae (Bengtsson, 1924)	Braconidae	Idiobiont ectoparasitoid of concealed hosts (*Megastigmus brevicaudis*, *Argyresthia conjugella*, and *Pontiana*)	*M. brevicaudis*	[5,34–36]
Bracon(Glabrobracon) pulcher (Bengtsson, 1924)	Braconidae	Ectoparasitoid of concealed hosts (*M. brevicaudis*, *A. conjugella*, and *Metzenera lapella*)	*M. brevicaudis*	[5,34–36]
Dicladocerus westwoodii Westwood, 1832	Eulophidae	Idiobiont ectoparasitoid (various Lepidoptera and Diptera)	*A. conjugella* [1]	[37,38]
Enytus neoapostata [2] (Horstmann, 1969)	Ichneumonidae	Koinobiont endoparasitoid (*Depressaria assimilella*)	This parasitoid not found by Ahlberg? [3]	[39,40]
Microgaster polita [4] Marshall, 1885	Braconidae	Koinobiont endoparasitoid (*A. conjugella*)	*A. conjugella*	[5,41,42]
Scambus calobatus Gravenhorst, 1829	Ichneumonidae	Idiobiont ectoparasitoid (various Lepidoptera, Coleoptera, and Hymenoptera)	*A. conjugella* [5]	[5,29,40,43]
Torymus aucupariae Rodzianko, 1908	Torymidae	Not known? (*Megastigmus*)	*M. brevicaudis* [6]	[5,37]

[1] In Ahlberg (1927), the name of *D. westwoodii* was *Diglyphus rugifrons* Thomson 1878 [37]. [2] At the time of the study, this species was in the genus *Diadegma*. [3] Ahlberg found *Angitia exareolata* (Ratz.). Horstmann 1969 [39] synonymized this with *Diadegma apostata* (now *Enytus apostata*, [44]), and noted some confusion on the identity of *apostata* specimens collected by Thomson. [4] At the time of study, the name of this species was *M. politus*. [5] In Ahlberg (1927), *S. calobatus* was in the genus *Epiurus*. [6] In Ahlberg (1927), *T. aucupariae* was in the genus *Syntomaspis* [37].

Moreover, most of the parasitoids are also known from other habitats than rowan trees (Table 1), and some of them are found on exophytic hosts. In particular, *S. calobatus* is reported to have a wide range of hosts and host habitats [34,43], including other parasitoids. *D. westwoodi* is also a habitat generalist and is reported as a parasitoid of various Lepidoptera that feed on conifer needles [38,45]. However, there is still a lot we do not know about these wasps. For example, *B. conjugellae* was recently reared from *Pontania* galls on *Salix* [35], after more than a hundred years with only rowanberry associations in the literature. Nevertheless, the conclusion from the Swedish study [5] also holds for the Norwegian one: *M. polita* is the most important parasitoid of the apple fruit moth, both because of its abundance and because it seems to be the only *A. conjugella* specialist. For *M. brevicaudis*, the little studied *T. aucupariae* may be the most important parasitoid.

Berry sampling has been the approach in all the studies on parasitoids of endophytic insects in rowanberries that we are aware of [5,16], including this one. Thus, nothing is known of any parasitoids specializing on *A. conjugella* larvae or pupae on the ground, or on the endophytic *M. brevicaudis* pupae in the spring. Predation of *A. conjugella* pupae on the ground has been briefly studied [46]. Egg parasitoids were looked for in the Swedish study without success [5], but on apples, *Trichogramma* has been reared from *A. conjugella* eggs in Latvia [47].

4.3. Prolonged Diapause

The rowanberry crop is rarely very low (with close to zero berries) for several years in a row. In a 22-year time series from the Norwegian apple fruit moth warning system, a very poor crop year occurred 5–6 times, but not for more than two years in a row [4]. Based on this, delayed emergence for 1–2 years should be relatively common in seed predators dependent on rowanberries, and longer delays rare. This was certainly the case for *M. brevicaudis*; in this species, one year of delay was almost as common as no delay. In a later study, a specimen of *M. brevicaudis* emerged 7 years after the berry crop [S.K. pers. obs.]. In contrast, only three percent of *A. conjugella* appeared delayed, and only for one or two years. Hanski [11] suggested that interspecific competition may lead to such quantitative differences in the distribution of diapause lengths, thereby easing coexistence.

The response of the two seed predators may also have been affected by the 9–10 months in a cold and dark storage room each year, but the same type of difference between a moth and a seed chalcid has been found in other studies, for example, between *Megastigmus strobilobius* Ratzeburg and the seed moth *Cydia strobilella* (L.) in spruce cones [11,14]. In the comprehensive Swedish study of rowanberry insects [5], prolonged diapause was noted in *M. brevicaudis*, but not in *A. conjugella*. In general, prolonged diapause is rather common in species of *Megastigmus* [15], while we have not found reports of it in other species of *Argyresthia* than *A. conjugella*.

Additionally, the seven parasitoid species differed in their pattern of delayed emergence: either they did not display it at all (two species), or to a small extent (two species), or as their most common strategy (three species). Explaining such patterns at the third trophic level in a food chain with masting at the first level is not easy [19]. An added complication is the possibility of parasitoids receiving physiological cues about prolonged diapause from their host, especially for koinobionts, while the host is still alive [20]. The delay pattern of the koinobiont *M. polita* was similar to that of its host, *A. conjugella*. The same was true for the pair of *T. aucupariae* and *M. brevicaudis*.

Prolonged diapause is costly, and in herbivores feeding on plant structures with intermittent occurrence, it only occurs in specialists that are not able to feed on anything else [48,49]. The same is probably true for parasitoids, meaning that species capable of exploiting hosts in other niches than rowanberries will be less inclined to prolong their diapause. In accordance with this, both parasitoid species were not found to delay emergence, *S. calobatus* and *D. westwoodi*, have a broad host spectrum (Table 1). As they were more abundant in the mapping samples than two of the species found in the delay study (Figure 1), their absence from the delay study probably was not due to a small sample size.

It should be noted that the two rarest species in the mapping samples, *B. pulcher* and *Enytis neoapostata* Horstmann, were among the three parasitoids with the most pronounced delay pattern (the third was *T. aucupariae*). More studies are needed on their biology and host preferences to understand their propensity for delayed emergence, taking into account the fact that standard rearing for one season will not give a complete picture. For example, if all the mapping samples in our study had been kept for five years, the relative abundance of these three parasitoids would have increased, possibly along with their geographical distribution. This is also true for the seed chalcid.

Improved knowledge on prolonged diapause in parasitoids and their hosts is important to understand the potential effects of climate change in temperate ecosystems [50]. This includes how the attacks of the apple fruit moth on apples will develop in the future.

Author Contributions: Conceptualization, S.K. and N.T.; methodology, S.K.; sample processing and identifications, S.K.; writing—original draft preparation, S.K.; writing—review and editing, N.T., K.W., T.H. and S.K.; visualization, K.W. and S.K.; project administration, S.K.; funding acquisition, S.K. and T.H. All authors have read and agreed to the published version of the manuscript.

Funding: This research received funding from The Agricultural Research Council of Norway (now The Research Council of Norway) in the 1970s and 80s. The apple fruit moth warning system is funded by the Norwegian Ministry of Agriculture and Food.

Data Availability Statement: The data presented in this study are available on request from the corresponding author.

Acknowledgments: T. Edland (deceased 1999) initiated and carried out the largest part of this study. We also thank D.M. Arnesen, A. Grøtli, B. Meyer, K. Nybøle, and H.M. Singh for their skillful and technical assistance, and the many agricultural advisors for sending rowanberry samples from their region.

Conflicts of Interest: The authors declare no conflict of interest. The funders had no role in the design of the study; in the collection, analyses, or interpretation of data; in the writing of the manuscript; or in the decision to publish the results.

References

1. Räty, M.; Caudullo, G.; de Rigo, D. Sorbus aucuparia in Europe: Distribution, habitat, usage and threats. In *European Atlas of Forest Tree Species*; San-Miguel-Ayanz, J., de Rigo, D., Caudullo, G., Houston Durrant, T., Mauri, A., Eds.; Publication Office of the European Union: Luxembourg, 2016; pp. 176–177. Available online: https://www.researchgate.net/publication/299471313_Sorbus_aucuparia_in_Europe_distribution_habitat_usage_and_threats (accessed on 20 January 2023).
2. Raspé, O.; Findlay, C.; Jaquemart, A.-L. *Sorbus aucuparia* L. Biological flora of the British Isles. *J. Ecol.* **2000**, *88*, 910–930. [CrossRef]
3. Koenig, W.D. A brief history of masting research. *Phil. Trans. R. Soc. B* **2021**, *376*, 20200423. [CrossRef] [PubMed]
4. Kobro, S.; Søreide, L.; Djønne, E.; Rafoss, T.; Jaastad, G.; Witzgall, P. Masting of rowan *Sorbus aucuparia* L. and consequences for the apple fruit moth *Argyresthia conjugella* Zeller. *Popul. Ecol.* **2003**, *45*, 25–30. [CrossRef]
5. Ahlberg, O. Rönnbärsmalen, *Argyresthia conjugella* Zell. En redogörelse för undersökningar åren 1921–1926. In *Meddelande från Centralanstalten för Försöksväsendet på Jordbruksområdet*; Lantbruksentomologiska Avdelingen: Stockholm, Sweden, 1927; Volume 324, pp. 1–127. (In Swedish with English Summary)
6. Jespersen, C.; Lomholt, O. De danske Frøvepse, *Megastigmus* Dalman, 1820 (Hymenoptera, Chalcidoidea, Torymidae). *Ent. Meddr.* **1983**, *50*, 111–118.
7. Sperens, U. Fruit production in *Sorbus aucuparia* L. (Rosaceae) and pre-dispersal seed predation by the apple fruit moth (*Argyresthia conjugella* Zell.). *Oecologia* **1997**, *110*, 368–373. [CrossRef]
8. Milliron, H.E. Taxonomic and biological investigations in the genus *Megastigmus* with particular reference to the taxonomy of the Nearctic species (Hymenoptera: Chalcidoidea; Callimomidae). *Am. Midl. Nat.* **1949**, *41*, 257–420. [CrossRef]
9. Roques, A.; Skrzypczyńska, M. Seed-infesting chalcids of the genus *Megastigmus* Dalman, 1820 (Hymenoptera: Torymidae) native and introduced to the West Palearctic region: Taxonomy, host specificity and distribution. *J. Nat. Hist.* **2003**, *37*, 127–238. [CrossRef]
10. Rouault, G.; Turgeon, J.; Candau, J.-N.; Roques, A.; Aderkas, P.v. Oviposition strategies of conifer seed chalcids in relation to host phenology. *Naturwissenschaften* **2004**, *91*, 472–480. [CrossRef]
11. Hanski, I. Four kinds of extra long diapause in insects: A review of theory and observations. *Ann. Zool. Fenn.* **1988**, *25*, 37–53.
12. Danks, H.V. *Insect Dormancy: An Ecological Perspective*; Biological Survey of Canada (Terrestrial Arthropods), National Museum of Natural Sciences: Ottawa, ON, Canada, 1987; pp. 175–187.
13. Bakke, A. Distribution of prolonged diapausing larvae in populations of *Laspeyresia strobilella* L. (Lep., Tortricidae) from spruce cones. *Norsk Entomol. Tidsskr.* **1971**, *18*, 89–93.
14. Annila, E. Fluctuations in cone and seed insect populations in Norway spruce. In *Communicationes Instituti Forestalis Fenniae (Finland)*; Metsäntutkimuslaitos, 1981; Volume 101, pp. 1–32. (In Finnish with English Summary)
15. Turgeon, J.J.; Roques, A.; Groot, P. Insect fauna of coniferous seed cones: Diversity, host plant interactions, and management. *Annu. Rev. Entomol.* **1994**, *39*, 179–212. [CrossRef]
16. Krämer, K. Zur Biologie der Ebereschenmotte (Argyresthia conjugella Zell.). *Anz. Für Schädlingskunde* **1960**, *33*, 102–107. [CrossRef]
17. Edland, T. Apple fruit moth (*Argyresthia conjugella* Zell.): Large attacks on apples are to be expected in 1971. *Gartneryrket* **1971**, *61*, 444–445. (In Norwegian)
18. Corley, J.C.; Capurro, A.F.; Bernstein, C. Prolonged diapause and the stability of host–parasitoid interactions. *Theor. Popul. Biol.* **2004**, *65*, 193–203. [CrossRef]
19. Satake, A.; Bjørnstad, O.N.; Kobro, S. Masting and trophic cascades: Interplay between rowan trees, apple fruit moth, and their parasitoid in southern Norway. *Oikos* **2004**, *104*, 540–550. [CrossRef]
20. Le Lann, C.; van Baaren, J.; Visser, B. Dealing with predictable and unpredictable temperatures in a climate change context: The case of parasitoids and their hosts. *J. Exp. Biol.* **2021**, *224*, jeb238626. [CrossRef]
21. Schøyen, W.M. *Beretning om Skadeinsekter og Plantesygdomme i 1898*; Grøndahl & Søns Bogtrykkeri: Kristiania, Norway, 1899; pp. 18–19. (In Norwegian)
22. Edland, T. Apple fruit moth (*Argyresthia conjugella* Zell.): Forecast of risk for attacks on apple. Methods and results. *Gartneryrket* **1974**, *64*, 524–533. (In Norwegian, English Summary)
23. Edland, T. Benefits of minimum pesticide use in insect and mite control in orchards. In *Techniques for Reducing Pesticide Use. Economic and Environmental Benefits*; Pimentel, D., Ed.; Wiley & Sons: New York, NY, USA, 1997; pp. 197–220.

24. Vips Platform. Available online: https://www.vips-landbruk.no/ (accessed on 24 February 2023).
25. Skog, T.-E.; Nordskog, B.; Eikemo, H.; Hole, H.; Schjøll, A.F.; Netland, J.; Trandem, N.; Rafoss, T. An open source technology platform for international collaboration and local adaptation of IPM tools. In *Book of Abstract*; IOBC VIII Workshop on Integrated Soft Fruit Production: Trentino, Italy, 2014; p. 24.
26. Hogstad, O.; Selås, V.; Kobro, S. Explaining annual fluctuations in breeding density of fieldfares *Turdus pilaris*—Combined influences of factors operating during breeding, migration and wintering. *J. Avian Biol.* **2003**, *34*, 350–354. [CrossRef]
27. Fox, A.D.; Kobro, S.; Lehikoinen, A.; Lyngs, P.; Väisänen, R.A. Northern Bullfinch *Pyrrhula p. pyrrhula* irruptive behaviour linked to rowanberry *Sorbus aucuparia* abundance. *Ornis Fenn.* **2009**, *86*, 51–60.
28. Økland, J. Utbredelse av noen ferskvannsmuslinger i Norge, og litt om European Invertebrate Survey. *Fauna* **1976**, *29*, 29–40.
29. Lampa, S. Rönnbärsmalen (*Argyresthia conjugella* Zell.). *Entomol. Tidskr.* **1906**, *27*, 1–13. Available online: https://archive.org/details/entomologisktids2728190607ento/page/n21/mode/2up (accessed on 1 February 2023).
30. Liston, A.; Prous, M.; Vårdal, H. A review of west palaearctic *Hoplocampa* species, focussing on Sweden (Hymenoptera, Tenthredinidae). *Zootaxa* **2019**, *4615*, 1–45. [CrossRef] [PubMed]
31. Thuróczy, C.K.; Hansen, L.O. *Systasis edlandi*—A new species of Ormocerinae (Hymenoptera, Pteromalidae) from Norway associated with rowan berries (*Sorbus aucuparia*) (Rosaceae). *Norw. J. Entomol.* **2015**, *62*, 205–209. Available online: http://urn.nb.no/URN:NBN:no-52321 (accessed on 27 February 2023).
32. Hawkins, B.A.; Askew, R.R.; Shaw, M.R. Influences of host feeding-niche and foodplant type on generalist and specialist parasitoids. *Ecol. Entomol.* **1990**, *15*, 275–280. [CrossRef]
33. Althoff, D.M. Does parasitoid attack strategy influence host specificity? A test with New World braconids. *Ecol. Entomol.* **2003**, *28*, 500–502. [CrossRef]
34. Yu, D.S.; van Achterberg, K.; Horstmann, K. World Ichneumonoidea 2004. Taxonomy, Biology, Morphology and Distribution. Taxapad 2005. DVD Version.
35. Kopelke, J.P.; Nymann, T.; Cazelles, K.; Gravel, D.; Vissault, S.; Roslin, T. Food-web structure of willow-galling sawflies and their natural enemies across Europe. *Ecology* **2017**, *98*, 1730. Available online: http://onlinelibrary.wiley.com/doi/10.1002/ecy.1832/suppinfo (accessed on 1 March 2023). [CrossRef]
36. Shaw, M.R.; Huddleston, T. Classification and biology of Braconid wasps (Hymenoptera: Braconidae). In *Handbooks for the Identification of British Insects*; Royal Entomological Society: London, UK, 1991; Volume 7, Part 11; pp. 54–58.
37. Noyes, J.S.; Universal Chalcidoidea Database. World Wide Web Electronic Publication 2019. Available online: http://www.nhm.ac.uk/chalcidoids (accessed on 1 February 2023).
38. Askew, R.R. Hymenoptera. 2. Chalcidoidea section b. In *Handbooks for the Identification of British Insects*; Royal Entomological Society: London, UK, 1968; Volume VIII, p. 18. Available online: https://www.royensoc.co.uk/wp--content/uploads/2022/01/Vol08_Part02b.pdf (accessed on 23 January 2023).
39. Horstmann, K. Typenrevision der europäischen Arten der Gattung *Diadegma* Foerster (syn. Angitia Holmgren) (Hymenoptera: Ichneumonidae). *Beitr. Ent. Bd.* **1969**, *19*, 413–472. [CrossRef]
40. Broad, G.R.; Shaw, M.R.; Fitton, M.G. Ichneumonid wasps (Hymenoptera: Ichneumonidae): Their classification and biology. In *Handbook for the Identification of British Insects*; Field Studies Council: London, UK, 2018; Volume 7, Part 12; pp. 146, 271.
41. Shaw, M.R. Notes on some European Microgastrinae (Hymenoptera: Braconidae) in the National Museums of Scotland, with twenty species new to Britain, new host data, taxonomic changes and remarks, and descriptions of two new species of *Microgaster* Latreille. *Entomol. Gaz.* **2012**, *63*, 173–201.
42. Gauld, I.; Bolton, B. *The Hymenoptera*; Oxford University Press: Oxford, UK, 1988; p. 217.
43. Shaw, M.R.; Jennings, M.T.; Quicke, D.L.J. The identity of *Scambus planatus* (Hartig, 1838) and *Scambus ventricosus* (Tschek, 1871) as seasonal forms of *Scambus calobatus* (Gravenhorst, 1829) in Europe (Hymenoptera, Ichneumonidae, Pimplinae, Ephialtini). *J. Hymenopt Res.* **2011**, *23*, 55–64. [CrossRef]
44. Scaramozzino, P.L.; Di Giovanni, F.; Loni, A.; Ricciardi, R.; Lucchi, A. Updated list of the insect parasitoids (Insecta, Hymenoptera) associated with *Lobesia botrana* (Denis & Schiffermüller, 1775) (Lepidoptera, Tortricidae) in Italy. 2. Hymenoptera, Ichneumonidae, Anomaloninae and Campopleginae. *ZooKeys* **2018**, *772*, 47–95. [CrossRef]
45. Ryan, R.B.; Denton, R.E. *Initial Release of Chrysocharis laricinellae and Dicladocerus Westwoodii for Biological Control of the Larch Casebearer in the Western United States*; US Department of Agriculture, Forest Service, Pacific Northwest Forest and Range Experiment Station: Portland, OR, USA, 1973; Volume 200, pp. 1–4. Available online: https://agris.fao.org/agris-search/search.do?recordID=US201303127920 (accessed on 1 February 2023).
46. Furenhed, S. Ground-Living Predators of the Apple Fruit Moth *Argyresthia conjugella* (Zell.). Master's Thesis, Sveriges Lantbruksuniversitet, Uppsala, Sweden, 2006.
47. Zilsparne, A. Phenologische Daten über der *Argyresthia conjugella* L. und auswendung von *Trichogramma* bei dessen bekämpfung im Jahre 1971. *Zool. Muz. Raksti* **1973**, *11*, 49–54.
48. Boivin, T.; Auger-Rozenberg, M.-A. Native fruit, cone and seed insects in the Mediterranean basin. In *Insects and Diseases of Mediterranean Forest Systems*; Paine, T.D., Lieutier, F., Eds.; Springer: Cham, Switzerland, 2016; pp. 47–88.

49. Denlinger, D. *Insect Diapause*; Cambridge University Press: Cambridge, UK, 2022; p. 38.
50. Tougeron, K.; Brodeur, J.; Le Lann, C.; Van Baaren, J. How climate change affects the seasonal ecology of insect parasitoids. *Ecol. Entomol.* **2019**, *45*, 167–181. [CrossRef]

forests

MDPI

Article

Population Dynamics of *Ips sexdentatus* (Börner) in the Czech Republic

Daniela Hlávková [1,2,*], Markéta Davídková [1,3], Jana Koudelková [2] and Petr Doležal [1,2,3,*]

1 Institute of Entomology, Biology Centre Czech Academy of Science, Branišovská 1160/31, 370 05 České Budějovice, Czech Republic
2 Faculty of Science, University of South Bohemia, Branišovská 1760, 370 05 České Budějovice, Czech Republic
3 Forestry and Game Management Research Institute, Forest Protection Service, Strnady 136, 156 00 Prague, Czech Republic
* Correspondence: hlavkova.daniela@entu.cas.cz (D.H.); dolezal@entu.cas.cz (P.D.)

Abstract: Recently, an outbreak of *Ips sexdentatus* (Börner, 1776) has caused considerable damage in the pine forests of the Czech Republic. As historical data on the biology of this pest are scarce due to its rare occurrence in recent decades, our work focused on monitoring flight activity and voltinism and investigating methods for monitoring its activity during the growing season. Observations were conducted from March to September 2021 and 2022 at three sites using 12 Theysohn traps with four types of pheromone lures (ACUMIPROTECT, ACUWIT, SEXOWIT and Pheagr IAC) together with data loggers to record weather conditions. The first beetles occurred in early May (daily mean temperatures above 13 °C). After the first egg laying stage, females re-emerged to establish a sister brood. The maximum flight activity appeared between late June and mid-July (daily mean temperatures about 20 °C), and the offspring occurred at the turn of June/July and in the first half of August. Since then, flight activity had a downward trend and quietened in September. According to the data, monitoring of *I. sexdentatus* should be conducted between May and September using the ACUMIPROTECT pheromone bait exhibiting the highest capturing efficacy. In future, however, the behavior of *I. sexdentatus* might alter due to climate change.

Keywords: monitoring; *Ips sexdentatus*; flight activity; voltinism; population dynamics

Citation: Hlávková, D.; Davídková, M.; Koudelková, J.; Doležal, P. Population Dynamics of *Ips sexdentatus* (Börner) in the Czech Republic. *Forests* **2024**, *15*, 961. https://doi.org/10.3390/f15060961

Academic Editor: Jun-Jun Liu

Received: 3 May 2024
Revised: 24 May 2024
Accepted: 28 May 2024
Published: 31 May 2024

1. Introduction

Climate change is triggering significant changes in the bionomy of many bark beetle species. The effects are particularly pronounced in primary spruce pests such as *Ips typographus* (Linnaeus, 1758) and *Ips duplicatus* (Sahlberg, 1836). However, recent observations have also revealed dramatic changes in bark beetles previously considered to be of less importance [1–4]. Under favorable conditions, epidemic population levels are quickly reached, causing significant damage to both weakened and healthy trees [5]. In fact, even endemic species can become major pests, exacerbating the severity of infestation [6]. While outbreaks of these beetles were relatively rare in the past [1], climate change is now creating optimal living conditions such as rapid changes in their voltinism, reproduction, development rate and overwintering success [7–11].

Such a species is *Ips sexdentatus* (Börner, 1776) (Supplementary Figure S1), which is distributed across warm regions, including Portugal, Central Europe, Turkey, and Russia [12,13]. While it causes enormous damage in the forests of Caucasian spruce in Turkey (*Picea orientalis* (L.)) [14], its reproductive success is limited in Central Europe [15]. However, rising temperatures are creating favorable conditions for it to thrive in areas where it was previously scarce or absent. In fact, several European countries have reported an increase in occurrence and damage caused by pine bark beetles, including Austria, the Czech Republic, and Bulgaria [16–19].

In the Czech Republic before the 1990s [20], *Ips sexdentatus* was rare, but since the dry periods documented in the 1990s and 2000s [21,22], it has been found regularly in the original refuges and it even proliferated in central and south Bohemia where it had not been observed before [13,18,23,24] (Supplementary Figure S2). This expansion is attributed to the absence of effective pest management strategies, which contributes to delayed detection, accelerates spread of *I. sexdentatus* [25], and boosts heavy infestation of trees [2].

In Turkey, *Ips sexdentatus* populations typically begin to swarm in late April or early May when daily maximum temperatures reach 18–20 °C [14,15,26]. The pioneering sex, the males, colonize in the area near the base of trees with a bark thickness of 5–15 mm [15,27,28], create a nuptial chamber, and produce the aggregation pheromone to attract other males and lure females [29–31]. After mating, females bore maternal tunnels and lay eggs [32]. Environmental conditions affect the developmental time and number of generations during the season [14,29]. The higher humidity and colder climate in northern Europe extend the development time, allowing only one generation to complete its development (univoltine population) [14,15,29]. Conversely, the low humidity and warm conditions of Central and Southern Europe shorten the development time, and so two to six generations can evolve [33,34]. The flight period of Turkish populations typically ends in early September [10].

Overall, the *I. sexdentatus* population dynamics are influenced by environmental conditions. Understanding these factors is crucial for developing effective monitoring and prevention methods for pest control in the Czech Republic. In our work, we therefore focused on (1) mapping flight activity during the season in three affected localities in the Czech Republic, (2) revealing the number of generations under local environmental conditions, and (3) providing a preliminary insight into the typology of available pheromone lures against *I. sexdentatus*.

2. Materials and Methods

2.1. Study Sites

This study was conducted at three sites over two consecutive years, 2021 and 2022. All sites were located on dry sandy-gravelly soils with moderate nutrient supply and low water retention capacity [34–36].

Locality1 was located near Brandýs nad Labem–Stará Boleslav (GPS: 50.2109169N, 14.7128108E; Supplementary Figures S2 and S3) in a clear-cut surrounded by a stand of 80–100-year-old Scots pines (*Pinus sylvestris* (L.)) on podzols [36,37] with an approximate mean plant density (MPD) of 506 trees per hectare. This locality is characterized by an altitude of about 187 m above sea level, a mean annual temperature (MAT) of 9.5 °C, and a mean annual precipitation (MAP) of 575 mm [38]. Locality1 was the most recent western edge of the distribution.

Locality2 was located near Šemíkovice in the Jevišovická Highlands (GPS: 49.0487050N, 16.0878489E; Supplementary Figures S2 and S3) in an 80-year-old Scots pine stand after extensive felling of spruce (*Picea abies* (L.)) on cambisols–luvisols [36,37] of approximately MPD 198 trees per hectare. The locality is characterized by an altitude of about 411 m above sea level, MAT of 8.9 °C, and MAP total of 482 mm [39]. Locality2 was the area where the species spread in the 1990s.

Locality3 was located near Bzenec in South Moravia (GPS: 48.9498897N, 17.2159839E; Supplementary Figures S2 and S3) in a clear-cut surrounded by a 100-year-old Scots pine stand on cambisols [36,37] of approximately MPD 482 trees per hectare. This locality is characterized by an altitude of about 216 m above sea level, MAT of 9.7 °C, and MAP total of 537 mm [38]. Locality3 was a location where *I. sexdentatus* had been observed before 1995 [39].

2.2. Air Temperature Measurements

From the beginning of March to the end of September in both years, data loggers (Comet system, Rožnov pod Radhoštěm, the Czech Republic) were installed in all localities to measure air temperature at 30 min intervals.

2.3. Flight Activity

Theysohn pheromone traps (RIDEX, Vrbno pod Pradědem, the Czech Republic) were set up in April 2021 and 2022. Twelve traps created a transect placed approximately 20 m from the forest edge and 2 m above the ground at a distance of 30 m (Supplementary Figure S4). A set of four different synthetic pheromone baits (Table 1) was distributed in triplicate at the sites and the same bait types were distributed at 80 m intervals. The positioning of the baits was not changed during the sampling dates. The pheromone baits were replaced every eight weeks according to the manufacturer's instructions. Samples were removed from the traps weekly and taken to the laboratory where they were separated, identified, and numbered. Observations were completed in September in both study years.

Table 1. An overview of the synthetic pheromone baits used at selected sites during the growing season in both years of the study.

The Lure Name	The Manufacturer	The Composition	Targeted Species
ACUMIPROTECT	SEDQ s.l., Barcelona, Spain	ipsenol, ipsdienol, verbenol, 2,6-di-terc-butyl-4-methylphenol	*Ips acuminatus*
ACUWIT	Witasek-Pflanzenschutz GmbH, Feldkirchen, Austria	ipsenol, ipsdienol	*Ips acuminatus*
Pheagr IAC	Scitech s.r.o., Prague, the Czech Republic	ipsdienol, ipsenol, verbenol, 2-methylbut-3-en-2-ol, 2,6-di-terc-butyl-4-methylfenol	*Ips acuminatus*
SEXOWIT	Witasek-Pflanzenschutz GmbH, Feldkirchen, Austria	2-fenylethanol, ipsdienol, α-pinene	*Ips sexdentatus*

2.4. Statistical Analysis

2.4.1. Temperatures

The temperature recordings were downloaded from the data loggers at the end of September. The daily minimum (Ti) and maximum (Tm) temperatures were detected. Average temperatures (Ta) were calculated as an average of temperatures (Ti and Tm) measured in interval among two consecutive midnights.

2.4.2. The Thermal Sum

The thermal sums for certain occasions correspond to an accumulation of degree days (DD) from March 1. The daily DD were counted as the sum of the daily Tm and Ti divided by two and subtracted from the temperature threshold for the development of each day [40]. The temperature threshold (TT) for *I. sexdentatus* flight is considered to be 20 °C [14].

2.4.3. Cumulative Percentage of Captured Beetles

The cumulative percentage of *I. sexdentatus* was calculated as the division of the cumulative number of beetles in traps trapped in a one-week interval by the total number of beetles found in the season. Then the result was multiplied by one hundred.

2.4.4. Analyses in R

The analyses were performed with the statistical language R (Posit®, Boston, MA, USA; version 4.2.2; [41]) on Windows 10 x64 (build 19045) using the packages 'Matrix' (version 1.5.1; [42]) and 'lme4' (version 1.1.31; [43]). First, the D'Agostino and Pearson omnibus normality test was applied to the data before other statistical tests were performed. Based on the result, parametric tests were used.

To test the effect of pheromone lures on catches, we fitted generalized linear mixed models for Poisson distributed data [44] using the package 'lme4' (version 1.1.31; [43]) and the package 'Matrix' (version 1.5.1; [42]). In the model, our locations and the position of the traps were included as a random effect (1 | locationlocality:trap position) to account for a nested design. The season and bait type were included as fixed-effects variables. ML (maximum likelihood estimation) [45] and BOBYQA optimizer [46] were used to optimize model output. We compared our model with simplified models that included only locality as a covariate to assess treatment significance. Post hoc comparisons between bait efficacy were also performed for parametric models with simple adjustment of *p* values (stepwise method) provided by the 'glht' function from the R package 'multcomp' (version 1.4-23; [47]). A mixed Poisson model (estimated with ML and BOBYQA optimizer) was used to estimate differences in the abundance of catches of *I. sexdentatus* due to bait type.

3. Results

3.1. Flight Activity

The first captured beetles were observed in early June 2021, when the accumulated thermal sum was 56.5 ± 46.6 DD, corresponding to 35.0 ± 1.0 days of positive daily heat sum, calculated from early March above the 11 °C threshold [48]. In 2022, the first records occurred at the beginning of May, when the thermal sum amounted to 63.5 ± 16.7 DD, after 30.3 ± 15.7 days with a positive daily thermal sum.

3.1.1. Locality1

In the first decade of May in 2021, Ta reached TT suitable for flight. However, due to the cold weather, the first adults of *I. sexdentatus* were not captured until early June when maximum temperatures regularly reached 22 °C. Flight activity peaked in early July. About two weeks later, catches increased again. Two further increases were recorded in mid-August and early September. In the following weeks, low temperatures prevailed, and flight activity dropped to zero, so that no more beetles were caught (Figure 1).

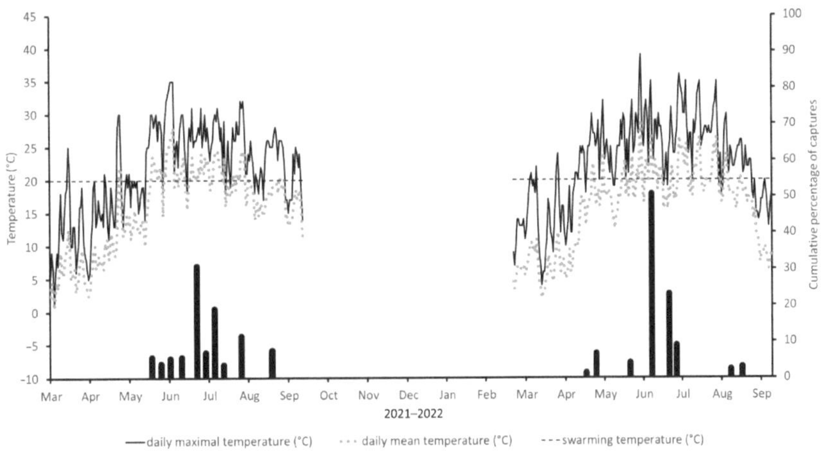

Figure 1. Cumulative percentage of captures of *Ips sexdentatus* at Locality1 in period 2021–2022. The dotted line corresponds to mean daily air temperature and the solid line to maximum daily air temperature. The horizontal (dashed) line corresponds to the temperature threshold of 20 °C for the flight activity of *I. sexdentatus*. The sampling dates are indicated on the *x*-axes.

In 2022, the Ta exceeded flight TT shortly at the start of May; however, during the whole season, the temperature oscillated around TT. The first swarming adults were recorded at the beginning of May. At this time, the average daily Ta reached 16 °C. In the following weeks, daily Ta dropped and fluctuated around 15 °C, which led to an interruption in flight activity (zero catches) until the end of June, when flight activity increased again together with temperature. Due to the cold weather with daily mean temperatures of only about 17.5 °C, in mid-July, catches dropped and remained low or at zero until early September when the last flying adults were caught (Figure 1).

3.1.2. Locality2

In the second week of 2021, the Ta reached TT values for flight. The swarming of *I. sexdentatus* began in early June after a six-week period in which the daily Tm rose to 31 °C, even though the Ta at that time was around 11 °C. The beetles flew out when the daily Ta exceeded 19 °C and the daily Tm reached 23.8 °C. At this point, in the following weeks, catches decreased with temperature, and an increase in flight activity was observed in mid-July when daily Ta exceeded 23 °C. A further increase in activity was recorded in early August. From this point onwards, the daily Ta remained below 20 °C, so that the catches showed a downward trend. The last flying adults were detected in early September (Figure 2).

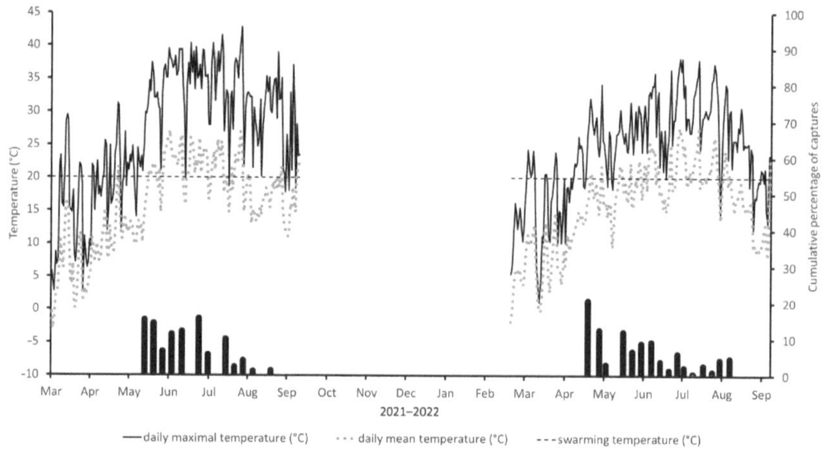

Figure 2. Cumulative percentage of captures of *Ips sexdentatus* at Locality2 in period 2021–2022. The dotted line corresponds to mean daily air temperature and the solid line to maximum daily air temperature. The horizontal (dashed) line corresponds to the temperature threshold of 20 °C for the flight activity of *I. sexdentatus*. The sampling dates are indicated on the *x*-axes.

In 2022, flight activity started about six weeks after a week in March with a Tm of about 24 °C and the second half of April with a daytime Tm of 20 °C. However, the Ta necessary for flight occurred no earlier than in the second half of May. The first bark beetles emerged in early May when daily Ta reached 17 °C and daily Tm reached 22.5 °C. Thereafter, the temperature decreased, and so only small numbers of beetles were caught until early June. The next time any flight activity was observed was in the second half of July and at the end of August. In September, the temperature dropped, and flight activity came to a standstill and no beetles were caught (Figure 2).

3.1.3. Locality3

In 2021, although the Tm was regularly above 20 °C, Ta oscillated only around 11 °C; however, the flight TT was exceeded for a short time in the second decade of May. Moreover, the week-long cold snap in mid-April probably led to a shift in spring swarming, which did not begin until early June with a maximum in the first June decade, when daily Ta fluctuated around 20 °C and daily Tm was 32 °C. Thereafter, temperatures dropped, and catches were low until mid-July. Since then, the number of beetles caught tended to decrease with temperature, and the last catches were monitored in the first half of September (Figure 3).

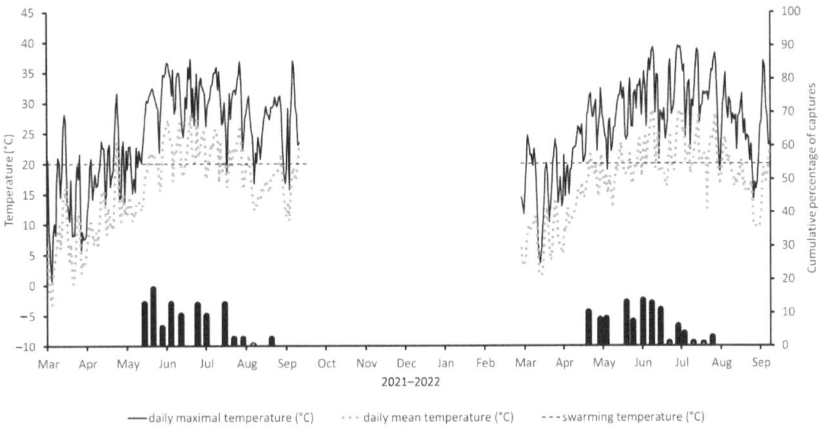

Figure 3. Cumulative percentage of captures of *Ips sexdentatus* at Locality3 in period 2021–2022. The dotted line corresponds to mean daily air temperature and the solid line to maximum daily air temperature. The horizontal (dashed) line corresponds to the temperature threshold of 20 °C for the flight activity of *I. sexdentatus*. The sampling dates are indicated on the *x*-axes.

In 2022, a warm week at the end of March (daily Tm up to 23 °C, Ta about 10 °C), followed by a drop in daily Tm to 3 °C, and a five-week period with a mean daily Ta of about 8 °C caused the start of swarming that year in early May, when daily Ta fluctuated around 18 °C and the maximum Ta reached 23.9 °C. Daily Ta was low in May and June, so the next catches were not recorded until the second half of June. In the second half of July and in August, flight activity remained at a minimum. Flight activity decreased thereafter and came to a halt in September (Figure 3).

3.2. The Efficiency of Pheromone Lures

A total of 5538 individuals of *I. sexdentatus* were caught at all our sites in both years (Table 2). From other representatives of the genus *Ips*, *I. typographus*, *I. duplicatus* and *I. acuminatus* were found in the traps in quantities of 19,832, 218, and 2768 individuals, respectively.

Table 2. The total number of *Ips sexdentatus* caught at our sites in 2021 and 2022 with four different pheromone baits.

		2021			2022	
Lure	**Locality1**	**Locality2**	**Locality3**	**Locality1**	**Locality2**	**Locality3**
ACUMIPROTECT	834	1526	403	406	320	834
ACUWIT	38	311	151	4	119	38
Pheagr IAC	20	352	89	18	35	20
SEXOWIT	26	319	117	9	77	26

3.3. Lure Attractiveness

The number of *I. sexdentatus* in the traps in 2021 and 2022 (df = 1, χ^2 = 1974.3, $p < 0.001$) differed significantly between the sites (Table 2; df = 1, χ^2 = 3614.4, $p < 0.001$). In general, the lowest number was caught at site 1 in 2021 (Table 2), and the highest number was observed at Locality2 in 2021 (Table 2). Comparing the abundance of first- and second-generation beetles caught, large differences can be observed at Locality1 (Figure 4). The number of beetles caught depending on the type of bait differed significantly for *I. sexdentatus* (df = 23, χ^2 = 2518.9, $p < 0.001$). The highest number of catches was observed in traps with ACP bait (3775 individuals), and conversely, the lowest number of catches was in traps with PH bait (563 individuals). However, the less effective baits SX, PH and AW had similar attractiveness values, with a significant difference observed between AW and PH ($p < 0.05$) (Figure 4). On the other hand, *I. typographus* and *I. duplicatus* responded most positively to PH (19,227 and 171 captured individuals, respectively). In addition, the number of catches was influenced by the minimum temperatures in the week prior to sampling (df = 4, $\chi^2 = 55.19$, $p < 0.001$).

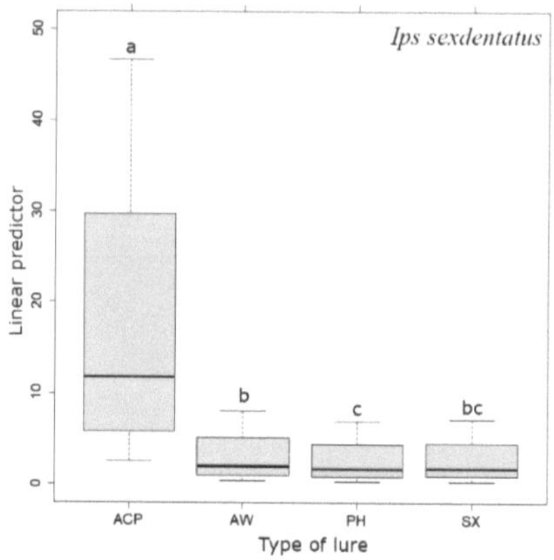

Figure 4. Differences in catching *Ips sexdentatus* with different pheromone lures. Different letters above the boxplots indicate statistically significant differences ($p < 0.05$). The type of bait is shown on the x-axes: ACP (ACUMIPROTECT), AW (ACUWIT), PH (Pheagr IAC) and SX (SEXOWIT). The linear predictor shown on the y-axes refers to a combination of the factors considered in the model (location, season, trap position).

4. Discussion

Compared to the bark beetle species on pine trees (e.g., *Tomicus piniperda* (Linnaeus, 1758), *Tomicus minor* (Hartig, 1834), *Orthotomicus erosus* (Wollaston 1857)), *Ips sexdentatus* is considered "the late swarmer", as the first active beetles are usually caught in June [49–51]. The flight activity of *I. sexdentatus* at the studied sites follows the pattern previously reported in northern Italy, Turkey, eastern Romania, and France [14,48,52–55]. In contrast, *I. sexdentatus* only emerged at our sites when daily temperatures exceeded 13 °C. Similarly to Turkish populations, mass flight occurred at temperatures above 20 °C [14]. The first increase in abundance occurred two to three weeks after the peak in the spring swarming (mid-July), which could indicate that females are re-emerge and establish a sister brood, similar to *I. typographus* and *I. duplicatus*; however, due to few observations, further studies are needed to confirm this theory. The second increase occurred in the first half of August, which indicates the emergence of offspring; therefore, it is certain that *I. sexdentatus* is univoltine in the Czech Republic. Flight activity lasts 3–4 months and begins when the mean daily temperature reaches 13 °C (early May), similarly to *I. typographus* and *I. duplicatus* [56]. Conversely, the flight activity of *I. sexdentatus* already ceased in August and ended in early September, about one month earlier than that of *I. duplicatus* and *I. typographus* [3,57–62]. This significantly shortens the period during which the pines are exposed to bark beetle infestation, which considerably reduces the damage. Although the current flight activity and population dynamics have been mapped, some changes could be observed in the future due to progressive warming. For example, the shift in spring swarming to an earlier date and a clearly recognizable increase in flight activity could be expected, similar to what has been predicted for *I. typographus* [63–66]; or permanent sister broods could develop, similar to *I. sexdentatus* in France [47]; and finally, additional generations could appear at the end of the growing season. Therefore, the monitoring period could be extended to capture *I. sexdentatus* in April and September. In addition, surveys could be increased at the time of female re-emergence and the emergence of offspring.

Pheromone baits are considered the most practical means of monitoring and controlling bark beetle outbreaks [67,68]. Our results show that their effectiveness varies considerably. Traps baited with ACP lures showed the highest attractiveness for both species studied. In general, the number of *I. sexdentatus* caught was about 4.7–8.1 times higher when ACP baits were used than when other baits (AW, PH or SX) were used. The high affinity of *I. sexdentatus* to ACP baits, which were originally developed for the capture of *I. acuminatus*, is not surprising as interspecific inhibition between *Ips* genera [69] towards pheromone compounds is eminent [29,31,70,71]. Thus, similar to *I. acuminatus*, *I. sexdentatus* is also attracted to the substances ipsenol, ipsdienol, and verbenol contained in ACP [72]. In contrast, our data contradict Knížek [18], who found PH and SX baits to be very effective for catching *I. sexdentatus*. Considering that the chirality of ipsdienol varies considerably between species and populations within the same species [73,74] and that the exclusion or inappropriate enantiomer drastically reduces the attractiveness of the aggregation pheromone [71], a plausible explanation for the low efficacy of the SX lure could be the use of an inappropriate optical enantiomer of ipsdienol in the mixture. Similarly, the low efficacy of PH could be due to the solvent 2-methylbut-3-en-2-ol (MB) [75]. However, MB is a dominant component of the aggregation pheromone of *I. typographus*, *I. duplicatus*, *O. erosus*, *Pityogenes* spp. or *Ips aminitus* (Eichhoff, 1871) [76–81], it is minimally attractive to pine bark beetles [71]. Therefore, the ACP bait is an optimal mixture of suitable enantiomers with potentially high attractiveness and thus a suitable tool for monitoring *I. sexdentatus*.

5. Conclusions

Characteristics of flight activity, such as spring emergence and re-emergence of females, sister broods and termination of flight, are crucial information for monitoring and mapping the population dynamics of *Ips sexdentatus*. Under the climatic conditions of the Czech Republic, the optimal survey period begins in early May, when the overwintering generation begins to swarm, and lasts until early September, when activity comes to ahalt. However, it is important to consider the possibility of extending the monitoring period due to global warming. It may also be useful to increase protection measures before the females re-emerge and establish a sister brood and before callows emerge in search of new hosts. Theysohn traps with ACP lures containing an optimal combination of the relevant enantiomers have proven to be a suitable tool for monitoring the population dynamics of *I. sexdentatus*.

Supplementary Materials: The following supporting information can be downloaded at: https://www.mdpi.com/article/10.3390/f15060961/s1, Figure S1: Six-toothed pine bark beetle Ips sexdentatus adult and damage caused on Scots pine. The scale bar for head (A) and back part of elytra (B) is 250 μm, and for maternal gallery (C) is 2.5 cm. Figure S2: Distribution map of Ips sexdentatus in the Czech Republic and location of experimental localities (✱). Distribution data were extracted from following literature sources: (✕) Knížek et al. [18]; (▲) Knížek and Zahradník [21]; (■) Knížek and Liška [22]; (○) Nature Conservation Agency of the Czech Republic [24]; (•) Jelínek [39]. Figure S3: A set of maps: (A) a map with the location of the Czech Republic in Europe; (B) a map with the location of the research area in the Czech Republic; and detailed maps with the location of the sites (C—Locality1; D—Locality2, E—Locality3). Figure S4: Illustrative photos of monitored localities. A—Locality1; B—Locality2; C—Locality3.

Author Contributions: Conceptualization, P.D.; methodology, P.D.; validation, P.D.; formal analysis, D.H.; investigation, P.D., M.D., J.K. and D.H.; resources, P.D.; data curation, D.H.; writing—original draft preparation, D.H. and P.D.; writing—review and editing, P.D. and D.H.; visualization, D.H.; supervision, P.D.; project administration, P.D.; funding acquisition, P.D. All authors have read and agreed to the published version of the manuscript.

Funding: This research was funded by the Forests of the Czech Republic, state enterprises [project 05/2019]. Petr Doležal and Markéta Davídková acknowledge support from the Ministry of Agriculture of the Czech Republic, institutional support MZE-RO0123.

Data Availability Statement: Data available on request.

Acknowledgments: Authors would like to thank to Jan Okrouhlík for consultation with statistical analyses and Pavel Doležal for his support in the field. Sincere thanks come to three anonymous reviewers for their comments that considerably improved the quality of this manuscript.

Conflicts of Interest: The authors declare no conflicts of interest.

References

1. Etxebeste, I.; Lencina, J.L.; Pajares, J. Saproxylic community, guild and species responses to varying pheromone components of a pine bark beetle. *Bull. Entomol. Res.* **2013**, *103*, 497–510. [CrossRef]
2. Foit, J.; Čermák, V. Colonization of disturbed Scots pine trees by bark- and wood-boring beetles. *Agric. For. Entomol.* **2014**, *16*, 184–195. [CrossRef]
3. Wermelinger, B.; Schneider Mathis, D.; Knížek, M.; Forster, B. Tracking the spread of the northern bark beetle (*Ips duplicatus* [Sahlb.]) in Europe and first records from Switzerland and Liechtenstein. *Alp. Entomol.* **2020**, *4*, 179–184. [CrossRef]
4. Holuša, J.; Foit, J.; Knížek, M.; Schovankova, J.; Lukášová, K.; Vanicka, H.; Trombik, J.; Kula, E. The bark beetles *Orthotomicus laricis* and *Orthotomicus longicollis* are not pests in Central Europe: A case study from the Czech Republic. *Bull. Insectology* **2019**, *72*, 253–260.
5. Guérard, N.; Dreyer, E.; Lieutier, F. Interactions between Scots pine, *Ips acuminatus* (Gyll.) and *Ophiostoma brunneo-ciliatum* (Math.): Estimation of the critical thresholds of attack and inoculation densities and effects on hydraulic properties in the stem. *Ann. For. Sci.* **2000**, *57*, 681–690. [CrossRef]
6. Trugman, A.T.; Anderegg, L.D.L.; Anderegg, W.R.L.; Das, A.J.; Stephenson, N.L. Why is tree drought mortality so hard to predict? *Trends Ecol. Evol.* **2021**, *36*, 520–532. [CrossRef] [PubMed]
7. Wermelinger, B.; Seifert, M. Analysis of temperature dependent development of the spruce bark beetle *Ips typographus* (L.) (Col., Scolytidae). *J. Appl. Entomol.* **1998**, *122*, 185–191. [CrossRef]
8. Régnière, J.; Bentz, B. Modeling cold tolerance in the mountain pine beetle, *Dendroctonus ponderosae*. *J. Insects Physiol.* **2007**, *53*, 559–572. [CrossRef]
9. Powell, J.A.; Bentz, B.J. Connecting phenological predictions with population growth rates for mountain pine beetle, an outbreak insect. *Landsc. Ecol.* **2009**, *24*, 657–672. [CrossRef]
10. Bentz, B.J.; Régnière, J.; Fettig, C.J.; Hansen, E.M.; Hayes, J.L.; Hicke, J.A.; Kelsey, R.G.; Negrón, J.F.; Seybold, S.J. Climate change and bark beetles of the Western United States and Canada: Direct and indirect effects. *BioScience* **2010**, *60*, 602–613. [CrossRef]
11. Weed, A.S.; Ayres, M.P.; Hicke, J.A. Consequences of climate change for biotic disturbances in North American forests. *Ecol. Monogr.* **2013**, *83*, 441–470. [CrossRef]
12. Jonášová, M.; Prach, K. Central-European Mountain spruce (*Picea abies* (L.) Karst.) forests: Regeneration of tree species after a bark beetle outbreak. *Ecol. Eng.* **2004**, *23*, 15–27. [CrossRef]
13. Knížek, M. Lýkožrout borový (*Ips sexdentatus*). *Lesn. Práce* **2020**, *99*, 14–17. (In Czech)
14. Özcan, G.E. Use of pheromone-baited traps for monitoring *Ips sexdentatus* (Boerner) (Coleoptera: Curculionidae) in oriental spruce stands. *Afr. J. Biotechnol.* **2011**, *10*, 16351–16360.
15. Bakke, A. Ecological studies on bark beetles (Coleoptera: Scolytidae) associated with Scots pine (*Pinus sylvestris* L.) in Norway with particular reference to the influence of temperature. *Medd. Fra Det Nor. Skogsforsoksvesen* **1968**, *21*, 443–602.
16. Barna, M.; Ferezliev, A.; Tsakov, H.; Mihál, I. Investigations of mature Scots pine stands in wind-throw areas in Norway spruce forests in Western Rhodopes. *Folia Oecol.* **2020**, *47*, 1–9. [CrossRef]
17. Georgieva, M.; Belilov, S.; Dimitrov, S.; Iliev, M.; Trenkin, V.; Mirchev, P.; Georgiev, G. Application of remote sensing data for assessment of bark beetle attacks in pine plantations in Kirkovo region, the Eastern Rhodopes. *Forests* **2022**, *13*, 620. [CrossRef]
18. Knížek, M.; Liška, J.; Véle, A. Efficacy of synthetic lures for pine bark beetle monitoring. *J. For. Sci.* **2022**, *68*, 19–25. [CrossRef]
19. Kovac, M.; Rigling, D.; Pernek, M. Ophiostomatales associated with Mediterranean pine engraver, *Orthotomicus erosus* (Coleoptera, Curculionidae) in Dalmatia, Croatia. *J. Fungi* **2022**, *8*, 788. [CrossRef]
20. Pfeffer, A. *Fauna ČSR: Kůrovci—Scolytoidea*, 1st ed.; Československá Akademie Věd: Prague, Czech Republic, 1995; p. 324. (In Czech)
21. Knížek, M.; Zahradník, P. Bark and wood boring beetles in the pine stands. In *Methodology of Forest Insect and Disease Survey in Central Europe, Proceedings of the Second Workshop of the IUFRO Working Party 7.03.10, Sion-Châteauneuf, Switzerland, 20–23 April 1999*; Forster, B., Knížek, M., Grodzki, W., Eds.; Swiss Federal Institute for Forest, Snow and Landscape Research: Sion-Châteauneuf, Switzerland, 1999; pp. 54–59.
22. Knížek, M.; Liška, J. Výskyt lesních škodlivých činitelů v roce 2020 a jejich očekávaný stav v roce 2021. In *Zpravodaj Ochrany Lesa, Supplementum*; Forestry and Game Management Research Institute: Jíloviště, Czech Republic, 2021; p. 76. Available online: https://www.vulhm.cz/files/uploads/2021/06/ZOL_Suppl_2021.pdf (accessed on 11 August 2022).
23. Liška, J.; Knížek, M.; Véle, A. Evaluation of insect pest occurrence in areas of calamitous mortality of Scots pine. *Cent. Eur. For. J.* **2021**, *67*, 85–90. [CrossRef]
24. Nature Conservation Agency of the Czech Republic. Available online: https://portal.nature.cz/w/druh-6250#/ (accessed on 18 May 2024).

25. Lubojacký, J.; Lorenc, F.; Samek, M.; Knížek, M.; Liška, J. Main problems in forest protection in the Czech Republic in 2021 and forecast for 2022. *Zprav. Ochr. Lesa* **2022**, *25*, 17–26. (In Czech)
26. Özcan, G.E. Assessment of *Ips sexdentatus* population considering the capture in pheromone traps and their damages under non-epidemic conditions. *Šumarski List* **2017**, *141*, 47–56. [CrossRef]
27. Bouhot, L.; Lieutier, F.; Debouzie, D. Spatial and temporal distribution of attacks by *Tomicus piniperda* L. and *Ips sexdentatus* Boern. (Col., Scolytidae) on *Pinus sylvestris*. *J. Appl. Entomol.* **1987**, *106*, 356–371. [CrossRef]
28. Markalas, S. Frequency and distribution of insect species on trunks in burnt pine forests of Greece. *Schweiz. Entomol. Ges.* **1997**, *70*, 57–61. [CrossRef]
29. Vité, J.P.; Bakke, A.; Renwick, J.A.A. Pheromones in Ips (Coleoptera: Scolytidae): Occurrence and production. *Can. Entomol.* **1972**, *104*, 1967–1975. [CrossRef]
30. Kohnle, U.; Meyer, M.; Kluber, J. Formulation of population attractant for the pine bark beetle, *Ips sexdentatus* (Col, Scolytidae). *Allg. Forst Jagdztg.* **1992**, *163*, 81–87.
31. Etxebeste, I.; Pajares, J.A. Verbenone protects pine trees from colonization by the six-toothed pine bark beetle, *Ips sexdentatus* Boern. (Col.: Scolytinae). *J. Appl. Entomol.* **2011**, *135*, 258–268. [CrossRef]
32. Jactel, H.; Lieutier, F. Effects of attack density on fecundity of the Scots pine beetle *Ips sexdentatus* Boern (Col.; Scolytidae). *J. Appl. Entomol.* **1987**, *104*, 190–204. [CrossRef]
33. Sierra, J.M.; Martín, A.B. Efectividad de trampas de feromona en la captura masiva de *Ips sexdentatus* Boern. (Coleoptera: Scolytidae), escoítido perforador de los pinos. *Bol. San. Veg. Plagas* **2004**, *30*, 745–752. (In Spanish)
34. Péter, S. Effect of Climate Change on Population Dynamics of Bark Beetles: Relationships between Temperature and Development Rate of *Ips sexdentatus* (Boern.). Ph.D. Dissertation, Université de Lorraine, Lorraine, Italy, 2014.
35. ÚVT, BENETA.cz. Taxonomický Klasifikační Systém půd České Republiky. Available online: https://klasifikace.pedologie.czu.cz (accessed on 17 April 2023).
36. Czech Geological Survey, Soil Map 1: 50 000. Available online: https://mapy.geology.cz/pudy/#; https://mapy.geology.cz/geo/ (accessed on 17 April 2023).
37. Forests of the Czech Republic, State Enterprises: Porostní Mapy—Informační Portál Stáří Lesních Porostů na Území České Republiky. Available online: https://geoportal.lesycr.cz/itc/?serverconf=default&wmcid=882 (accessed on 17 April 2023).
38. METEOBLUE (2006–2023): Pozorované Historické Údaje o Klimatu a počasí. Available online: www.meteoblue.com (accessed on 18 April 2023).
39. Jelínek, J. Check List of Czechoslovak Insects IV (Coleoptera). *Folia Heyrovskyana* **1993**, *1*, 5–172.
40. Bessin, R.; Villanueva, R. *Predicting Insect Development Using Insect Degree Days*; Cooperative Extension Service University of Kentucky College of Agriculture, Food and Environment: Lexington, KY, USA, 2019; Available online: https://entomology.ca.uky.edu/ef123 (accessed on 18 April 2023).
41. R Core Team. R: A Language and Environment for Statistical Computing. R Foundation for Statistical Computing, Vienna, Austria. Available online: https://www.R-project.org/ (accessed on 18 April 2023).
42. Bates, D.; Maechler, M.; Jagan, M. Matrix: Sparse and Dense Matrix Classes and Methods. R Package Version 1.5-1. 2022. Available online: https://CRAN.R-project.org/package=Matrix (accessed on 17 April 2023).
43. Bates, D.; Maechler, M.; Bolker, B.; Walker, S. Fitting Linear Mixed-Effects Models Using lme4. *J. Stat. Softw.* **2015**, *67*, 1–48. [CrossRef]
44. Bolker, B.M.; Brooks, M.E.; Clark, C.J.; Geange, S.W.; Poulsen, J.R.; Stevens, M.H.H.; White, J.-S.S. Generalized linear models: A practical guide for ecology and evolution. *Trends Ecol. Evol.* **2009**, *24*, 127–135. [CrossRef] [PubMed]
45. Mullen, K.M.; Ardia, D.; Gil, D.L.; Windover, D.; Cline, J. DEoptim: An R Package for Global Optimization by Differential Evolution. *J. Stat. Softw.* **2011**, *40*, 1–26. [CrossRef]
46. Powell, M.J.D. *The BOBYQA Algorithm for Bound Constrained Optimization without Derivatives*; Technical Report NA2009/06; Department of Applied Mathematics and Theoretical Physics: Cambridge, UK, 2009.
47. Hothorn, T.; Bretz, F.; Westfall, P. Simultaneous inference in general parametric models. *Biom. J.* **2008**, *50*, 346–363. [CrossRef] [PubMed]
48. Pineau, X.; David, G.; Peter, Z.; Sallé, A.; Baude, M.; Lieutier, F.; Jactel, H. Effect of temperature on the reproductive success, developmental rate and brood characteristics of *Ips sexdentatus* (Boern.). *Agric. For. Entomol.* **2016**, *19*, 23–33. [CrossRef]
49. Långström, B. Life cycles and shoot-feeding of pine shoot beetles. *Stud. For. Suec.* **1983**, *163*, 1–29.
50. Tribe, G.D. Phenology of *Pinus radiata* log colonization and reproduction by the European bark beetle *Orthotomicus erosus* (Wollaston) (Coleoptera: Scolytidae) in the south-western Cape Province. *J. Entomol. Soc. South. Afr.* **1990**, *53*, 117–126.
51. Mendel, Z.; Boneh, O.; Shenhar, Y.; Riov, J. Diurnal flight patterns of *Orthotomicus erosus* and *Pityogenes calcaratus* in Israel. *Phytoparasitica* **1991**, *19*, 23–31. [CrossRef]
52. Lozzia, G.C. Monitoring and control of *Ips sexdentatus* Boerner using synthetic pheromones. *DiPSA* **1998**, *27*, 71–84.
53. Isaia, G.; Manea, A.; Paraschiv, M. Study on the effect of pheromones on the bark beetles of the Scots pine. *Bull. Transilv. Univ. Bras.* **2010**, *3*, 67–72.

54. Sarikaya, O.; Avci, M.; Yildirim, S. Flight activity and biology of *Ips sexdentatus* Boerner in black pine (*Pinus nigra* Arnold) forests of Isparta, Turkey. In Proceedings of the International Science Conference, Forests in the Future—Sustainable Use, Risks and Challenges, Belgrade, Serbia, 4–5 October 2012; Rakonjac, L., Ed.; Institute of Forestry: Belgrade, Serbia, 2012; pp. 595–604.
55. López, S.; Goldarazena, A. Flight dynamics and abundance of *Ips sexdentatus* (Coleoptera: Curculionidae: Scolytinae) in different sawmills from Northern Spain: Differences between local *Pinus radiate* (Pinales: Pinaceae) and Southern France incoming *P. pinaster* timber. *Psyche A J. Entomol.* **2012**, *2012*, 145930. [CrossRef]
56. Holuša, J.; Lukášová, K.; Lubojacký, J. Comparison of seasonal flight activity of *Ips typographus* and *Ips duplicatus*. *Sci. Agric. Bohem.* **2012**, *43*, 109–115.
57. Bakke, A.; Austara, O.; Pettersen, H. Seasonal flight activity and attack pattern of *Ips typographus* in Norway under epidemic conditions. *Medd. Norsk I Skogfor* **1977**, *33*, 257–268.
58. Harding, S.; Ravn, H. Seasonal activity of *Ips typographus* L. (Col., Scolytidae) in Denmark. *Z. Angew. Entomol.* **1985**, *99*, 123–131. [CrossRef]
59. Faccoli, M.; Stergulc, F. *Ips typographus* (L.) pheromone trapping in south Alps: Spring catches determine damage thresholds. *J. Appl. Entomol.* **2004**, *128*, 307–311. [CrossRef]
60. Baier, P.; Pennerstorfer, J.; Schopf, A. PHENIPS – a comprehensive phenology model of *Ips typographus* (L.) (Col., Scolytinae) as a tool for hazard rating of bark beetle infestation. *For. Ecol. Manag.* **2007**, *249*, 171–186. [CrossRef]
61. Chen, G.; Zhang, Q.-H.; Wang, Y.; Liu, G.-T.; Zhou, X.; Niu, J.; Schlyter, F. Catching *Ips duplicatus* (Sahlberg) (Coleoptera: Scolytidae) with pheromone-baited traps: Optimal trap type, colour, height and distance to infestation. *Pest. Manag. Sci.* **2009**, *66*, 213–219. [CrossRef] [PubMed]
62. Öhrn, P.; Långström, B.; Lindelöw, Å.; Björklund, N. Seasonal flight patterns of *Ips typographus* in southern Sweden and thermal sums required for emergence. *Agric. For. Entomol.* **2014**, *16*, 147–157. [CrossRef]
63. Lobinger, G. Die Lufttemperatur als limitierender Faktor für die Schwärmaktivität zweier rindenbrütender Fichtenborkenkäferarten, *Ips typographus* L. und *Pityogenes chalcographus* L. (Col., Scolytidae). *Anz. Für Schädlingskunde Pflanzenschutz Umweltschutz* **1994**, *67*, 14–17. [CrossRef]
64. Jönsson, A.M.; Appelberg, G.; Harding, S.; Bärring, L. Spatio-temporal impact of climate change on the activity and voltinism of the spruce bark beetle, *Ips typographus*. *Glob. Chang. Biol.* **2009**, *15*, 486–499. [CrossRef]
65. Jönsson, A.M.; Harding, S.; Krokene, P.; Lange, H.; Lindelöw, Å.; Økland, B.; Raven, H.P.; Schroeder, L.M. Modelling the potential impact of global warming on *Ips typographus* voltinism and reproductive diapause. *Clim. Chang.* **2011**, *109*, 695–718. [CrossRef]
66. Jakoby, O.; Lischke, H.; Wermelinger, B. Climate change alters elevational phenology patterns of the European spruce bark beetle (*Ips typographus*). *Glob. Chang. Biol.* **2019**, *25*, 4048–4063. [CrossRef] [PubMed]
67. Vité, J.P.; Baader, E. Present and future use of semiochemicals in pest management of bark beetles. *J. Chem. Ecol.* **1990**, *16*, 3031–3041. [CrossRef] [PubMed]
68. Bakke, A. Using Pheromones in the Management of Bark Beetle Outbreaks. Forest insect guilds: Patterns of interaction with host trees. In *Report No. NE-153*; Baranchikov, Y.N., Mattson, W.J., Hain, F.P., Payne, T.L., Eds.; Department of Agriculture, Forest Service, Northeastern Forest Experiment Station: Radnor, PA, USA, 1991.
69. Byers, J.A. Chemical ecology of bark beetles. *Experientia* **1989**, *45*, 271–283. [CrossRef]
70. Francke, W.; Pan, M.L.; Bartels, J.; König, W.A.; Vité, J.P.; Krawielitzki, S.; Kohnle, U. The odour bouquet of three pine engraver beetles (*Ips* spp.). *J. Appl. Entomol.* **1986**, *101*, 453–461. [CrossRef]
71. Birgersson, G.; Dalusky, M.J.; Espelie, K.E.; Berisford, C.W. Pheromone production, attraction, and interspecific inhibition among four species of Ips bark beetles in the southeastern USA. *Psyche A J. Entomol.* **2012**, *2012*, 532652. [CrossRef]
72. Paiva, M.R.; Pessoa, M.F.; Vité, J.P. Reduction in the pheromone attractant response of *Orthotomicus erosus* (Woll.) and *Ips sexdentatus* Boern. (Col., Scolytidae). *J. Appl. Entomol.* **1988**, *106*, 198–200. [CrossRef]
73. Seybold, S. The Role of Chirality in the Olfactory-Directed Aggregation Behavior of Pine Engraver Beetles in the Genus *Ips* (Coleoptera: Scolytidae). Ph.D. Dissertation, University of California, CSU, Berkeley, CA, USA, 1992.
74. Ivarsson, P. The Pheromone Systems of the Spruce Bark Beetles, *Ips duplicatus* and *I. typographus*: Biosynthesis and Regulations. Ph.D. Dissertation, Göteborg University, Göteborg, Sweden, 1995.
75. Doležal, P.; Davídková, M.; Vovesný, P.; Drašar, P. Pheromone dispenser ACUMIPROTECT for mass trapping of the sharp-dentated bark beetle, *Ips acuminatus* (Coleoptera; Curculionidae). *Chem. Listy* **2023**, *117*, 13–16. [CrossRef]
76. Bakke, A.; Froyen, P.; Skattebol, L. Field response to a new pheromonal compound isolated from *Ips typographus*. *Naturwissenschaften* **1977**, *64*, 98–99. [CrossRef]
77. Giesen, H.; Kohnle, U.; Vité, J.P.; Pan, M.-L.; Francke, W. Aggregationspheromon des mediterranen Kiefernborkenkäfers *Ips* (*Orthotomicus*) *erosus*. *J. Appl. Entomol.* **1984**, *98*, 95–97. [CrossRef]
78. Baader, E.J. *Pityogenes* spp. (Col., Scolytidae): Investigations on semiochemicals and their application in forest protection. *J. Appl. Entomol.* **1989**, *107*, 1–31. [CrossRef]
79. Zuber, M. Racemat und Enantiomere von Ipsdienol zur Anlockung von *Ips amitinus* (Eich.) (Col., Scolytidae). *Anz. Für Schädlingskunde Pflanzenschutz Umweltschutz* **1994**, *67*, 92–93. [CrossRef]

80. Seybold, S.J.; Huber, D.P.W.; Lee, J.C.; Graves, A.D.; Bohlmann, J. Pine monoterpenes and pine bark beetles: A marriage of convenience for defense and chemical communication. *Phytochem. Rev.* **2006**, *5*, 143–178. [CrossRef]
81. Duduman, M.-L.; Beránková, K.; Jakuš, R.; Hradecký, J.; Jirošová, A. Efficiency and sustainability of *Ips duplicatus* (Coleoptera: Curculionidae) pheromone dispensers with different designs. *Forests* **2022**, *13*, 511. [CrossRef]

forests

MDPI

Article

Ecological Traits of Three Species of *Xiphydria* Woodwasps from Japan: Host Tree Species and Eggs, Symbiotic Fungi, and Mucus in Their Bodies

Ryu Takagi * and Hisashi Kajimura *

Laboratory of Forest Protection, Graduate School of Bioagricultural Sciences, Nagoya University, Nagoya 464-8601, Japan
* Correspondence: takagi.ryu.z5@s.mail.nagoya-u.ac.jp (R.T.); kajimura@agr.nagoya-u.ac.jp (H.K.)

Abstract: Woodwasps (Siricidae and Xiphydriidae) inhabit living, weakened, or freshly dead trees and their larvae feed on sapwood. Siricidae have been intensively researched for pest management. In contrast, the ecology of Xiphydriidae remains largely unknown. In the present study, we collected xiphydriid woodwasp adults and dissected female adults to elucidate the cornerstone ecology of this family and compared these findings with those of siricid woodwasps. The findings provide new host records for these species and indicate that their host ranges span multiple families. Notably, all *Xiphydria* species had female-biased sex ratios. All adult females had gourd-shaped eggs, similar to those found in gall wasps (Cynipidae), which contrast with the oval-shaped eggs of Siricidae. Slit-like mycangia were located at the base of the ovipositor, with pairs of fungal masses composed of hyphal fragments or spores directly positioned below the seventh sternum, differing structurally from the pouch-like mycangia in Siricidae. Mucus reservoirs and secretory glands were found in the terminal abdominal segments, similar to Siricidae. Mucus in *X. annulitibia* and *X. ogasawarai* was colorless and transparent, as reported in Siricidae, whereas *X. eborata* exhibited deep wine-red mucus, which is the world's first discovery in all dissected species of Siricidae and Xiphydriidae.

Keywords: woodwasp; Xiphydriidae; host tree species; female-biased sex ratio; gourd-shaped egg; mycangia; symbiotic fungi; colored mucus; comparative morphology

Academic Editors: Katarína Pastirčáková and Rostislav Zemek

Received: 13 December 2024
Revised: 28 January 2025
Accepted: 29 January 2025
Published: 1 February 2025

Citation: Takagi, R.; Kajimura, H. Ecological Traits of Three Species of *Xiphydria* Woodwasps from Japan: Host Tree Species and Eggs, Symbiotic Fungi, and Mucus in Their Bodies. *Forests* **2025**, *16*, 264. https://doi.org/10.3390/f16020264

1. Introduction

Woodwasps (Siricidae and Xiphydriidae) inhabit living, weakened, or freshly dead trees and their larvae feed on the sapwood [1]. Female adults insert ovipositors into tree trunks and lay eggs in the sapwood [2,3]. Hatched larvae bore into the sapwood, where they pupate and then emerge as adults [1,4,5]. In Siricidae, the subfamily Siricinae (e.g., *Sirex*, *Urocerus*, *Xoanon*, *Xeris*) utilizes coniferous trees [6,7], and the subfamily Tremecinae (e.g., *Tremex*) utilizes broadleaf trees [8]. In contrast, all Xiphydriidae species utilize only broadleaf trees [8]. During oviposition, female adults drill single or multiple holes in the sapwood, where they lay eggs [2,3]. Simultaneously, spores of symbiotic fungi and slime-like secretions (mucus) stored in their bodies are injected into the sapwood together with the eggs [2]. They carry symbiotic fungal spores in specialized storage organs called mycangia or intersegmental sacs [9–11]. Siricinae and Tremecinae have symbiotic relationships with basidiomycete fungi *Amylostereum* spp. and *Cerrena* spp., respectively [7,12]. In Xiphydriidae, there are symbiotic relationships with the ascomycete fungi *Daldinia* spp. [13]. In contrast, *Xeris spectrum* L. of Siricinae has no symbiotic fungi and utilizes trees that have already been colonized by other woodwasps [14]. The fungi inoculated into trees colonize sapwood, which is a food resource for larvae [4] and is thought to be responsible for transporting and providing nutrients, such as nitrogen, to larvae [4] or providing some enzymes for the digestion of woody tissue [15,16]. Mucus is a transparent secretion primarily composed of protein polysaccharides [17] and contains several enzymes, such as laccase, which contribute to the degradation of tree tissues [18,19], and proteins such as noctilisin, which are toxic to trees [20,21]. These components, particularly in species that utilize living trees, weaken living trees [22,23] and stimulate the growth of symbiotic fungi [22], thereby helping woodwasp larvae overcome tree defenses [5]. However, mucus storage has also been reported in some woodwasp species that do not utilize living trees [11,24,25], and the role of mucus in these species remains unclear.

Several siricid species cause economic losses in forestry. For example, *Sirex noctilio* Fabricius attacks *Pinus* spp. trees by injecting arthrospores of the symbiotic fungus *Amylostereum areolatum* (Fr.) Boid. and phytotoxic mucus during oviposition [2,23]. Fungal hyphae cause xylem desiccation and blockage, and mucus causes leaf discoloration and wilting [22,26]. This combination causes devastating damage to trees, leading to their decline or death [26]. *S. noctilio*, native to Europe, invaded Australia and caused extensive dieback damage to pine tree plantations [27]. It subsequently invaded South Africa [28], and in recent years, it has been found in North America [29] and China [18]. In Japan, *Urocerus japonicus* Smith has been noted for oviposition on living Japanese cedar, *Cryptomeria japonica* (L.f.) D.Don, and Japanese cypress, *Chamaecyparis obtusa* (Siebold et Zucc.) Endl. [7,30]. Although this woodwasp does not kill trees, injection of the mycangial fungus *Amylostereum laevigatum* (Fr.) Boid. discolors woody tissue and decreases timber value [7,10,30]. Knowledge regarding woodwasps, such as host tree species, oviposition behavior, and the presence of symbiotic fungi and mucus, is largely biased towards Siricidae, especially species inhabiting living trees. Thus, to comprehensively understand the ecology of woodwasps and advance pest management strategies, focusing on the species inhabiting dead trees is essential.

Xiphydriid woodwasps inhabit broadleaf trees immediately after death [31,32] and thus, unlike the siricid species mentioned above, are not considered pests. This family includes over 150 species worldwide [33]. Since it has received less attention than Siricidae, for most of the species, life history, including ecological traits such as host tree species, is still unknown. Additionally, their symbiotic fungi and mucus, which are also closely related to their life history, are confirmed to be present in only a few species [11,25]. Understanding the life history of Xiphydriidae can fill critical knowledge gaps in the life history of woodwasps, which will contribute to clarifying their diversity in the utilization of wood substrates with a continuum from healthy to declining and dead trees. Xiphydriidae is a taxonomic sister group to Siricidae [34,35], and its ecology is similar but different from that previously reported for *S. noctilio* and *U. japonicus*. Ecological comparisons between these families can provide important insights not only into the evolutionary processes of woodwasp–fungi–tree relationships but also into potential applications for strategies of forest management and conservation.

We aims to investigate the entire life history of Xiphydriid woodwasps and compare it with that of siricid woodwasps, seeking to uncover evolutionary and ecological differences between these taxonomic groups as our ultimate goal. In this study, we exhaustively collected adults of xiphydriid species that emerged from deciduous broadleaf tree species and monitored their occurrence periods. We also measured their body size, dissected the females to examine the number and morphology of eggs, and confirmed the presence of mycangia and mucus.

2. Materials and Methods

2.1. Collection of Study Logs and Adult Woodwasps

Tree collection was conducted in a mixed stand of coniferous and broadleaf trees in Inabu Town, Toyota City, Aichi Prefecture, central Japan (35°11′ N, 137°10′ E, elevation 980 m). In July 2021 and June and July 2022, six species of deciduous broadleaf trees were felled: *Padus grayana* (Maxim.) C. K. Schneid., *Alnus hirsuta* (Spach) Fisch. ex Rupr., *Carpinus tschonoskii* Maxim., *Carpinus japonica* Blume, *Castanea crenata* Siebold et Zucc., and *Acer sieboldianum* Miq. The felled trees were laid on the forest floor for nine to 10 months to allow woodwasp oviposition (Figure S1a). In April 2022 and April 2023, they were divided into 1 m long logs and stored within outdoor mesh insectaries for each tree species (Figure S1b). From May to October 2022 to 2024, xiphydriid adults emerging from the logs were collected, and their fresh body weight, head width, and ovipositor length were measured (Table 1).

Table 1. Summary of body size and fecundity of emerged adults in three woodwasp species.

Woodwasp Species	Emerged Year	Host Tree Species	Sex (n)	Fresh Body Weight (mg)	Head Width (mm)	Ovipositor Length (mm)	Number of Eggs
Xiphydria eborata	2023	*Padus grayana*	♀(7)	137.8 ± 24.5 (107.3–178.6)	3.3 ± 0.1 (3.1–3.5)	7.1 ± 0.3 (6.8–7.7)	139 ± 28 (104–182)
			♂(1)	87	3.2	-	-
		Alnus hirsuta	♀(2)	173.6 ± 14.7 (163.2–184.0)	3.5 ± 0.1 (3.4–3.6)	7.9 ± 0.4 (7.6–8.2)	226 ± 21 (212–241)
Xiphydria annulitibia	2023	*Carpinus tschonoskii*	♀(20)	34.5 ± 4.8 (24.5–41.4)	2.1 ± 0.2 (1.6–2.4)	3.2 ± 0.1 (2.9–3.4)	169 ± 20 (125–202)
Xiphydria ogasawarai	2023	*Carpinus japonica*	♀(25)	45.6 ± 9.6 (24.1–71.8)	2.3 ± 0.2 (1.9–2.8)	4.6 ± 0.5 (3.9–5.6)	238 ± 43 (158–328)
			♂(2)	15.7 ± 2.2 (14.1–17.2)	1.8 ± 0.1 (1.7–1.9)	-	-
		Acer sieboldianum	♀(107)	40.6 ± 10.8 (20.2–67.5)	2.2 ± 0.2 (1.7–2.7)	4.7 ± 0.5 (3.4–5.6)	217 ± 48 (94–357)
			♂(4)	13.1 ± 4.5 (8.4–18.2)	1.7 ± 0.1 (1.6–1.8)	-	-
	2024	*Carpinus japonica*	♀(8)	76.7 ± 13.7 (58.1–102.4)	2.7 ± 0.2 (2.4–3.0)	6.0 ± 0.3 (5.8–6.6)	380 ± 74 (270–488)
			♂(1)	16.7	1.7	-	-
		Acer sieboldianum	♀(18)	66.2 ± 22.9 (31.2–111.0)	2.6 ± 0.3 (2.0–3.1)	5.7 ± 0.6 (4.5–6.9)	316 ± 73 (195–429)
			♂(6)	18.8 ± 5.5 (11.3–26.7)	1.8 ± 0.2 (1.5–2.0)	-	-

Body size and fecundity are shown as mean ± SD (minimum–maximum).

2.2. Sample Pre-Treatment Before Dissection

The measured female adults were stored at 15 °C in 1.5 mL tubes. Of the stored adults, the individuals living at the time of the experiment were used as samples and, as pre-treatment for dissection, were transferred at 4 °C for 10 min to paralyze them. For further paralysis, the head was dipped in 99.5% ethanol. They were then surface-washed in sterile water for 1 min.

2.3. Dissection, Internal Observation, and Fungal Isolation of Female Adults

Adult females were dissected to confirm the presence of mycangia and mucus and to isolate ovaries and symbiotic fungi, as described previously [11]. The surface-washed samples were pinned onto a sterile paraffin wax plate. Their abdomens were then dissected from the ventral side using sterile needles and microscissors under a stereomicroscope (SZ2-1LST, OLYMPUS, Tokyo, Japan). Their ovaries were collected from the body cavities, and each egg was separated in distilled water and counted (Table 1). Mycangial fungi were separated with sterile needles and inoculated on PDA medium (Difco, Detroit, MI, USA) in Petri dishes (9 cm in diameter) to which 100 ppm streptomycin sulfate (Meiji, Tokyo, Japan) was added to inhibit bacterial contamination. Fungal cultures were incubated at 25 °C in the dark. Fungi from some individuals were stained with lactophenol cotton blue (ourself-formulated) and prepared for observation under an optical microscope (BX53, OLYMPUS, Tokyo, Japan).

2.4. Statistical Analysis

The relationships between fresh body weight and (1) head width, (2) ovipositor length, and (3) number of eggs in female adults were analyzed in the respective pairs of parameters for each woodwasp species using a linear regression model. Slopes of the regression lines obtained for each species were then compared, and their values were tested for significant differences. All analyses were performed using R 4.3.0 [36], and the comparisons of the slope were conducted using the emmeans package.

3. Results

3.1. Adult Woodwasps Emerged from Study Logs

Xiphydria eborata Konow, *X. annulitibia* Takeuchi, and *X. ogasawarai* Matsumura (Figure 1) emerged from the logs of *P. grayana* and *A. hirsuta*, *C. tschonoskii*, and *C. japonica* and *A. sieboldianum*, respectively (Table 2). However, no woodwasps were found on *C. crenata*. Neither *X. eborata* nor *X. annulitibia* emerged in 2022, one year after tree felling, but both emerged two years later, in mid-May 2023 (Table 2 and Figure 2a).

(**a**) (**b**) (**c**)

Figure 1. Female adults of *Xiphydria eborata* (**a**), *X. annulitibia* (**b**), and *X. ogasawarai* (**c**).

In 2024, only one adult (female) of *X. annulitibia* emerged on 24 June (Table 2). In contrast, *X. ogasawarai* emerged from late June to late September 2023 (Table 2 and Figure 2a) and from mid-May to early July 2024 (Table 2 and Figure 2b) in the first two consecutive years after tree felling. The sex ratio was female-biased in all the species: 10 females and one male in *X. eborata*, 251 females and 19 males in *X. ogasawarai*, and 28 females and no male in *X. annulitibia* (Table 2).

In addition to the three woodwasp species, several wood-boring insects were collected in the sample trees. Longhorn beetles (Cerambycidae) emerged from all tree species, weevils (Mesoptiliinae) emerged from *C. japonica* and *A. sieboldianum*, and ambrosia beetles (Scolytinae) emerged from *A. sieboldianum*.

Table 2. Annual number of adult woodwasps emerged from study logs.

Study Logs			Woodwasp Species	2022			2023			2024		
Tree Species	Date of Tree-Felling	No. of Logs		No. of Adults		Period of Emergence	No. of Adults		Period of Emergence	No. of Adults		Period of Emergence
				♀	♂		♀	♂		♀	♂	
Padus grayana	21 July 2021	10	*Xiphydria eborata*	0	0	-	8	1	10–22 May	0	0	-
Alnus hirsuta	30 June 2021	8	*Xiphydria eborata*	0	0	-	2	0	17–18 May	0	0	-
Carpinus tschonoskii	24 June 2021	9	*Xiphydria annulitibia*	0	0	-	27	0	17–26 May	1	0	June 24
Castanea crenata	17 June 2021	8	-	0	0	-	0	0	-	0	0	-
Padus grayana	16 June 2022, 13 July 2022	11	-	-	-	-	0	0	-	0	0	-
Carpinus japonica	16 June 2022, 13 July 2022	9	*Xiphydria ogasawarai*	-	-	-	29	3	10 July–5 September	39	3	12 May–1 July
Acer sieboldianum	16 June 2022, 14 July 2022	9	*Xiphydria ogasawarai*	-	-	-	120	4	19 June–13 September	63	9	10 May–8 July

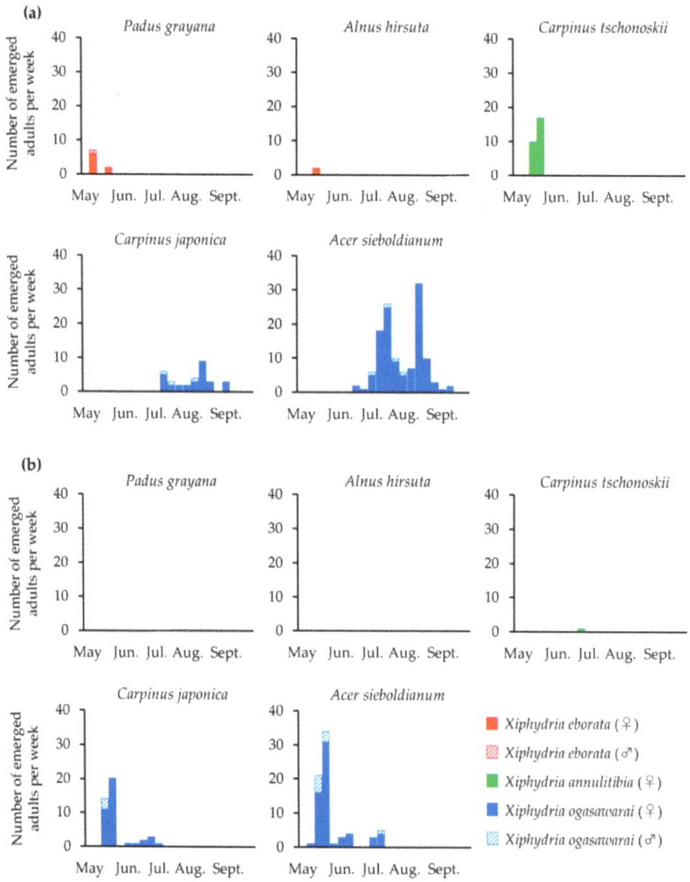

Figure 2. Seasonal occurrence trend of emerged adults in three woodwasp species in 2023 (**a**) and 2024 (**b**).

3.2. Body Size and Fecundity of Adult Woodwasps

Overall, *X. eborata* exhibited the largest body size, *X. annulitibia* exhibited the smallest body size, and male adults were smaller than female adults (Table 1). There were significant positive correlations between fresh body weight and head width (Figure 3a) and between fresh body weight and ovipositor length (Figure 3b) in adult females of all species. The slope of the regression line between fresh body weight and head width for *X. eborata* was significantly smaller than that for the other species (Figure 3a). Additionally, for the regression between fresh body weight and ovipositor length, *X. ogasawarai* had the largest slope, with a significant difference between *X. ogasawarai* and *X. eborata* (Figure 3b).

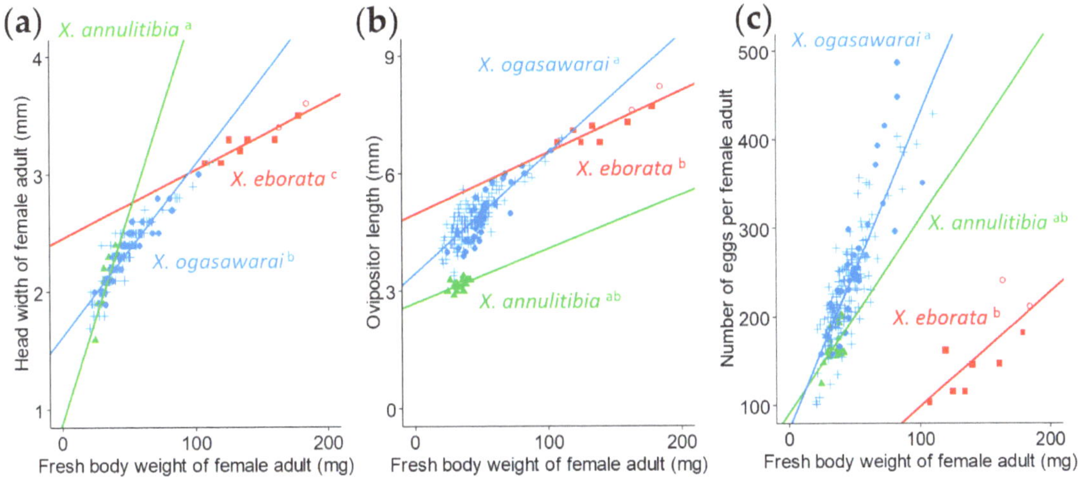

Figure 3. Relation between fresh body weight and head width (**a**), ovipositor length (**b**), and number of eggs (**c**) in three woodwasp species, *Xiphydria eborata*, *Xiphydria annulitibia*, and *Xiphydria ogasawarai*, emerged from felled deciduous trees. *Alnus hirsuta* (○); *Padus grayana* (■); *Carpinus tschonoskii* (▲); *Carpinus japonica* (●); *Acer sieboldianum* (+). Regression equations and Pearson's correlation coefficients for each line: (**a**) *X. eborata*: $y = 0.0058230x + 2.4625690$, $r = 0.929$ ($p = 0.0002882$); *X. annulitibia*: $y = 0.035276x + 0.913680$, $r = 0.872$ ($p = 0.0000005363$); *X. ogasawarai*: $y = 0.0145946x + 1.6300464$, $r = 0.917$ ($p < 2 \times 10^{-16}$). (**b**) *X. eborata*: $y = 0.015744x + 4.983579$, $r = 0.882$ ($p = 0.001631$); *X. annulitibia*: $y = 0.013733x + 2.706501$, $r = 0.473$ ($p = 0.03516$); *X. ogasawarai*: $y = 0.030519x + 3.480342$, $r = 0.830$ ($p < 2 \times 10^{-16}$). (**c**) *X. eborata*: $y = 1.2947x - 30.2170$, $r = 0.756$ ($p = 0.0185$); *X. annulitibia*: $y = 2.1814x + 93.3853$, $r = 0.533$ ($p = 0.01563$); *X. ogasawarai*: $y = 3.5311x + 76.9107$, $r = 0.852$ ($p < 2 \times 10^{-16}$). Different small letters in the legend indicate significant differences in the slopes of the regression lines (comparisons of slopes, $p < 0.05$).

There were also significant positive correlations between fresh body weight and the number of eggs (fecundity) (Figure 3c) in female adults of all species. *X. eborata* had the smallest slope, whereas *X. ogasawarai* had the largest, with a significant difference between these species (Figure 3c).

3.3. Eggs, Symbiotic Fungi, and Mucus in Female Adults

Ovaries, mycangia, and mucus were found in the abdomen of adult females of all species (Figure S2). All eggs consisted of oval egg bodies and elongated projections (Figure 4a–c). All egg bodies were lined towards the ovipositor (Figure S3). Egg sizes in the ovaries were almost the same, but interspecific variation was distinct; *X. eborata* had the largest eggs, and *X. annulitibia* had the smallest eggs.

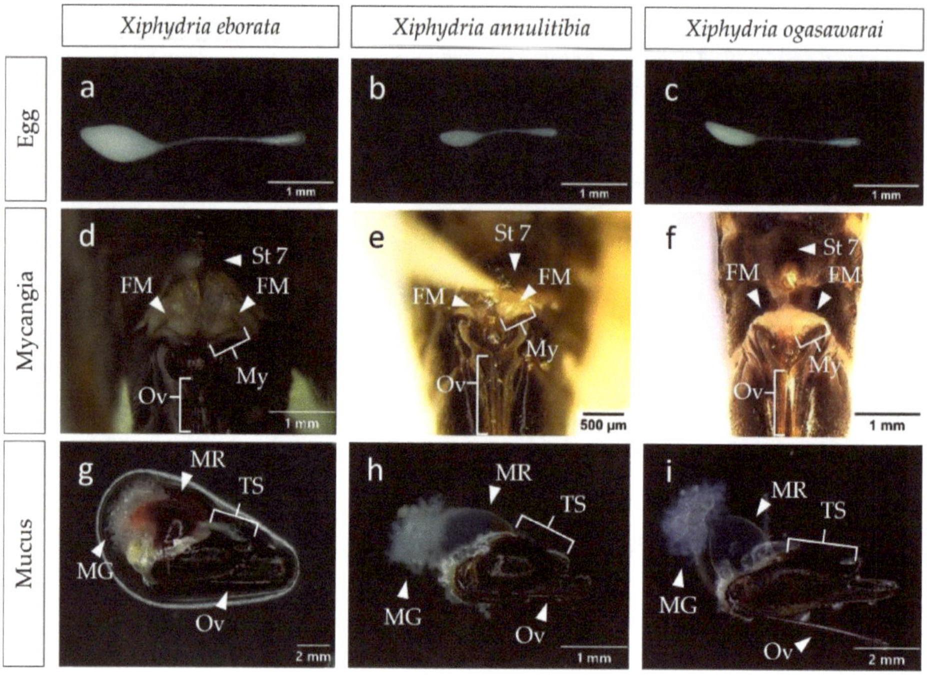

Figure 4. Comparative morphology of an egg (**a–c**), mycangia (**d–f**), and mucus (**g–i**) in three woodwasp species, *Xiphydria eborata*, *Xiphydria annulitibia*, and *Xiphydria ogasawarai*. St7: the 7th sternum; FM: fungal mass; My: mycangia; Ov: ovipositor; MR: mucus reservoir; MG: mucus grand; TS: terminal segment of the abdomen.

Mycangia were at the base of the ovipositor below the 7th sternum and had slit-like structures filled with fungal masses (Figure 4d–f and Figure S3). The masses were placed directly on a slit and were composed of mycelial fragments or spores. *X. eborata* had mycelial fragments of 1.7–5.1 µm width (Figure 5a). *X. annulitibia* and *X. ogasawarai* had spores of 2.0–6.8 µm width and 8.4–27.1 µm length (Figure 5c) and 2.3–13.4 µm width and 6.5–30.9 µm length (Figure 5e), respectively. Spores of both species comprised 1–3 cells. Mycangial fungi produced single colonies on PDA, and their traits varied among the woodwasp species. Fungal colonies of *X. eborata* had felt-like and white mycelia, filled a 9 cm Petri dish in 10–12 days, and then turned concentrically brown to black from the inside (Figure 5b). *X. annulitibia* also had white but cottony mycelia, grew slowly to half of the dish in 10 days, and did not discolor (Figure 5d). *X. ogasawarai* initially showed white and felt-like mycelia, expanded quickly, filled the dish within 6–8 days, and turned grey to black, similar to *X. eborata* (Figure 5f).

The mucus reservoir was located in the terminal segment of the abdomen. The glandular organ was attached to the reservoir, and its terminal was connected to the base of the ovipositor. The mucus of *X. annulitibia* and *X. ogasawarai* were colorless and transparent (Figure 4h,i), as in previous reports including Siricidae, but surprisingly, that of *X. eborata* exhibited a deep wine-red color (Figure 4g).

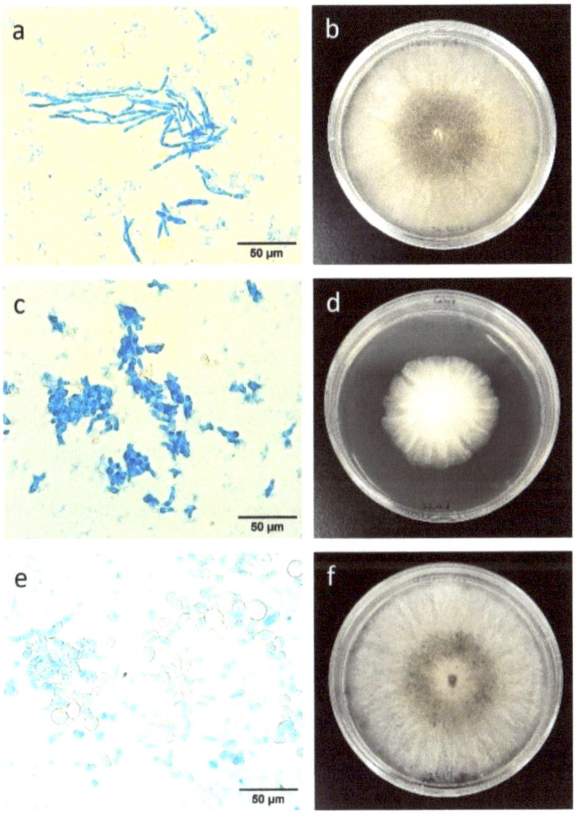

Figure 5. Mycelial fragments (**a**) and spores (**c**,**e**) stored in mycangia and their colonies (**b**,**d**,**f**) cultured on PDA at 25 °C in darkness for 10 days. (**a**,**b**) *Xiphydria eborata*; (**c**,**d**) *Xiphydria annulitibia*; (**e**,**f**) *Xiphydria ogasawarai*. Mycangial fungi were stained with lactophenol cotton blue.

4. Discussion and Conclusions

In the present study, three species of xiphydriid woodwasps emerged from five species of deciduous broadleaf trees, excluding *C. crenata*; *X. eborata* emerged from *P. grayana* and *A. hirsuta*, *X. annulitibia* emerged from *C. tschonoskii*, and *X. ogasawarai* emerged from *C. japonica* and *A. sieboldianum*. *X. eborata* is a cryptic species of *X. camelus* that confuses host tree records, and only *Betula grossa* Siebold et Zucc. and *Juglans mandshurica* Maxim. are definite for *X. eborata* [37], although *A. hirsuta* was reported as a host for *X. camelus* [8]. Therefore, our results have added two tree species belonging to different genera to this record. *X. ogasawarai* is primarily collected from *Acer* spp. and utilizes *Juglans mandshurica* Maxim. subsp. *sachalinensis* (Komatsu) Kitam., *Pterocarya rhoifolia* Siebold et Zucc., and *Aesculus turbinata* Blume [38]. Previous studies have confirmed the emergence of *X. ogasawarai* from *A. sieboldianum* [11], which aligns with our results. In addition to this, the present study identifies *C. japonica* as a new host tree species for this species. More notably, *X. annulitibia*, which is distributed in Japan, Korea, and the Russian Far East, has no known host tree species [39]. Therefore, we found *C. tschonoskii* for the first time. Additionally, multiple xiphydriid species never utilize the same tree species even though they belong to the same genus, *Carpinus*. This suggests that they coexist harmoniously with different host tree preferences. Our results also revealed extensive variation in the emergence periods of these three woodwasp species, spanning one to two years after tree felling. Previous studies have suggested the emergence of *X. ogasawarai* to be delayed along a latitudinal gradient [11], as well as the water content of the wood to influence oviposition preference and larval boring activities in *Sirex noctilio* [4,5]. These findings suggest that environmental conditions such as temperature and humidity affect the behavior and life cycle of Xiphydriidae, which should be considered in future studies.

Fewer than females, one male *X. eborata* emerged from *P. grayana*. Six and thirteen males of *X. ogasawarai* emerged from *C. japonica* and *A. sieboldianum*, respectively, without much difference in the sex ratio between host tree species. These results suggest female-biased sex ratios, which are typical of Hymenoptera. Conversely, Siricidae exhibited a male-to-female ratio of 1–2:1, except for 0:1 *Urocerus sah* Mocsáry and 0.8:1 *Urocerus fantoma* Fabricius [6,40,41]. Those of *S. noctilio* and *U. japonicus* were 32:1 and 2–4:1, respectively, depending on their habitat and host tree status [40,42,43]. Female-biased sex ratios are often observed in insects that undergo local mating or inbreeding [44,45]. Generally, Hymenoptera reproduces through arrhenotokous or haplodiploidy, which is characterized by the development of unfertilized eggs into haploid male adults and fertilized eggs into diploid female adults [44,46]. These reproductive patterns are adaptively advantageous for minimizing the number of males as follows, assuming parasitoid wasps: (1) arrhenotokous female adults (foundresses) oviposit a cluster of eggs within the host; (2) the sex ratio is biased towards females, with at least one male present in the offspring group; (3) siblings develop together within the host, with male adults losing first; (4) male adults gain multiple mating opportunities, sequentially mating with later-emerging females; and (5) female adults, having stored sperm, disperse and search for new hosts. It is unclear whether Xiphydriidae has this ideal cycle, but we have shown that local oviposition and the early emergence of males fit it. Furthermore, in this study, *X. annulitibia* emerged only as females, and no males were found in previous studies, suggesting parthenogenesis [39]. Another possibility for the observed female-biased sex ratio of emerging adults in Xiphydriidae is that the mortality rate in eggs, larvae, pupae, and adults of males may be higher than that of females in the galleries of wood. Therefore, future research should be focused on obtaining the life table by counting the number of males and females during oviposition and following development, as well as ascertaining whether Xiphydriidae has a local mating system.

We confirmed that all dissected female adults of the three xiphydriid species had eggs, mycangia, and mucus in their abdomens, similar to Siricidae [24,47]. Eggs of any species were "gourd-shaped" (i.e., oval egg bodies with elongated projections). The projection is a unique structure that has not been found in the oval-shaped eggs of Siricidae [3]. A similar structure was reported in eggs of Cynipoidea, especially gall wasps, where the projection from the anterior end of the egg body is called the peduncle (pedicel) and is thought to help large eggs pass through ovipositors [48]. The egg body first enters the narrow egg canal, followed by the peduncle [48,49]. When the canal strongly presses against the body, the egg content flows into the peduncle. Once the body leaves the ovipositor, the content flows back into it [49]. In cynipoid parasitoid wasps, such as those of the family Figitidae, eggs have an inconspicuous peduncle or no distinct transition point between the body and peduncle, whereas eggs of cynipoid gall wasps have a long peduncle and sharp transition points [48]. The projections of xiphydriid wasps appear to be morphologically similar to the peduncle of gall wasps, which may result from the adaptation of both wasps to lay their eggs within the plant tissue through a narrow ovipositor. Compared to Siricidae, which have simple oval-shaped eggs, Xiphydriidae pay higher costs for egg production but may reduce the risk of egg bursting during oviposition.

Adult females of *X. eborata* had noticeably larger eggs than those of *X. annulitibia* and *X. ogasawarai*, although no egg size variation occurred within individuals or species. There were significant positive correlations between body size (fresh body weight) and the fecundity (number of eggs) of female adults, as in *Sirex nitobei* Matsumura (Siricidae) [50]. Additionally, interspecific comparisons revealed an interesting implication that the rate of increase in fecundity relative to body size varied among the three xiphydriid species. The slope of the regression line was the smallest for *X. eborata* and largest for *X. ogasawarai*, with a significant difference between them. There is a tradeoff between the number of eggs (offspring) and their size because of limited resource availability [51]. Thus, our findings suggest that in egg production, *X. eborata* invests resources (energy) in size, whereas *X. ogasawarai* invests in number. In ectoparasitoid wasps, the egg size is associated with nutrient availability and protection from desiccation [52]. However, the effect of egg size on Xiphydriidae survival requires further investigation. Moreover, studies are needed to determine how many and where the eggs from the ovaries are actually laid into host trees.

In the present study, female adults of all Xiphydriidae species had paired mycangia at the base of the ovipositor. In Siricidae, the mycangial location is similar, but its structure is pouch-like and composed of intersegmental membranes that project into the body cavity [10,24,47]. In contrast, the mycangia of Xiphydriidae have a slit-like structure below the sternum, and fungal masses of hyphal fragments or spores are present on the slit, as reported for *X. ogasawarai* [11]. For *X. eborata* and *X. annulitibia*, we first identified the mycangia. Additionally, Siricidae larvae possess a fungus-harboring structure called the hypopleural organ on the body surface [53,54], whereas Xiphydriidae larvae lack this structure [55]. These differences in mycangia and hypopleural organs suggest that the mechanisms for acquiring and maintaining symbiotic fungi differ between the two families.

All fungal masses in the mycangia formed single colonies with no morphological differences among individuals of the same woodwasp species or between the tree species from which they emerged. However, differences in morphology and radial growth rate were observed between woodwasp species. This suggests that each female woodwasp carries a single fungal species that is specific to the woodwasp species. Previous studies have indicated that Xiphydriidae are symbiotic with fungi of the genus *Daldinia* [13] and that the mycangial fungi of *X. eborata* and *X. ogasawarai* have similar colony morphologies [56]. However, in this study, *X. annulitibia* showed a different colony morphology from those of the other two species, implying a deviation from typical *Daldinia* characteristics [56]. This suggests that *X. annulitibia* has a symbiotic relationship with fungi that do not belong to *Daldinia*. To test this hypothesis, we will conduct molecular identification and phylogenetic analyses of the isolated fungi in the following paper.

Mucus reservoirs and secretory glands were found within the body cavity of the abdominal terminal segment in all woodwasp species, consistent with results for Siricidae [24,47]. Regarding secretion color, the mucus of *X. annulitibia* and *X. ogasawarai* are colorless and transparent, as reported previously [24,54]. Surprisingly, the mucus of *X. eborata* has a deep wine-red color. To our knowledge, this is the first report of colored mucus in woodwasps, including Siricidae. In siricid woodwasps attacking living trees, the mucus contains toxic proteins that kill trees and enzymes that degrade tree tissues [18,19,21]. Thus, its function depends on the oviposition and reproduction strategies of the woodwasp species. Considering our results showing distinct host tree preferences among the three xiphydriid species, differences in secretion color might reflect variations in the utilization of host tree species as larval food sources. To support these predictions about the structure and function of the mucus of Xiphydriidae discovered in this study, we will address its chemical analysis and bioassays as a first step. Such studies can provide critical insights into the life history of Xiphydriidae and the role of mucus in woodwasps.

In conclusion, we examined the ecological traits of three Xiphydriidae species from Japan and compared them among species, including Siricidae. Our results, particularly the clear differences in egg shape, mycangial structure, and mucus coloration, contribute significantly to elucidating the ecology and evolutionary background of both families in terms of pest management.

Supplementary Materials: The following supporting information can be downloaded at: https://www.mdpi.com/article/10.3390/f16020264/s1, Figure S1: Felled tree and study logs of *Acer sieboldianum*; Figure S2: Ventral view of female adult of *Xiphydria ogasawarai* before (a) and after (b) dissection, and an egg (c), mycangia (d), mucus (e) found in abdomen; Figure S3: Dorsal view of female adult of *Xiphydria ogasawarai* before (a) and after (b) removal of ovary and mucus.

Author Contributions: Conceptualization, H.K. and R.T.; methodology, H.K. and R.T.; software, R.T.; validation, H.K.; formal analysis, R.T.; investigation, R.T.; resources, H.K.; data curation, R.T. and H.K.; writing—original draft preparation, R.T.; writing—review and editing, H.K.; visualization, R.T.; supervision, H.K.; project administration, H.K.; funding acquisition, H.K. All authors have read and agreed to the published version of the manuscript.

Funding: This research was funded by the Kakubun-Suzuki Environmental Foundation (Grant No. 22-1-10) for H.K.

Data Availability Statement: All data are included in the text and the Supplementary Materials.

Acknowledgments: We thank Akihiko Shinohara (National Museum of Nature and Science) for identifying the woodwasp species. We also thank Wataru Toki and members of the Laboratory of Forest Protection (LFP) of Nagoya University for their constructive suggestions on our work. Additionally, we thank the LFP members and Norio Yamaguchi and Honami Takabe of Nagoya University Forest for their help with log collection.

Conflicts of Interest: The authors declare no conflicts of interest.

References

1. Gauld, I.; Bolton, B. *The Hymenoptera*; Oxford University Press: New York, NY, USA, 1988; pp. 122–125.
2. Coutts, M.P.; Dolezal, J.E. Emplacement of fungal spores by woodwasp, *Sirex noctilio*, during oviposition. *For. Sci.* **1969**, *15*, 412–416.
3. Spradbery, J.P. The oviposition biology of siricid woodwasps in Europe. *Ecol. Entomol.* **1977**, *2*, 225–230. [CrossRef]
4. Madden, J. Egg and larval development in the woodwasp, *Sirex noctilio* F. *Aust. J. Zool.* **1981**, *29*, 493–506. [CrossRef]
5. Ryan, K.; Hurley, B. Life history and biology of *Sirex noctilio*. In *The Sirex Woodwasp and Its Fungal Symbiont: Research and Management of a Worldwide Invasive Pest*; Slippers, B., de Groot, P., Wingfield, M.J., Eds.; Springer: Dordrecht, The Netherlands, 2012; pp. 15–30. [CrossRef]
6. Spradbery, J.P.; Kirk, A.A. Aspects of the ecology of siricid woodwasps (Hymenoptera: Siricidae) in Europe, North Africa and Turkey with special reference to biological control of *Sirex noctilio* F in Australia. *Bull. Entomol. Res.* **1978**, *68*, 341–359. [CrossRef]
7. Tabata, M.; Miyata, H.; Maeto, K. Siricid woodwasps and their fungal symbionts in Asia, specifically those occurring in Japan. In *The Sirex Woodwasp and Its Fungal Symbiont: Research and Management of a Worldwide Invasive Pest*; Slippers, B., de Groot, P., Wingfield, M.J., Eds.; Springer: Dordrecht, The Netherlands, 2012; pp. 95–102. [CrossRef]
8. Okutani, T. Food plants of Japanese Symphyta (I). *Jpn. J. Appl. Entomol. Zool.* **1967**, *11*, 43–49. [CrossRef]
9. Terashita, T. A basidiomycets symbiotic to a siricid in Japan. *J. Jpn. For. Soc.* **1970**, *52*, 313–316. [CrossRef]
10. Tabata, M.; Abe, Y. *Amylostereum laevigatum* associated with the Japanese horntail, *Urocerus japonicus*. *Mycoscience* **1997**, *38*, 421–427. [CrossRef]
11. Kajimura, H. Discovery of mycangia and mucus in adult female xiphydriid woodwasps (Hymenoptera: Xiphydriidae) in Japan. *Ann. Entomol. Soc. Am.* **2000**, *93*, 312–317. [CrossRef]
12. Pažoutová, S.; Šrůtka, P. Symbiotic relationship between *Cerrena unicolor* and the horntail *Tremex fuscicornis* recorded in the Czech Republic. *Czech Mycol.* **2007**, *59*, 83–90. [CrossRef]
13. Pažoutová, S.; Šrůtka, P.; Holuša, J.; Chudíčková, M.; Kolařík, M. Diversity of xylariaceous symbionts in *Xiphydria* woodwasps: Role of vector and a host tree. *Fungal Ecol.* **2010**, *3*, 392–401. [CrossRef]
14. Fukuda, H.; Hijii, N. Reproductive strategy of a woodwasp with no fungal symbionts, *Xeris spectrum* (Hymenoptera: Siricidae). *Oecologia* **1997**, *112*, 551–556. [CrossRef] [PubMed]
15. Kukor, J.J.; Martin, M.M. Acquisition of digestive enzymes by siricid woodwasps from their fungal symbiont. *Science* **1983**, *220*, 1161–1163. [CrossRef] [PubMed]

16. Thompson, B.M.; Grebenok, R.J.; Behmer, S.T.; Gruner, D.S. Microbial symbionts shape the sterol profile of the xylem-feeding woodwasp, *Sirex noctilio*. *J. Chem. Ecol.* **2013**, *39*, 129–139. [CrossRef] [PubMed]
17. Wong, L.K.; Crowden, R.K. Preliminary studies on mucus secretion of wood wasp, *Sirex noctilio* F. I. Physicochemical and biochemical properties. *Aust. J. Biol. Sci.* **1976**, *29*, 21–32. [CrossRef]
18. Li, D.; Shi, J.; Lu, M.; Ren, L.; Zhen, C.; Luo, Y. Detection and identification of the invasive *Sirex noctilio* (Hymenoptera: Siricidae) fungal symbiont, *Amylostereum areolatum* (Russulales: Amylostereacea), in China and the stimulating effect of insect venom on laccase production by *A. areolatum* YQL03. *J. Econ. Entomol.* **2015**, *108*, 1136–1147. [CrossRef]
19. Gao, C.; Ren, L.; Wang, M.; Wang, Z.; Fu, N.; Wang, H.; Wang, X.; Ao, T.; Du, W.; Zheng, Z.; et al. Proteo-transcriptomic characterization of *Sirex nitobei* (Hymenoptera: Siricidae) venom. *Toxins* **2021**, *13*, 562. [CrossRef]
20. Wang, T.; Zhao, M.; Rotgans, B.A.; Ni, G.; Dean, J.F.D.; Nahrung, H.F.; Cummins, S.F. Proteomic analysis of the venom and venom sac of the woodwasp, *Sirex noctilio*—Towards understanding its biological impact. *J. Proteom.* **2016**, *146*, 195–206. [CrossRef]
21. Bordeaux, J.M.; Lorenz, W.W.; Johnson, D.; Badgett, M.J.; Glushka, J.; Orlando, R.; Dean, J.F.D. Noctilisin, a venom glycopeptide of *Sirex noctilio* (Hymenoptera: Siricidae), causes needle wilt and defense gene responses in pines. *J. Econ. Entomol.* **2014**, *107*, 1931–1945. [CrossRef]
22. Coutts, M. Mechanism of pathogenicity of *Sirex noctilio* on *Pinus radiata*. II. Effects of *S. noctilio* mucus. *Aust. J. Biol. Sci.* **1969**, *22*, 1153–1161. [CrossRef]
23. Bordeaux, J.; Dean, J. Susceptibility and response of pines to *Sirex noctilio*. In *The Sirex Woodwasp and Its Fungal Symbiont: Research and Management of a Worldwide Invasive Pest*; Slippers, B., de Groot, P., Wingfield, M.J., Eds.; Springer: Dordrecht, The Netherlands, 2012; pp. 31–50. [CrossRef]
24. Kuramitsu, K.; Ishihara, T.; Sugita, A.; Yooboon, T.; Lustig, B.; Matsumori, Y.; Yamada, H.; Kinoshita, N. The attraction of *Tremex apicalis* (Hymenoptera, Siricidae, Tremecinae) and its parasitoid *Ibalia japonica* (Hymenoptera, Ibaliidae) to the fungus *Cerrena unicolor*. *J. Hymenopt. Res.* **2019**, *68*, 37–48. [CrossRef]
25. Šrůtka, P.; Pažoutová, S.; Kolařík, M. *Daldinia decipiens* and *Entonaema cinnabarina* as fungal symbionts of *Xiphydria* wood wasps. *Mycol. Res.* **2007**, *111*, 224–231. [CrossRef] [PubMed]
26. Coutts, M. The mechanism of pathogenicity of *Sirex noctilio* on *Pinus radiata*. I. Effects of symbiotic fungus *Amylostereum* sp. (Thelophoraceae). *Aust. J. Biol. Sci.* **1969**, *22*, 915–924. [CrossRef]
27. Bedding, R.A.; Akhurst, R.J. Use of the nematode *Deladenus siridicola* in the biological control of *Sirex noctilio* in Australia. *Aust. J. Entomol.* **1974**, *13*, 129–135. [CrossRef]
28. Tribe, G. The woodwasp *Sirex noctilio* Fabricius (Hymenoptera: Siricidae), a pest of *Pinus* species, now established in South Africa. *Afr. Entomol.* **1995**, *3*, 215–217.
29. De Groot, P.; Nystrom, K.; Scarr, T. Discovery of *Sirex noctilio* (Hymenoptera: Siricidae) in Ontario, Canada. *Great Lakes Entomol.* **2006**, *39*, 5. [CrossRef]
30. Fukuda, H.; Maeto, K. Discoloration injury of *Cryptomeria japonica* (L.f.) D. Don and *Chamaecyparis obtusa* (Sieb. et Zucc.) Endl. caused by woodwasps and the associated fungi: Constructing of control methods against the injury. *J. Jpn. For. Soc.* **2001**, *83*, 161–168. [CrossRef]
31. Smith, D.R. The xiphydriid woodwasps of North America (Hymenoptera: Xiphydriidae). *Trans. Am. Entomol. Soc.* **1976**, *102*, 101–131.
32. Deyrup, M. A maple wood wasp, *Xiphydria maculata*, and its insect enemies (Hymenoptera, Xiphydriidae). *Great Lakes Entomol.* **1984**, *17*, 17–28.
33. Shinohara, A.; Smith, D.R. Japanese Xiphydriidae (Hymenoptera) in the collection of the National Museum of Natural History, Washington, DC. *Jpn. J. Syst. Entomol.* **2020**, *26*, 23–28.
34. Peters, R.S.; Krogmann, L.; Mayer, C.; Donath, A.; Gunkel, S.; Meusemann, K.; Kozlov, A.; Podsiadlowski, L.; Petersen, M.; Lanfear, R.; et al. Evolutionary history of the Hymenoptera. *Curr. Biol.* **2017**, *27*, 1013–1018. [CrossRef]
35. Wutke, S.; Blank, S.M.; Boevé, J.L.; Faircloth, B.C.; Koch, F.; Linnen, C.R.; Malm, T.; Niu, G.; Prous, M.; Schiff, N.M.; et al. Phylogenomics and biogeography of sawflies and woodwasps (Hymenoptera, Symphyta). *Mol. Phylogenet. Evol.* **2024**, *199*, 108144. [CrossRef] [PubMed]
36. R Core Team. *R: A Language and Environment for Statistical Computing*; R Foundation for Statistical Computing: Vienna, Austria, 2023; Available online: https://www.R-project.org/ (accessed on 11 December 2024).
37. Shinohara, A.; Kameda, Y. DNA barcodes and morphology revealed three species masquerading in *Xiphydria camelus* of authors (Hymenoptera, Xiphydriidae) in Northeast Asia: *X. eborata* sp. rev. and *X. albopicta* sp. nov. *Zootaxa* **2019**, *4612*, 171–186. [CrossRef] [PubMed]
38. Shinohara, A. The *Xiphydria palaeanarctica* group from Japan (Hymenoptera, Xiphydriidae). *Zootaxa* **2019**, *4608*, 531–542. [CrossRef] [PubMed]
39. Shinohara, A.; Hara, H.; Smith, D.R. The *Xiphydria annulitibia* group in northeastern Asia (Hymenoptera, Xiphydriidae). *Zootaxa* **2020**, *4755*, 375–389. [CrossRef]

40. Inada, T. Effects of *Cryptomeria japonica* and *Chamaecyparis obtusa* thinning season and bucking method on the emergence of *Urocerus japonicus* adults. *J. Jpn. For. Soc.* **2003**, *85*, 95–99. [CrossRef]
41. Matsumoto, T.; Sato, S. Oviposition preference and reproductive success of *Xeris spectrum* on logs inoculated with a fungal symbiont *Amylostereum laevigatum* of wood wasps. *J. Jpn. For. Res.* **2015**, *97*, 238–242. [CrossRef]
42. Iede, E.T.; Penteado, S.R.C.; Schaitza, E.G. *Sirex noctilio* problem in Brazil—Detection, evaluation and control. In Proceedings of a Conference: Training in the Control of Sirex Noctilio by Use of Natural Enemies, Colombo, Brazil, 4–9 November 1996; Iede, E., Shaitza, E., Penteado, S., Reardon, R., Murphy, T., Eds.; USDA Forest Service (FHTET 98-13): Morgantown, WV, USA, 1998.
43. Queffelec, J.; Wooding, A.L.; Greeff, J.M.; Garnas, J.R.; Hurley, B.P.; Wingfield, M.J.; Slippers, B. Mechanisms that influence sex ratio variation in the invasive hymenopteran *Sirex noctilio* in South Africa. *Ecol. Evol.* **2019**, *9*, 7966–7973. [CrossRef]
44. Hamilton, W.D. Extraordinary sex ratios: A sex ratio theory for sex linkage and inbreeding has new implications in cytogenetics and entomology. *Science* **1967**, *156*, 477–488. [CrossRef]
45. Hardy, I.C.W. Sex-ratio and mating structure in the parasitoid Hymenoptera. *Oikos* **1994**, *69*, 3–20. [CrossRef]
46. Heimpel, G.E.; de Boer, J.G. Sex determination in the Hymenoptera. *Annu. Rev. Entomol.* **2008**, *53*, 209–230. [CrossRef]
47. Spradbery, J.P. A comparative study of phytotoxic effects of siricid woodwasps on conifers. *Ann. Appl. Biol.* **1973**, *75*, 309–320. [CrossRef]
48. Vårdal, H.; Sahlén, G.; Ronquist, F. Morphology and evolution of the cynipoid egg (Hymenoptera). *Zool. J. Linn. Soc.* **2003**, *139*, 247–260. [CrossRef]
49. Bronner, R. Anatomy of the ovipositor and oviposition behavior of the gall wasp *Diplolepis rosae* (Hymenoptera, Cynipidae). *Can. Entomol.* **1985**, *117*, 849–858. [CrossRef]
50. Fukuda, H.; Kajimura, H.; Hijii, N. Fecundity of the woodwasp, *Sirex nitobei* Matsumura, in relation to its body size. *J. Jpn. For. Soc.* **1993**, *75*, 405–408. [CrossRef]
51. Smith, C.C.; Fretwell, S.D. The optimal balance between size and number of offspring. *Am. Nat.* **1974**, *108*, 499–506. [CrossRef]
52. Mayhew, P.J.; Heitmans, W.R.B. Life history correlates and reproductive biology of *Laelius pedatus* (Hymenoptera: Bethylidae) in the Netherlands. *Eur. J. Entomol.* **2000**, *97*, 313–322. [CrossRef]
53. Stillwell, M.A. Hypopleural organs of woodwasp larva *Tremex columba* (L.) containing fungus *Daedalea unicolor* Bull ex Fries. *Can. Entomol.* **1965**, *97*, 783–784. [CrossRef]
54. Boros, C.B. The Relationship Between the Woodwasp, *Sirex noctilio* F. and the Wood-Rot Fungus *Amylostereum* sp. Ph.D. Thesis, University of Adelaide, Adelaide, Australia, 1968.
55. Francke-Grosmann, H. Ectosymbiosis in wood-inhabiting insects. In *Symbiosis*; Henry, S.M., Ed.; Academic Press: London, UK, 1967; pp. 141–205.
56. Stadler, M.; Læssøe, T.; Fournier, J.; Decock, C.; Schmieschek, B.; Tichy, H.V.; Peršoh, D. A polyphasic taxonomy of *Daldinia* (Xylariaceae). *Stud. Mycol.* **2014**, *77*, 1–143. [CrossRef]

forests

MDPI

Article

Species Composition and Seasonal Abundance of Predatory Mites (Acari: Phytoseiidae) Inhabiting *Aesculus hippocastanum* (Sapindaceae)

Michal Kopačka and Rostislav Zemek *

Institute of Entomology, Biology Centre CAS, 37005 České Budějovice, Czech Republic
* Correspondence: rosta@entu.cas.cz

Abstract: Species of the family Phytoseiidae (Acari: Mesostigmata) inhabit a wide range of herbs, shrubs, and trees. Horse chestnut, *Aesculus hippocastanum*, is an important ornamental tree in Europe and, in addition to its importance for pollinators, it can serve as a host plant of these predatory mites. Little is, however, known about the factors affecting spatiotemporal variability in the density of phytoseiids on *A. hippocastanum* in an urban environment. We therefore assessed the species composition and the spatial and seasonal variability in the abundance of Phytoseiidae species in the city of České Budějovice, South Bohemia, Czech Republic. Compound-leaf samples were randomly collected from horse chestnut tree branches at eight sites, five times during the vegetation season in 2013. The mites were collected by washing technique and mounted on slides for identification. In total, 13,903 specimens of phytoseiid mites were found, and eight species were identified: *Amblyseius andersoni, Euseius finlandicus, Kampimodromus aberrans, Neoseiulella tiliarum, Phytoseius macropilis, Paraseiulus talbii, Paraseiulus triporus,* and *Typhlodromus* (*Typhlodromus*) *pyri*. The predominant species was *E. finlandicus* (96.25%). The mean number of mites per compound leaf was 2.53, 10.40, 23.54, 11.59, and 9.27 on the sampling dates in each month between May and September, respectively. The results further revealed that the mite abundance varied significantly among sampling sites and that it was negatively related to percentage of greenery area, intensity of greenery care, distance to water body, and density and age of horse chestnut trees, while it was positively related to air pollution index. The importance of leaf micromorphology for the attractiveness of *A. hippocastanum* to Phytoseiidae is discussed.

Keywords: horse chestnut tree; diversity; population dynamics; mite density; spatial variability; city parks

Citation: Kopačka, M.; Zemek, R. Species Composition and Seasonal Abundance of Predatory Mites (Acari: Phytoseiidae) Inhabiting *Aesculus hippocastanum* (Sapindaceae). *Forests* **2023**, *14*, 942. https://doi.org/10.3390/f14050942

Academic Editor: Zoltán Elek

Received: 2 March 2023
Revised: 25 April 2023
Accepted: 30 April 2023
Published: 3 May 2023

1. Introduction

The horse chestnut, *Aesculus hippocastanum* L. (Sapindales: Sapindaceae), is native to the mountains of the Balkans in southeast Europe. The biggest native populations are in mainland Greece (Thessaly and Pindus Mountains), but its native range also includes populations in Albania, the Republic of Macedonia, Serbia, and eastern Bulgaria [1]. This tree species has been planted in Europe since the seventeenth century primarily for ornamental purposes. Many *A. hippocastanum* trees are currently grown in city parks and urban forests for their beauty, providing shade and reducing the urban heat island effect. Moreover, extracts from the various parts of this tree have been widely used in herbal medicine [2]. The horse chestnut trees are also planted in forests in order to increase and diversify food resources because seeds are eaten by deer and wild boar [1], though the presence of various metabolites makes the seeds a less attractive food source [3].

Aesculus hippocastanum attracted researchers' attention when the horse chestnut leafminer, *Cameraria ohridella* Deschka and Dimic (Lepidoptera: Gracillariidae), became an invasive pest and spread to all countries of Europe [4–6]. Larvae of this moth inflict substantial leaf damage, and together with the horse chestnut leaf blotch caused by *Guignardia aesculi* (Peck) V.B. Stewart

(Botryosphaeriales: Botryosphaeriaceae), cause a decline in horse chestnut trees' aesthetical value [7]. Although ornamental function and providing shade are the main reasons why growing *A. hippocastanum* is so popular in city parks, other functions of this tree species might be less known. For example, horse chestnut provides pollen and nectar for pollinators [8]. *Aesculus hippocastanum* also serves as a habitat for many organisms, including natural enemies such as predatory mites of the family Phytoseiidae (Acari: Mesostigmata).

A survey of phytoseiid mites on deciduous trees and bushes conducted in Finland in 1989–1991 revealed the highest density of mites to be on *A. hippocastanum*, with 1063 mites per 100 leaves on average and a maximum of 14.4 mites per leaf in a single sample [9,10]. Similar results were reported for South Bohemia, Czech Republic, where the population density of phytoseiids ranged between 1 and 28, with a mean number of mites per compound leaf of 10.5 [11]. The predominant species in both Finland and in the Czech Republic was *Euseius finlandicus* (Oudemans), which represented more than 90% of all phytoseiids found [9,11,12]. In Greece, the number of phytoseiids was found to be lower, ranging between 0 and 16, with a mean density of 4.2 mites per compound leaf; another species, *Kampimodromus aberrans* (Oudemans), was nearly as abundant as *E. finlandicus* [11].

Although the above data provide basic information on mite abundance, the spatiotemporal variability in the density of phytoseiids on *A. hippocastanum* in an urban environment has not yet been studied. The objectives of the present study were therefore to investigate the species composition of phytoseiid mites inhabiting horse chestnut trees and seasonal changes in their abundance within a medium-sized city. Since various factors affect populations of both herbivorous and phytoseiid mites, as demonstrated, e.g., by Fidelis et al. [13] and Tixier et al. [14], we also analyzed which factors might have a significant effect on the abundance of Phytoseiidae. For example, we were interested as to whether greenery maintenance has any effect, as it was shown to in the case of horse chestnut leaf miner parasitoids [15], or if mite density is affected by air pollution [16].

2. Materials and Methods

2.1. Sampling Sites

The research was carried out in the town of České Budějovice, Czech Republic (48°59′ N, 14°29′ E; 386 m above sea level). The town, which is predominantly surrounded by forests and agricultural fields, has about 260 hectares of open public green space inhabiting 534 horse chestnut trees [17]. Eight sampling sites were defined within the cadastral area of the town (Figure 1), which differed in several parameters (Table 1).

Table 1. Characteristics of *Aesculus hippocastanum* sampling sites within České Budějovice.

Site Name	Geographical Coordinates of Geometric Centre	Greenery		Distance to Water [2] (m)	*A. hippocastanum*	
		Area (%)	Care [1]		Density (ha^{-1})	Age [3] (years; $\bar{x} \pm SE$)
City center	48.9744136 N, 14.4770594 E	9.02	3	69	1.40	74.99 ± 1.86
Šumava and Máj estate	48.9841631 N, 14.4407714 E	31.69	2	660	0.22	39.02 ± 3.19
Vltava estate	48.9962106 N, 14.4519647 E	36.96	2	360	0.43	41.79 ± 3.49
Třebotovice and Kaliště village	48.9612606 N, 14.5651828 E	6.67	1	700	0.17	17. 82 ± 6.86
Rožnov estate	48.9601436 N, 14.4763944 E	9.86	2	110	0.19	70.49 ± 4.43
Pražské předměstí estate	48.9860569 N, 14.4674681 E	29.84	2	338	0.40	49.74 ± 2.75
Stromovka park	48.9670208 N, 14.4551158 E	40.00	2	102	0.30	43.53 ± 4.13
Nádražní street	48.9779942 N, 14.4866511 E	7.69	0	920	1.38	77.74 ± 3.02

[1] Intensity of greenery care: 0—no mowing or raking, 1—mowing 1–2 times per season, 2—mowing 3–4 times per season, 3—mowing >4 times per season. [2] The shortest distance from geometric center to larger water body (river, pond, or stream). [3] Tree age was estimated by the method proposed by Jura [18].

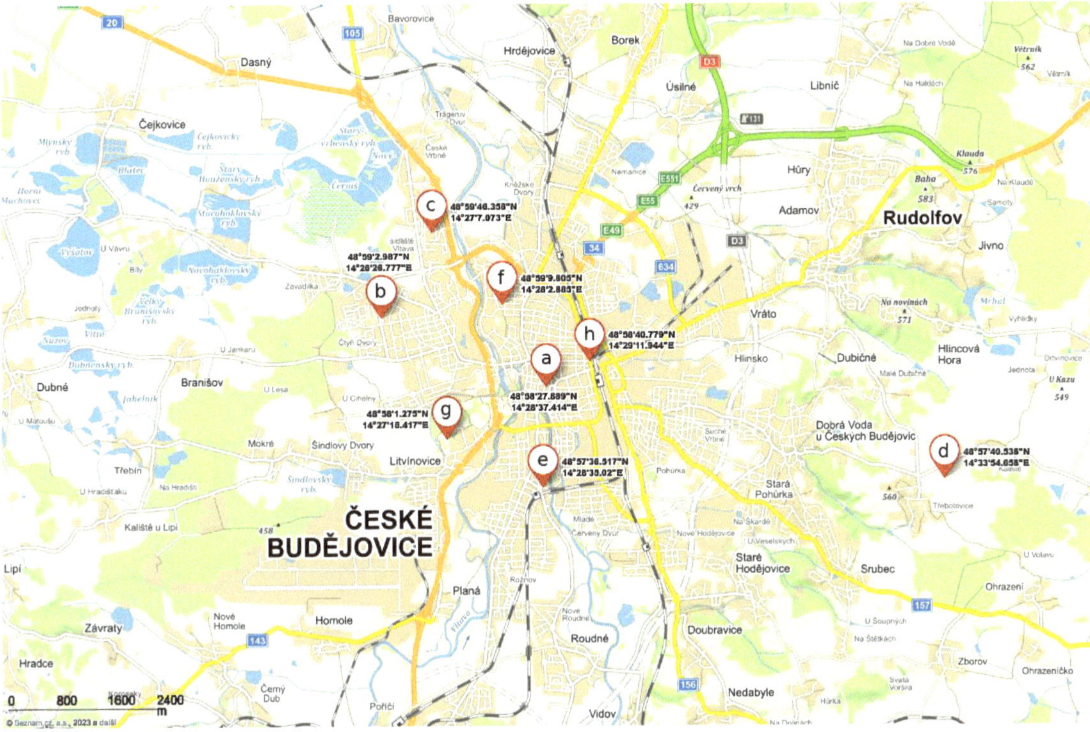

Figure 1. The map of České Budějovice with marked sites where *Aesculus hippocastanum* leaf samples were collected. (**a**) City center; (**b**) Šumava and Máj estate; (**c**) Vltava estate; (**d**) Třebotovice and Kaliště village; € Rožnov estate; (**f**) Pražské předměstí estate; (**g**) Stromovka park; (**h**) Nádražní street. Source: Mapy.cz.

Weather data are shown in Figure 2. The highest daily mean temperature and sum of precipitation were found in July and June, respectively (Table 2). In addition, data on air pollution provided by Czech Hydrometeorological Institute were obtained for each sampling site from Geographical Information System of South Bohemia [19] (Table S1).

Table 2. Mean temperature and sum of precipitations in České Budějovice for the period May–September 2013.

Month	Temperature in °C (Mean ± SE)	Precipitations in mm
May	12.95 ± 3.17	84.5
June	16.95 ± 4.96	190.0
July	20.35 ± 2.53	74.2
August	18.93 ± 3.77	59.5
September	13.80 ± 2.98	35.2

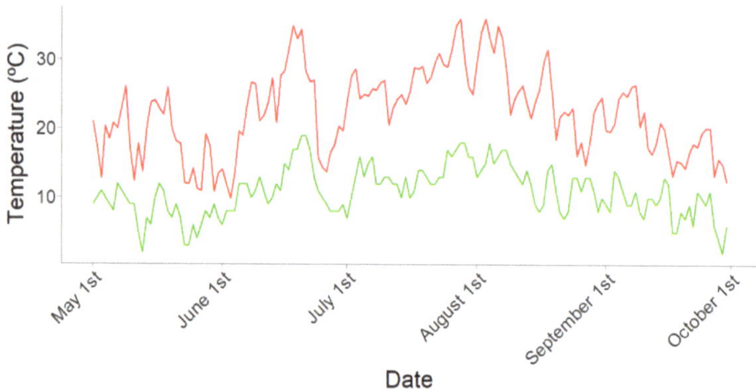

(a)

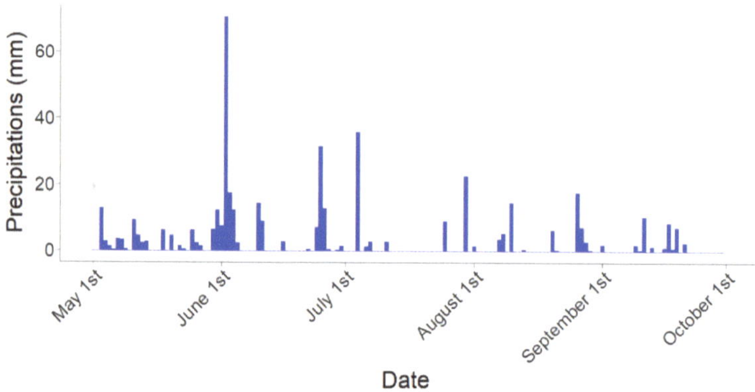

(b)

Figure 2. The maximum (red line) and minimum (green line) daily temperatures (**a**) and daily precipitations (**b**) in České Budějovice during sampling season in 2013.

2.2. Sampling of Horse Chestnut Leaves

The population of Phytoseiidae was assessed on horse chestnut leaves that were collected randomly five times during the vegetation season in 2013. The sampling dates were from 16 to 24 May, from 13 to 21 June, from 11 to 19 July, from 11 to 22 August, and from 9 to 19 September. The sampling took place only when there had been no rain for at least 48 h prior to sampling. A randomly selected compound leaf collected from *A. hippocastanum* tree up to 2.5 m above the ground represented the sample unit. In total, 30 compound leaves were collected per sampling site, each leaf from a different tree except at sites with fewer than 30 trees. The sampled trees were selected evenly across the whole site. The leaves at each individual site were all sampled on the same day, placed individually into polyethylene bags, and transported in a portable cool box to the laboratory where the leaves were stored at a temperature of 4 °C for a maximum of 24 h before they were processed.

2.3. Collection and Identification of Phytoseiid Mites

The mites were collected from individual leaves using the washing technique [20]. After taking the photographs, the horse chestnut leaf was held by the petiole, and individual leaflets were cut off by pruning shears into a glass jar (volume 700 mL) containing 350 mL of 85% ethanol. The jar was closed carefully and shaken vigorously for two minutes. Afterwards, each leaflet was removed by tweezers and washed with ethanol using a plastic wash bottle. Particular attention was paid to trichomes and domatia during washing.

The contents of the glass jar were poured through a plastic funnel into a glass dividing funnel with a Teflon® stopcock. The empty glass jar and the plastic funnel were washed with 85% ethanol, which was poured into the dividing funnel immediately. All invertebrates settled at the bottom of the dividing funnel within 5 min. Then, the Teflon® stopcock was opened, and approximately 50 mL of ethanol sample containing all invertebrates was poured off into a small glass vial with a plastic plug. The material was stored in this vial until microscope slide preparation and identification.

Each vial with preserved mites was placed in a paper holder to prevent spillage, and the ethanol in the vials was gradually transferred onto a watch glass using a plastic Pasteur pipette. The sample on the watch glass was inspected using a dissection microscope (Technival, Carl Zeiss, Jena, Germany). All mites were removed with a wire loop and mounted on temporal microscope slides in lactic acid. The mites were identified using a Peraval Interphaco microscope (Carl Zeiss, Jena, Germany) and the morphological identification keys [21–25].

2.4. Data Presentation and Statistical Analysis

The coefficient of constancy (C) [26] was used to indicate the frequency of different species in the studied localities:

$$C(\%) = \frac{N_a}{N} \times 100 \tag{1}$$

where N_a is the number of samples with species a, and N is the total number of samples. The species were classified as accidental (C < 25%), accessory (C = 25%–50%), constant (C = 50%–75%), or euconstant (C > 75%) [26].

The abundance of phytoseiid mites was expressed as the number of mites per compound *A. hippocastanum* leaf and analyzed by a generalized linear model using a Poisson distribution and log-link function. The following predictors were used: percentage of greenery area, intensity of greenery care (mowing/raking, ranging from 0 to 3), distance to water body, density and age of horse chestnut trees, temperature, precipitations, and overall air pollution index. The latter was calculated as a sum of indexes of individual pollutants:

$$I_t = \sum_i^n I_i \tag{2}$$

where I_i is the air pollution index of pollutant i calculated as:

$$I_i = \frac{C_i}{Co_i} \tag{3}$$

where C_i and Co_i are the monitoring value and its quality standard limit in ng/m^3 or μg/m^3, respectively [27].

The analysis was performed in SAS® Studio (Release 3.81, Enterprise Edition, SAS Institute Inc., Cary, NC, USA) using the GLM procedure (PROC GENMOD) of SAS/STAT® module [28]. Since the data represented repeated measurements and thus could not be regarded as independent, the generalized estimating equation (GEE) approach [29] was used to account for within-subject correlations, using option statement REPEATED in GENMOD procedure. p values < 0.05 were considered statistically significant.

3. Results

3.1. Species Composition

A total of 13,903 phytoseiid mites belonging to eight species were identified. The species composition was as follows: *E. finlandicus* 96.25% (C = 82.33%), *Typhlodromus (Typhlodromus) pyri* Scheuten 1.35% (C = 10.00%), *Amblyseius andersoni* (Chant) 0.80% (C = 4.50%), *Neoseiulella tiliarum* (Oudemans) 0.73% (C = 5.00%), *K. aberrans* 0.58% (C = 5.83%), *Phytoseius (Ph.) macropilis* (Banks) 0.22% (C = 1.00%), *Paraseiulus (Pa.) triporus*

(Chant & Yoshida-Shaul) 0.06% (*C* = 0.33%), and *Paraseiulus* (*Pa.*) *talbii* (Athias-Henriot) 0.01% (*C* = 0.08%). According to the coefficient of constancy, all species can be considered accidental except *E. finlandicus*, which is a euconstant species.

3.2. Mite Abundance and Seasonal Dynamics

The mean abundance of phytoseiid mites per compound horse chestnut leaf across all sites was 2.53 ± 0.29 (*n* = 240), 10.40 ± 0.51 (*n* = 240), 23.54 ± 0.91 (*n* = 240), 11.59 ± 0.49 (*n* = 240), and 9.27 ± 0.49 (*n* = 240) in May, June, July, August, and September, respectively. The overall annual average abundance was 11.47 ± 0.32 mites per leaf (*n* = 1200). With the exception of Stromovka park, mite abundance culminated in July (Figure 3) and thus correlated with the mean temperature. The highest mite density was observed in the Vltava estate site, while the lowest was in Stromovka park. The GLM analysis revealed that differences among sampling sites can be associated with all considered predictor variables with the exception of precipitations (Table 3). While the effects of greenery area, care intensity, distance to water, tree density, and age on mite abundance were negative, the statistical analysis showed that the mite abundance was positively related to temperature and air pollution index. Hence, e.g., Třebotovice and Kaliště village with the lowest pollution reported (Table S2) was inhabited by a lower density of Phytoseiidae throughout the season compared to the heavily polluted Nádražní street (Figure 3).

Table 3. Generalized linear model (GLM) with Poisson distribution and log-link function showing the relations between mite abundance and predicting variables.

Parameter	Estimate	Standard Error	95% Confidence Limits		Z	*p*
Intercept	−4.9278	0.5132	−5.9336	−3.9220	−9.60	<0.0001
Greenery area	−0.0462	0.0026	−0.0512	−0.0411	−17.93	<0.0001
Care intensity	−0.1688	0.0208	−0.2096	−0.1280	−8.11	<0.0001
Distance to water	−0.0002	0.0001	−0.0003	−0.0000	−2.26	00.0239
Tree density	−0.5152	0.0160	−0.5465	−0.4838	−32.20	<0.0001
Tree age	−0.0317	0.0030	−0.0375	−0.0258	−10.64	<0.0001
Temperature	0.1941	0.0183	0.1583	0.2300	10.62	<0.0001
Precipitations	−0.0009	0.0014	−0.0037	0.0019	−0.63	00.5257
Air pollution index	1.5798	0.0978	1.3881	1.7716	16.15	<0.0001

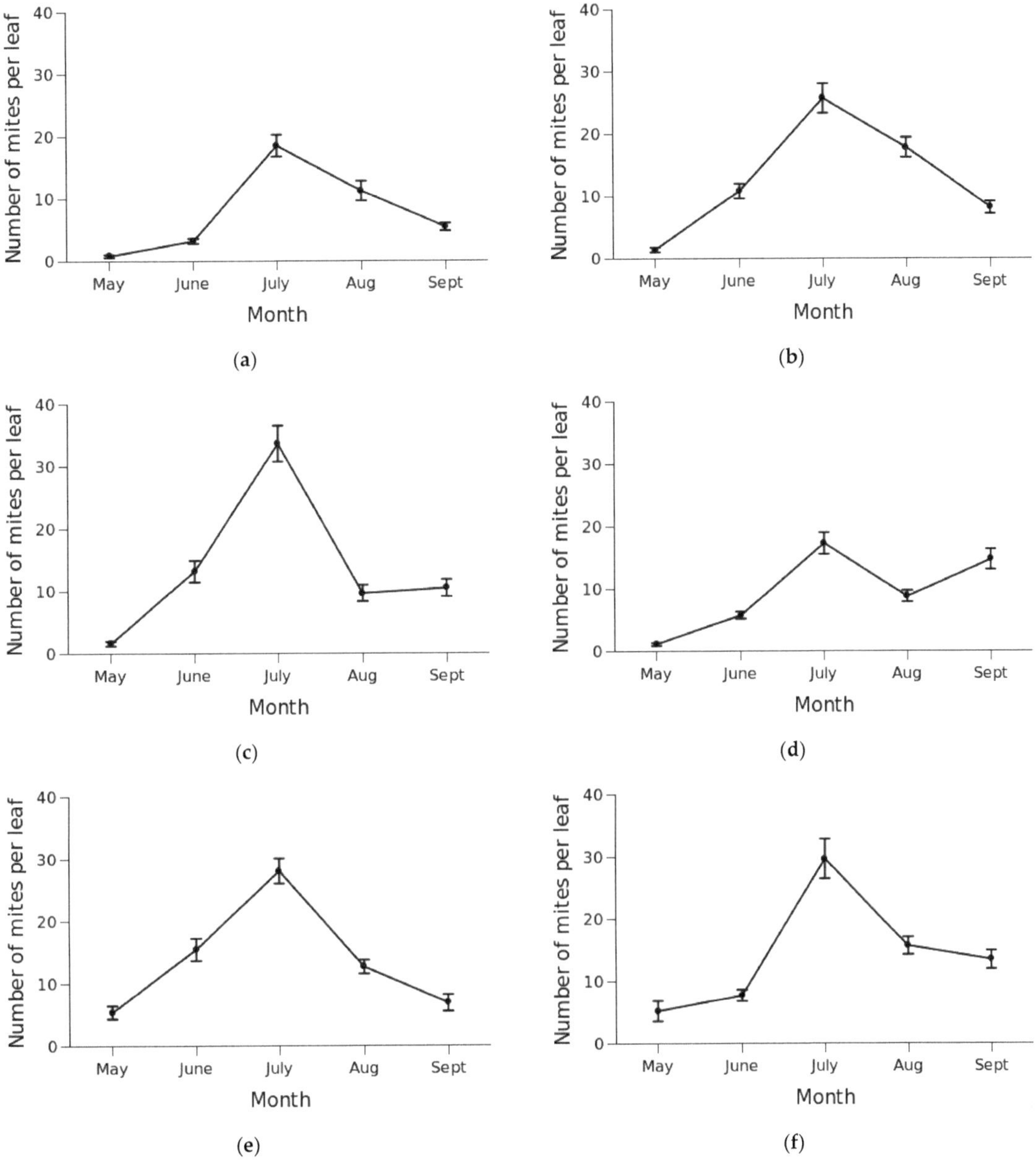

Figure 3. *Cont.*

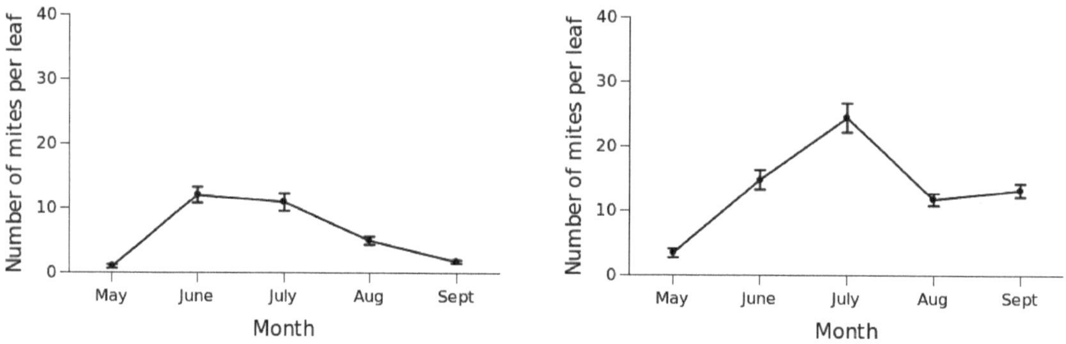

(g) (h)

Figure 3. The mean abundance of phytoseiid mites per *Aesculus hyppocastanum* compound leaf during vegetation period at eight sampling sites in České Budějovice. (**a**) City center; (**b**) Šumava and Máj estate; (**c**) Vltava estate; (**d**) Třebotovice and Kaliště village; (**e**) Rožnov estate; (**f**) Pražské předměstí estate; (**g**) Stromovka park; (**h**) Nádražní street. Vertical bars indicate standard error of the mean ($n = 30$).

4. Discussion

The Phytoseiidae family represents a very important group of predatory mites that inhabit, in addition to herbaceous plants, many species of deciduous trees and shrubs [9,10,30–42]. Six species of phytoseiids were identified on horse chestnut in city parks in Prague, Czech Republic, by Kabíček and Řeháková [12]: *E. finlandicus*, *Galendromus longipilus* (Nesbitt), *K. aberrans*, *Neoseiulella aceri* (Collyer), *N. tiliarum*, and *T.* (*T.*) *pyri*. The species richness, however, varied among the investigated sites, from one to four species.

The species composition of the present study confirmed the presence of these species, except *G. longipilus* and *N. aceri*, which were not found in any sample in České Budějovice. However, we found *A. andersoni*, *Ph. macropilis*, *Pa. triporus*, and *Pa. talbii*. Thus, eight species were found on *A. hippocastanum* in total. Six of them had also been found on horse chestnut in Hungary [43,44], while *Pa. talbii* and *Ph. macropilis*, which were identified in the present study, have not yet been reported on horse chestnut leaves. In Greece, where horse chestnut is an autochthonous tree, only four phytoseiid species were found, namely, *E. finlandicus*, *K. aberrans*, *Pa. talbii*, and *T.* (*T.*) *pyri* [11].

The most abundant species found in the present study was *E. finlandicus*, representing approximately 96% of all phytoseiid mites. This confirms that *E. finlandicus* is predominant in species complexes of phytoseiid mites in Europe on horse chestnut [9,12,38]. In Greece, however, the percentage of *E. finlandicus* on *A. hippocastanum* was found to be only 48%, and the second most dominant species, representing approximately 43% of all determined species of Phytoseiidae, was *K. aberrans* [11]. *Euseius finlandicus* is also the predominant species on other deciduous trees [9,10,32,35,36,38,45]. For example, in a survey by Kabíček and Povondrová [35], *E. finlandicus* represented more than 95% of all phytoseiid mites identified in leaf samples collected from various deciduous trees in a park in Prague.

The results of the present study revealed that one *A. hippocastanum* leaf can host 2.5–23.5 phytoseiid mites on average. These abundance values agree with those of previous studies conducted in Finland [9,10] and the Czech Republic [12]. The latter authors reported the highest population density of phytoseiids on horse chestnut trees in Prague to be 3.3 mites per leaflet on average [12]. Because *A. hippocastanum* compound leaves usually have five to seven leaflets [1], the above value could correspond to 19.8 mites per average compound leaf. This result matches that of our third sampling in July. In contrast, the density of phytoseiid mites in Greece was lower, with 4.2 mites per compound leaf on average [11].

Relatively high abundance of Phytoseiidae on horse chestnut compared to other deciduous tree species [32,33,35] confirms that *A. hippocastanum* is a favorable host plant species for phytoseiid generalists in urban environments. We would expect that it is due to the presence of prey availability. Although phytophagous mites are a common food source for most Phytoseiidae [31], we found only a few phytophagous mites in our samples. Among the mites infesting the horse chestnut tree are *Eotetranychus pruni* (Oudemans) (Acari: Tetranychidae) [46], *Aculus hippocastani* (Fockeu), and *Shevtchenkella carinatus* (Nalepa) (Acari: Eriophyidae) [47]. However, according to Tuovinen and Rokx [9], prey density does not seem to have any significant effect on the presence and density of *E. finlandicus* or *P. macropilis*. The number of phytoseiids on leaves is influenced by leaf surface characteristics rather than by food availability [48]. Horse chestnut trees have few glandular trichomes with a mean height of 84 μm located only on the midrib surface on the adaxial epidermis; nonglandular trichomes with lengths ranging from 116–436 μm were observed on the lower leaf surface, where they were located on the midrib and lateral veins as well as in the vein axils [49]. The density of leaf trichomes was reported to be 9.96/mm^2, which is relatively high compared to that in other tree species [50]. The relatively high phytoseiid density on *A. hippocastanum* can thus be explained by the favorable micromorphology of its leaves. A positive effect of leaf trichomes and domatia occurrence on the abundance of predatory mites has been well documented in many studies [32,33,48,51–61]. For example, no mites were found on tree species with glabrous leaves such as *Betula pendula* Ehrh. or *Populus tremula* L. with nonraised veins and no domatia [32,33]. While domatia mainly provide phytoseiid mites with shelter and act as protection from either natural enemies or abiotic stress [53], leaf pubescence also increases the capture and retention of pollen and fungal spores that serve as alternative foods [62]. We often observed many pollen grains on *A. hippocastanum* leaves. Kugler [63] estimated that this tree species itself can produce 42 million pollen grains from a single inflorescence. The *A. hippocastanum* pollen was also found to have a very high nutritional quality for phytoseiid mites [64,65]. Because the highest pollen concentration occurs during May [66], its availability probably facilitated the significant increase in phytoseiid mite density in June and July observed in our study. The other food resources such as extrafloral nectar or fungi are also considered to be important for generalist phytoseiid mites [67,68], and their role in nutrition of mites inhabiting horse chestnut needs to be investigated, too.

The different locality conditions may determine the variability of phytoseiid population densities on horse chestnut. Results of the present study showed that mite abundance can be affected by various factors. It is interesting that air pollution had a positive effect on density of phytoseiids. Air pollution of České Budějovice is relatively low as only benzo(a)pyrene concentration exceeded the limit (Table S2). The highest overall pollution index was found at the Nádražní street site, which is a narrow green belt between a busy road and railway line. Horse chestnut trees are thus likely to accumulate heavy metals in this site [69]. Some plants use this accumulation as a defense against herbivorous arthropods [70], and therefore a negative effect on predatory mites could be expected. Seniczak et al. reported that a high concentration of heavy metals was harmful to mites [16]. On the other hand, deposition of atmospheric nitrogen originating from vehicle exhaust can increase herbivore populations through effects on host quality [71]. Similarly, the use of de-icing salts has been implicated in the increased nutritive value of leaves on trees in street habitats and, in turn, in elevated populations of spider mites [72]. Moreover, Nádražní street is the only site without mowing and removing of leaf litter, measures which are recommended to control *C. ohridella* [73]. The highest greenery care is performed in the City center site resulting in low damage inflicted by *C. ohridella* [7,17]. However, this greenery open space maintenance was found to have a negative impact on abundance of hymenopteran parasioids and parasitism rate [15,17], so it could also negatively affect density of Phytoseiidae. Indeed, the abundance of mites in the City center site was lower than that in Nádražní street.

5. Conclusions

The present study showed that the horse chestnut tree grown in urban environment is a favorable host tree for phytoseiid mites. More than 90% of specimens collected were identified as *E. finlandicus*. Two species, *Pa. talbii* and *Ph. macropilis,* were recorded on horse chestnut for the first time. The other five species found were *A. andersoni*, *K. aberrans*, *N. tiliarum*, *Pa. triporus*, and *T. (T.) pyri*. A high abundance of Phytoseiidae confirmed that *A. hippocastanum* can serve as a good host plant of these predatory mites and should be planted in proximity to vineyards, orchards, or other crops for better control of spider mites and eriophyiid mites. The mite density varied significantly among sampling sites, which can be attributed to different site conditions. Whether air pollution has a positive effect, while removing leaf litter and mowing have a negative effect, on conservation of predatory mites will require more investigation.

Supplementary Materials: The following supporting information can be downloaded at: https://www.mdpi.com/article/10.3390/f14050942/s1, Table S1: Concentrations of air pollutants (weight per m^3 of air) in *Aesculus hippocastanum* sampling sites. Symbols in brackets indicate weight units. Data representing five-year averages of 2009–2013 were obtained from Geographical Information System of South Bohemia [19]; Table S2: Air pollution index for individual atmospheric pollutants and overall index of air pollution for *Aesculus hippocastanum* sampling sites. Individual indexes were calculated from measurements shown in Table S1 and concentration limits indicated in brackets using Equation (2). The concentration limits are from Geographical Information System of South Bohemia [19].

Author Contributions: Conceptualization, R.Z. and M.K.; methodology, R.Z. and M.K.; data curation, M.K.; writing—original draft preparation, M.K.; writing—review and editing, M.K. and R.Z.; visualization, M.K and R.Z.; supervision, R.Z. All authors have read and agreed to the published version of the manuscript.

Funding: This research was funded by the Czech Academy of Sciences, grant number RVO: 60077344.

Data Availability Statement: The data presented in this study are available on request from the corresponding author.

Acknowledgments: The authors thank Georgios Th. Papadoulis for his help with the identification of phytoseiid mites, Jiunn Luh Tan for his advice on data analysis, Marco del Olmo Linares for drawing graphs, and Barbora Kozelková for her technical assistance.

Conflicts of Interest: The authors declare no conflict of interest. The funders had no role in the design of the study; in the collection, analyses, or interpretation of data; in the writing of the manuscript; or in the decision to publish the results.

References

1. Thomas, P.A.; Alhamd, O.; Iszkuło, G.; Dering, M.; Mukassabi, T.A. Biological Flora of the British Isles: *Aesculus hippocastanum*. *J. Ecol.* **2019**, *107*, 992–1030. [CrossRef]
2. Ulbricht, C.; Tiffany, N.; Boon, H.; Ulbricht, C.; Basch, E.; Bent, S.; Barrette, E.P.; Smith, M.; Sollars, D.; Dennehy, C.; et al. Horse Chestnut: A Multidisciplinary Clinical Review. *J. Herb. Pharmacother.* **2002**, *2*, 71–85. [CrossRef]
3. Shimada, T. Nutrient Compositions of Acorns and Horse Chestnuts in Relation to Seed-Hoarding: Nutrient Composition of Seeds. *Ecol. Res.* **2001**, *16*, 803–808. [CrossRef]
4. Augustin, S.; Guichard, S.; Heitland, W.; Freise, J.; Svatoš, A.; Gilbert, M. Monitoring and Dispersal of the Invading Gracillariidae *Cameraria Ohridella*. *J. Appl. Entomol.* **2009**, *133*, 58–66. [CrossRef]
5. Kirichenko, N.; Augustin, S.; Kenis, M. Invasive Leafminers on Woody Plants: A Global Review of Pathways, Impact, and Management. *J. Pest Sci.* **2019**, *92*, 93–106. [CrossRef]
6. Musolin, D.L.; Kirichenko, N.I.; Karpun, N.N.; Aksenenko, E.V.; Golub, V.B.; Kerchev, I.A.; Mandelshtam, M.Y.; Vasaitis, R.; Volkovitsh, M.G.; Zhuravleva, E.N.; et al. Invasive Insect Pests of Forests and Urban Trees in Russia: Origin, Pathways, Damage, and Management. *Forests* **2022**, *13*, 521. [CrossRef]
7. Kopačka, M.; Nachman, G.; Zemek, R. Seasonal Changes and the Interaction between the Horse Chestnut Leaf Miner *Cameraria ohridella* and Horse Chestnut Leaf Blotch Disease Caused by *Guignardia aesculi*. *Forests* **2021**, *12*, 952. [CrossRef]
8. Somme, L.; Moquet, L.; Quinet, M.; Vanderplanck, M.; Michez, D.; Lognay, G.; Jacquemart, A.-L. Food in a Row: Urban Trees Offer Valuable Floral Resources to Pollinating Insects. *Urban Ecosyst.* **2016**, *19*, 1149–1161. [CrossRef]

9. Tuovinen, T.; Rokx, J.A.H. Phytoseiid Mites (Acari: Phytoseiidae) on Apple Trees and in Surrounding Vegetation in Southern Finland. Densities and Species Composition. *Exp. Appl. Acarol.* **1991**, *12*, 35–46. [CrossRef]
10. Tuovinen, T. Influence of Surrounding Trees and Bushes on the Phytoseiid Mite Fauna on Apple Orchard Trees in Finland. *Agric. Ecosyst. Environ.* **1994**, *50*, 39–47. [CrossRef]
11. Kopačka, M.; Stathakis, T.I.; Broufas, G.; Papadoulis, G.T.; Zemek, R. Diversity and Abundance of Phytoseiidae (Acari: Mesostigmata) on Horse Chestnut (*Aesculus hippocastanum* L.) in an Urban Environment: A Comparison between Greece and the Czech Republic. *Acarologia* **2018**, *58*, 83–90. [CrossRef]
12. Kabíček, J.; Řeháková, M. Phytoseiid Mite Community on *Aesculus Hippocastanum* in the Parks. *Acta Fytotech. Zootech.* **2004**, *7*, 114–115.
13. Fidelis, E.G.; Farias, E.d.S.; Joaquim Da Silva Junior, R. Contribution of Natural Enemies and Weather to the Population Buildup of *Raoiella Indica* on Coconut. *Syst. Appl. Acarol.* **2022**, *27*, 8. [CrossRef]
14. Tixier, M.-S.; Arnaud, A.; Douin, M.; Kreiter, S. Effects of Agroforestry on Phytoseiidae Communities (Acari: Mesostigmata) in Vineyards. A Synthesis of a 10-Year Period of Observations. *Acarologia* **2015**, *55*, 361–375. [CrossRef]
15. Volter, L.; Prenerová, E.; Weyda, F.; Zemek, R. Changes in the Parasitism Rate and Parasitoid Community Structure of the Horse Chestnut Leafminer, *Cameraria Ohridella* (Lepidoptera: Gracillariidae), in the Czech Republic. *Forests* **2022**, *13*, 885. [CrossRef]
16. Seniczak, S.; Dabrowski, J.; Długosz, J. Effect of Copper Smelting Air Pollution on the Mites (Acari) Associated with Young Scots Pine Forests Polluted by a Copper Smelting Works at Giogów, Poland. I. Arboreal Mites. *Water Air Soil Pollut.* **1997**, *94*, 71–84. [CrossRef]
17. Kopacka, M.; Zemek, R. Spatial Variability in the Level of Infestation of the Leaves of Horse Chestnut by the Horse Chestnut Leaf Miner, *Cameraria ohridella* (Lepidoptera: Gracillariidae) and in the Number of Adult Moths and Parasitoids Emerging from Leaf Litter in an Urban Environment. *Eur. J. Entomol.* **2017**, *114*, 42–52. [CrossRef]
18. Jura, S. Určování Stáří Stromů. *Silva Bohem.* **2001**, *1*, 19–21.
19. Mapa Znečištění Ovzduší. Available online: https://geoportal.kraj-jihocesky.gov.cz/portal/mapy/zivotni-prostredi (accessed on 15 February 2023).
20. Zacharda, M.; Pultar, O.; Muška, J. Washing Technique for Monitoring Mites in Apple Orchards. *Exp. Appl. Acarol.* **1988**, *5*, 181–183. [CrossRef]
21. Beglyarov, G.A. Opredelitel chiscnych klescej fitoseiid (Parasitiformes, Phytoseiidae) fauny SSSR. *Info Bjull IOBC* **1981**, *3*, 1–45.
22. Miedema, E. Survey of Phytoseiid Mites (Acari: Phytoseiidae) in Orchards and Surrounding Vegetation of Northwestern Europe, Especially in the Netherlands. Keys, Descriptions and Figures. *Neth. J. Plant Pathol.* **1987**, *93*, 1–63. [CrossRef]
23. Chant, D.A.; Shaul, E.Y. A World Review of the *Soleiger* Species Group in the Genus *Typhlodromus* Scheuten (Acarina: Phytoseiidae). *Can. J. Zool.* **1982**, *60*, 3021–3032. [CrossRef]
24. Chant, D.A.; Yoshida-Shaul, E. A World Review of the *Tiliarum* Species Group in the Genus *Typhlodromus* Scheuten (Acari: Phytoseiidae). *Can. J. Zool.* **1989**, *67*, 1006–1046. [CrossRef]
25. Karg, W. Acari (Acarina), Milben, Parasitiformes (Anactinochaeta), Cohors Gamasina Leach, Raubmilben. In *Tierwelt Deutschlands und der angrenzenden Meeresteile nach ihren Merkmalen und nach ihrer Lebensweise*, 2nd ed.; VEB Gustav Fischer Verlag: Jena, Germany, 1993.
26. Dajoz, R. *Introduction to Ecology*; Hodder and Stoughton: London, UK, 1977; ISBN 978-0-340-16254-5.
27. Air Pollution Index Online Calculator. Available online: https://drr.ikcest.org/app/sf020 (accessed on 27 February 2023).
28. SAS Institute. *SAS/STAT 14.3: User's Guide*; SAS Institute: Cary, NC, USA, 2017.
29. Liang, K.-Y.; Zeger, S.L. Longitudinal Data Analysis Using Generalized Linear Models. *Biometrika* **1986**, *73*, 13–22. [CrossRef]
30. Ragusa di Chiara, S.; Papaioannou-Souliotis, P.; Tsolakis, H.; Tsagkarakou, N. Acari Fitoseidi (Parasitiformes, Phytoseiidae) Della Grecia Associati a Piante Forestali a Diverse Altitudini. *Boll. Zool. Agr. Bachicol.* **1995**, *27*, 85–91.
31. McMurtry, J.A.; Croft, B.A. Life-Styles of Phytoseiid Mites and Their Roles in Biological Control. *Annu. Rev. Entomol.* **1997**, *42*, 291–321. [CrossRef]
32. Kabíček, J. Broad Leaf Trees as Reservoirs for Phytoseiid Mites (Acari: Phytoseiidae). *Plant Prot. Sci.* **2003**, *39*, 65–69. [CrossRef]
33. Kabíček, J. Intraleaf Distribution of the Phytoseiid Mites (Acari, Phytoseiidae) on Several Species of Wild Broad Leaf Trees. *Biologia* **2005**, *60*, 523–528.
34. Kabíček, J. Linden Trees Are Favourable Host Plants for Phytoseiid Generalists in Urban Environment. *Balt. For.* **2019**, *25*, 1. [CrossRef]
35. Kabíček, J.; Povondrová, K. Phytoseiid Mite Communities on Urban Deciduous Trees. *Acta Fytotech. Zootech.* **2004**, *7*, 119–121.
36. Omeri, I. Phytoseiid Mites (Parasitiformes, Phytoseiidae) on Plants in Trostyanets Dendrological Park (Ukraine). *Vestn. Zool.* **2009**, *43*, e-7–e-14. [CrossRef]
37. Barbar, Z. Occurrence, Population Dynamics and Winter Phenology of Spider Mites and Their Phytoseiid Predators in a Citrus Orchard in Syria. *Acarologia* **2014**, *54*, 409–423. [CrossRef]
38. Grabovska, S.L.; Kolodochka, L.A. Species Complexes of Predatory Phytoseiid Mites (Parasitiformes, Phytoseiidae) in Green Urban Plantations of Uman' (Ukraine). *Vestn. Zool.* **2014**, *48*, 495–502. [CrossRef]
39. Vieira de Souza, I.; Argolo, P.S.; Gondim Júnior, M.G.C.; de Moraes, G.J.; Bittencourt, M.A.L.; Oliveira, A.R. Phytoseiid Mites from Tropical Fruit Trees in Bahia State, Brazil (Acari, Phytoseiidae). *ZK* **2015**, *533*, 99–131. [CrossRef]

40. Stojnić, B.; Mladenović, K.; Marčić, D. Spider Mites and Predatory Mites (Acari: Tetranychidae, Phytoseiidae) on Stone Fruit Trees (*Prunus* Spp.) in Serbia. *Int. J. Acarol.* **2018**, *44*, 322–329. [CrossRef]
41. Grabovska, S.L.; Mykolaiko, I.I.; Mykolaiko, V.P. Structure Patterns of Phytoseiid Mite Communities in Urban Plant Associations. *Ukr. J. Ecol.* **2017**, *7*, 179–186. [CrossRef]
42. Grabovska, S.L.; Mykolaiko, I.I. Mites of Phytoseiidae (Acari, Parasitiformes) in Urban Vegetative Plantations. *Ukr. J. Ecol.* **2017**, *7*, 216–222. [CrossRef]
43. Ripka, G. Checklist of the Phytoseiidae of Hungary (Acari: Mesostigmata). *Folia Entomol. Hung.* **2006**, *67*, 229–260.
44. Ripka, G. New Data to the Knowledge on the Phytoseiid Fauna in Hungary (Acari: Mesostigmata). *Acta Phytopathol. Entomol. Hung.* **1998**, *33*, 395–405.
45. Komlovszky, S.Z.I.; Jenser, G. The frequent occurrence of the predatory mites *Amblyseius finlandicus* Oudemans and *Phytoseius plumifer*. *Növényvédelem* **1987**, *23*, 193–201.
46. Gyenis, K.; Pénzes, B.; Hegyi, T. Phytophagous and predatory mites on the horse chestnut tree. *Növényvédelem* **2005**, *41*, 143–148.
47. Ripka, G.; de Lillo, E. New Data to the Knowledge on the Eriophyoid Fauna in Hungary (Acari: Eriophyoidea). *Folia Entomol. Hung.* **1997**, *58*, 147–157.
48. Karban, R.; English-Loeb, G.; Walker, M.A.; Thaler, J. Abundance of Phytoseiid Mites on *Vitis* Species: Effects of Leaf Hairs, Domatia, Prey Abundance and Plant Phylogeny. *Exp. Appl. Acarol.* **1995**, *19*, 189–197. [CrossRef]
49. Weryszko-Chmielewska, E.; Haratym, W. Leaf Micromorphology of *Aesculus Hippocastanum* L. and Damage Caused by Leaf-Mining Larvae of *Cameraria Ohridella* Deschka and Dimić. *Acta Agrobot.* **2012**, *65*, 25–34. [CrossRef]
50. Muhammad, S.; Wuyts, K.; Samson, R. Atmospheric Net Particle Accumulation on 96 Plant Species with Contrasting Morphological and Anatomical Leaf Characteristics in a Common Garden Experiment. *Atmos. Environ.* **2019**, *202*, 328–344. [CrossRef]
51. Overmeer, W.P.J.; Zon, A.Q. *The Preference of Amblyseius Potentillae (Garman) (Acarina: Phytoseiidae) for Certain Plant Substrates*; Griffiths, D., Bowman, C., Eds.; Ellis Horwood Limited, Chichester: Chichester, UK, 1984; pp. 591–596.
52. O'Dowd, D.J.; Willson, M.F. Leaf Domatia and Mites on Australasian Plants: Ecological and Evolutionary Implications. *Biol. J. Linn. Soc.* **1989**, *37*, 191–236. [CrossRef]
53. Walter, D.E.; O'Dowd, D.J. Leaves with Domatia Have More Mites. *Ecology* **1992**, *73*, 1514–1518. [CrossRef]
54. Kreiter, S.; Tixier, M.-S.; Croft, B.A.; Auger, P.; Barret, D. Plants and Leaf Characteristics Influencing the Predaceous Mite *Kampimodromus Aberrans* (Acari: Phytoseiidae) in Habitats Surrounding Vineyards. *Environ. Entomol.* **2002**, *31*, 648–660. [CrossRef]
55. Matos, C.H.C.; Pallini, A.; Chaves, F.F.; Schoereder, J.H.; Janssen, A. Do Domatia Mediate Mutualistic Interactions between Coffee Plants and Predatory Mites? *Entomol. Exp. Appl.* **2006**, *118*, 185–192. [CrossRef]
56. Loughner, R.; Goldman, K.; Loeb, G.; Nyrop, J. Influence of Leaf Trichomes on Predatory Mite (*Typhlodromus Pyri*) Abundance in Grape Varieties. *Exp. Appl. Acarol.* **2008**, *45*, 111–122. [CrossRef]
57. Loughner, R.; Wentworth, K.; Loeb, G.; Nyrop, J. Leaf Trichomes Influence Predatory Mite Densities through Dispersal Behavior. *Entomol. Exp. Appl.* **2010**, *134*, 78–88. [CrossRef]
58. Loughner, R.; Wentworth, K.; Loeb, G.; Nyrop, J. Influence of Leaf Trichomes on Predatory Mite Density and Distribution in Plant Assemblages and Implications for Biological Control. *Biol. Control* **2010**, *54*, 255–262. [CrossRef]
59. O'Connell, D.M.; Lee, W.G.; Monks, A.; Dickinson, K.J.M. Does Microhabitat Structure Affect Foliar Mite Assemblages? *Ecol. Entomol.* **2010**, *35*, 317–328. [CrossRef]
60. Schmidt, R.A. Leaf Structures Affect Predatory Mites (Acari: Phytoseiidae) and Biological Control: A Review. *Exp. Appl. Acarol.* **2014**, *62*, 1–17. [CrossRef] [PubMed]
61. Barba, P.; Loughner, R.; Wentworth, K.; Nyrop, J.P.; Loeb, G.M.; Reisch, B.I. A QTL Associated with Leaf Trichome Traits Has a Major Influence on the Abundance of the Predatory Mite *Typhlodromus Pyri* in a Hybrid Grapevine Population. *Hortic. Res.* **2019**, *6*, 87. [CrossRef] [PubMed]
62. Roda, A.; Nyrop, J.; English-Loeb, G. Leaf Pubescence Mediates the Abundance of Non-Prey Food and the Density of the Predatory Mite *Typhlodromus Pyri*. *Exp. Appl. Acarol.* **2003**, *29*, 193–211. [CrossRef]
63. Kugler, H. *Blütenökologie*; Gustav Fischer Verlag: Stuttgart, Germany, 1970.
64. Goleva, I.; Zebitz, C.P.W. Suitability of Different Pollen as Alternative Food for the Predatory Mite Amblyseius Swirskii (Acari, Phytoseiidae). *Exp. Appl. Acarol.* **2013**, *61*, 259–283. [CrossRef] [PubMed]
65. Goleva, I.; Rubio Cadena, E.C.; Ranabhat, N.B.; Beckereit, C.; Zebitz, C.P.W. Dietary Effects on Body Weight of Predatory Mites (Acari, Phytoseiidae). *Exp. Appl. Acarol.* **2015**, *66*, 541–553. [CrossRef]
66. Weryszko-Chmielewska, E.; Tietze, M.; Michońska, M. Ecological Features of the Flowers of *Aesculus Hippocastanum* L. and Characteristics of *Aesculus* L. Pollen Seasons under the Conditions of Central-Eastern Poland. *Acta Agrobot.* **2012**, *65*, 61–68. [CrossRef]
67. Pekas, A.; Wäckers, F.L. Multiple Resource Supplements Synergistically Enhance Predatory Mite Populations. *Oecologia* **2017**, *184*, 479–484. [CrossRef]
68. Pozzebon, A.; Loeb, G.M.; Duso, C. Role of Supplemental Foods and Habitat Structural Complexity in Persistence and Coexistence of Generalist Predatory Mites. *Sci. Rep.* **2015**, *5*, 14997. [CrossRef] [PubMed]
69. Yilmaz, R.; Sakcali, S.; Yarci, C.; Aksoy, A.; Ozturk, M. Use of *Aesculus Hippocastanum* L. as a Biomonitor of Heavy Metal Pollution. *Pak. J. Bot.* **2006**, *38*, 1519–1527.

70. Godinho, D.P.; Branquinho, C.; Magalhães, S. Intraspecific Variability in Herbivore Response to Elemental Defences Is Caused by the Metal Itself. *J. Pest. Sci.* **2023**, *96*, 797–806. [CrossRef]
71. Raupp, M.J.; Shrewsbury, P.M.; Herms, D.A. Ecology of Herbivorous Arthropods in Urban Landscapes. *Annu. Rev. Entomol.* **2010**, *55*, 19–38. [CrossRef]
72. Kropczynska, D.; Van De Vrie, M.; Tomczyk, A. Bionomics of *Eotetranychus Tiliarium* and Its Phytoseiid Predators. *Exp. Appl. Acarol.* **1988**, *5*, 65–81. [CrossRef]
73. Kehrli, P.; Bacher, S. Date of Leaf Litter Removal to Prevent Emergence of *Cameraria ohridella* in the Following Spring. *Entomol. Exp. Appl.* **2003**, *107*, 159–162. [CrossRef]

forests

MDPI

Article

Mites Associated with the European Spruce Bark Beetle *Ips typographus* (Linnaeus, 1758) in Europe, with New Evidence for the Fauna of Serbia

Marija Milosavljević [1,*], Mara Tabaković-Tošić [1], Milan Pernek [2], Ljubinko Rakonjac [3], Aleksandar Lučić [4], Saša Eremija [3] and Michal Rindos [5,6,*]

1 Department of Plant Protection, Institute of Forestry, Kneza Višeslava 3, 11030 Belgrade, Serbia
2 Division for Forest Protection and Game Management, Croatian Forest Research Institute, Cvjetno naselje 41, 10450 Jastrebarsko, Croatia
3 Department of Forest Establishment, Silviculture and Ecology, Institute of Forestry, Kneza Višeslava 3, 11030 Belgrade, Serbia
4 Department of Genetics, Plant Breeding, Seed and Nursery Production, Institute of Forestry, Kneza Višeslava 3, 11030 Belgrade, Serbia
5 Institute of Entomology, Biology Centre of the Czech Academy of Sciences, Branisovska 31, 37005 Ceske Budejovice, Czech Republic
6 Faculty of Forestry and Wood Sciences, Czech University of Life Sciences Prague, Kamycka 1176, 16500 Prague 6, Czech Republic
* Correspondence: m.milosavljevic04@gmail.com (M.M.); rindom00@prf.jcu.cz (M.R.)

Citation: Milosavljević, M.; Taboković-Tošić, M.; Pernek, M.; Rakonjac, L.; Lučić, A.; Eremija, S.; Rindos, M. Mites Associated with the European Spruce Bark Beetle *Ips typographus* (Linnaeus, 1758) in Europe, with New Evidence for the Fauna of Serbia. *Forests* **2022**, *13*, 1586. https://doi.org/10.3390/f13101586

Academic Editors: Katarína Pastirčáková and Rostislav Zemek

Received: 31 August 2022
Accepted: 19 September 2022
Published: 28 September 2022

Publisher's Note: MDPI stays neutral with regard to jurisdictional claims in published maps and institutional affiliations.

Abstract: Various biotic and abiotic factors are the cause of the decline of coniferous forests throughout Europe. Trees weakened by unfavorable weather conditions create an ideal environment for a possible outbreak of bark beetles. The damage caused by bark beetles costs billions of dollars worldwide every year. Extreme climate events are responsible for the enormous forest losses in Tara National Park in the last ten years, leading to a massive bark beetle infestation. The understanding of the diversity and role of mites as biological control agents is still insufficient. In this study, we summarize the current knowledge on the diversity of mites associated with *Ips typographus* L. in Europe and provide information on the diversity of these mites in Serbia. *Paraleius leontonychus*, *Uroobovella ipidis*, *Dendrolaelaps quadrisetus*, *Histiostoma piceae*, and *Trichouropoda polytricha* were detected for the first time in Serbia. Moreover, the occurrence of *Paraleius leontonychus* represents the southernmost occurrence of this species.

Keywords: Acari; Coleoptera; Curculionidae; Scolytinae; biodiversity; Tara National Park

1. Introduction

Biotic and abiotic disturbances associated with climate change currently pose the greatest threat to coniferous forests in Europe [1–4]. In Serbia, a sharp decline in coniferous forests has been observed since 2012 [5–10]. The main reason for this decline was droughts that occurred during growing seasons and caused damage of 116,290 m^3 with a percentage of 89% dead spruces and fir [11].

Tara National Park is located in the western part of Serbia, part of the Dinaric Alps. With an area of 19,175 ha and about 1013 plant species, Tara National Park is the third most diverse area in Serbia [12,13]. Forest ecosystems cover almost 80% of the total land area. The most common forest community is mixed forests of beech, fir, and spruce (*Piceo-Fago-Abietetum* Čolić, 1965), which are also known for the occurrence of the endemic and endangered Serbian spruce *Picea omorika* (Pančić) Purk. [14]. Tara National Park Enterprise data estimate the loss of about 6.4% (112,879.3 m^3 of wood) of the forest area in the last 10 years due to extreme climatic events and subsequent pest calamities [15].

Bark beetles are a major threat to physiologically weakened coniferous forests [16]. The most abundant is the European spruce bark beetle, *Ips typographus* Linnaeus, 1758

(Coleoptera, Curculionidae, Scolytinae). Research on natural enemies and potential control agents for this species in Serbia has so far been limited to hymenopteran parasitoids and entomopathogenic fungal associations [17–19]. However, interactions between mites and bark beetles have been overlooked. Previously, more than 250 mite species associated with bark beetles have been described worldwide [20]. Their interactions fall into three main categories: predatory, mite-fungus symbiosis, and phoretic [21]. The most common type of interaction between mites and bark beetles is phoresis. Phoresis may also be considered harmful to bark beetles by limiting their movement and potential dispersal [22]. In addition, previous studies have shown that mites are able to transmit fungal spores in beetle galleries, e.g., *Proctolaelaps scolyti* (Evans, 1958) and *Tarsonemus crassus* (Schaarschmidt, 1959), which contribute to the transmission of *O. novo-ulmi* Brasier, 1991 to elms [23] or spores of *Ophiostoma* sp. by mites of the genus *Histiostoma* Kramer, 1876 [24]. In addition, phoretic mites are potential vectors of entomopathogenic fungi such as *Beauveria bassiana* (Bals.-Criv.) Vuill. (1912) [25]. Studies on the diversity of phoretic mites associated with *I. typographus* have already been conducted in several European countries (see Table S1).

The main objective of this study was to summarize the knowledge about the mites associated with *I. typographus* reported so far from Europe and to provide a view of the diversity of these mites in Serbia.

2. Materials and Methods

2.1. Study Area

Our research was conducted at six selected sites in the Tara National Park in Serbia (Figure 1). Site No. 1 is located in the southwestern part of the National Park (43°53′25.22″ N, 19°30′43.87″ E) at an altitude of 1143 m.a.s.l. It is covered with mixed stands of spruce, fir, and beech (*Piceo-Abieti-Fagetum drymetosum* association). Site No. 2 was located near site No. 1 (43°53′24.63″ N 19°30′34.21″ E), at an altitude of 1155 m.a.s.l., with the same forest stands. Site No. 3 (43°53′49.49″ N 19°29′46.08″ E) was located in the western part, at an altitude of 1196 m.a.s.l. It is covered with mixed stands of spruce, fir, and beech (*Piceo-Abieti-Fagetum typicum*). Site No. 4 (43°55′33.85″ N 19°25′16.60″ E) is located near No. 3, with the same forest stands and at a similar altitude (1158 m.a.s.l.). Sites No. 5 (43°55′22.09″ N 19°27′13.90″ E) and No. 6 (43°56′15.19″ N 19°28′19.66″ E) are adjacent to each other and belong to the same forest stands as No. 3 and 4, with an average altitude of 1000 m.a.s.l.

2.2. Sampling Methods

In Tara National Park, cross-barrier pheromone traps (THEYSOHN®) have been used to monitor bark beetle populations since 2014 [26]. For our research, we used the same pheromone traps equipped with IT Ecolure pheromones. We collected the captured adults of *I. typopgraphus* during the first flight in April 2016. The collected adults were packed in plastic boxes and brought to the laboratory of the Institute of Forestry in Belgrade. We examined the bark beetles under a stereomicroscope, counted the mites, and placed them in Eppendorf tubes containing ethanol and lactic acid before preparation [27]. The mites were sorted into morphospecies using the key published by Moser and Bogenschütz [28] and identified by Milan Pernek. For the assessment of species dominance and frequency of occurrence in the bark beetle samples, we followed the statistical method used by other authors [29–33] (Table 1). The dominance parameters were eudominant (>30%), dominant (15.01%–30%), subdominant (7.01%–15%), recedent (3.01%–7%), and subrecedent (<3%); the frequency parameters were euconstant (>50%), constant (30.01%–50%), subconstant (15.01%–30%), accessory species (5.01%–15%) and accidental occurrence (<5%).

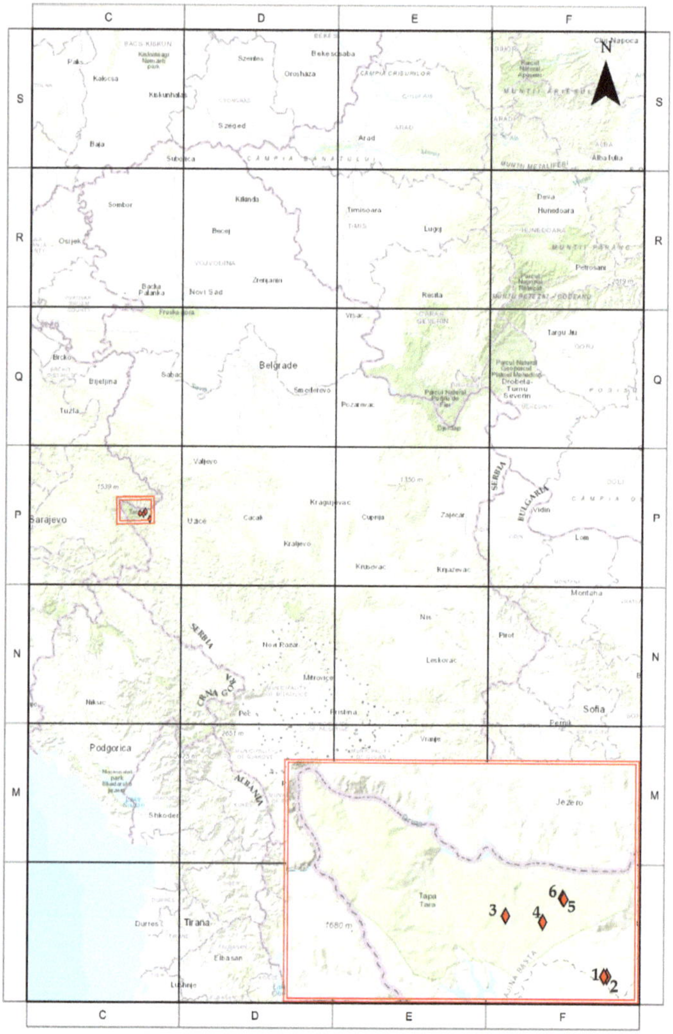

Figure 1. Location of the study areas in Tara National Park, Serbia.

Table 1. The mite species found on the captured adults of *I. typographus*, their total number, dominance, frequency in the samples, and location on the body of the bark beetles.

Species	Number	Dominance (%)	Frequency (%)	Location
Paraleius leontonychus (Berlese, 1910)	267	Dominant (18.98)	Eucostant (66.66)	Thorax, Elytral declivity, Head
Uroobovella ipidis (Vitzthum,1923)	191	Subdominant (13.57)	Eucostant (66.66)	Thorax, Elytral declivity, Head
Dendrolaelaps quadrisetus (Berlese 1920)	374	Eudominant (26.58)	Eucostant (83.33)	Under elytra, Elytral declivity
Histiostoma piceae (Scheucher, 1957)	10	Subrecedent (0.71)	Accidental (16.67)	Elytral declivity
Trichouropoda polytricha (Vitzthum, 1923)	565	Eudominant (40.16)	Eucostant (100.00)	Thorax, Elytral declivity, Head

2.3. Photographing Methods

Before photographing, we made permanent preparations for the identified mites according to the protocol of Saito et al. [34]. Photographing was performed with the Olympus BX63 fluorescence microscope and cellScens Dimension software. For photographing the adults of *I. typographus* with the mites, we used Olympus SZX16 with Olympus SDF Plapo 1XPF lens and PROMICRA 3-5CP camera together with QuickPHOTO MICRO 3.2 software.

3. Results

The data on species used to summarise current knowledge on mites associated with *I. typographus* in Europe were obtained from various literature sources (see Table S1). We found data about 97 species from 12 countries (Bulgaria, Croatia, the Czech Republic, Finland, Georgia, Germany, Poland, Romania, Russia, Slovakia, Sweden, and Turkey). The countries with the highest reported diversity of mites were Russia (53 species), Germany (25 species), and Finland (22 species). The lowest number of mite species was reported from Croatia (3 species). The most common mite species in Europe were *Dendrolaelaps quadrisetus* (Berlese, 1920), detected in 11 countries, *Trichouropoda polytricha* (Vitzthum, 1923)—found in 10 countries, *Uroobovella ipidis* (von Vitzthum, 1923)—documented in 9 countries, *Proctolaelaps fiseri* (Samsinak, 1960)—found in 8 countries, and *Histiostoma piceae* (Scheucher, 1957), detected in 5 countries.

We also examined 4093 adults of *I. typographus* collected in Tara National Park, from which we obtained 1407 mites (Table 1). All five species were found for the first time in Serbia (Figure 2).

3.1. List of the Species Newly Recorded in Serbia

1. Order: Oribatida
   Family: Scheloribatidae Jacot, 1935
   Genus: *Paraleius* Travé, 1960
   Species: *P. leontonychus* (Berlese, 1910)
   Figure 2a
   Distribution: Holarctic

   Note: The species is often also listed in the literature as *Siculobata leontonycha* (Berlese, 1910). *P. leontonychus* is a fungivorous (or detritivorous) species of oribatid mite with strong hooked claws helping it to reach the bark beetle galleries [35]. This phoretic species was previously found on various European bark beetle species, such as *Cryphalus abietis, Dryocoetes autographus, Hylurgops palliatus, Ips amitinus/I. sexdentatus/I. typographus, Pityogenes chalcographus, Pityokteines curvidens/P. spinidens/P. vorontzowi, Scolytus multistriatus*, and *Tomicus minor/T. piniperda* [36–39].

2. Order: Mesostigmata
   Family: Urodinychidae Berlese, 1917
   Genus: *Uroobovella* Berlese, 1903
   Species: *U. ipidis* (Vitzthum, 1923)
   Figure 2b
   Distribution: Europe and western Asia

   Note: The deutonymphs of *U. ipidis* are phoretic and were previously found on various European bark beetle species, such as *I. typographus/I. sexdentatus, Hylastes cunicularius, Pityokteines curvidens/P. spinidens/P. vorontzowi*, and *Pityogenes chalcographus* [28,32,37–46].

3. Order Mesostigmata
   Family: Digamasellidae Evans, 1957
   Genus: *Dendrolaelaps* Halbert, 1915
   Species: *D. quadrisetus* (Berlese, 1920)
   Figure 2c

Distribution: Holarctic and Neotropical region

Note: Deutonymphs of *D. quadrisetus* are phoretic and positioned mainly under the elytra of bark beetles. However, Khaustov et al. [47] observed that deutonymphs of this species increase their size by feeding on the eggs of *I. typographus*.

D. quadrisetus was previously found on many European bark beetle species, e.g., *Crypturgus cinereus, Dryocoetes autographus, Hylastes opacus, Hylesinus varius, Hylurgops glabratus/H. palliatus, Ips acuminatus/I. amitinus/I. cembrae/I. sexdentatus/I. typographus, Pityogenes chalcographus, Pityokteines curvidens/P. spinidens/P. vorontzowi, Polygraphus poligraphus, Scolytus intricatus/S. ratzeburgii, Tomicus minor/T. piniperda,* and *Xyleborus cryptographus* [24,28,37,38,41,44,48–52].

4. Order: Sarcoptiformes
Family: Histiostomatidae Berlese, 1897
Genus: *Histiostoma* Kramer, 1876
Species: *H. piceae* Scheucher, 1957
Figure 2d
Distribution: Palearctic region

Note: The deutonymphs of this species are phoretic and can also carry hyperphoretic plant pathogenic fungi (e.g., *Ophiostoma* spp.). *H. piceae* has previously been found mainly in the subelytral spaces of *I. typographus*. However, they have also previously been found in other bark beetle species, such as *Dryocoetes hectographus, Ips cembrae, Pityogenes chalcographus,* and *Pityokteines curvidens/P. spinidens/P. vorontzowi* [24,28,37,38,40,41,45,53–55].

5. Order: Mesostigmata
Family: Trematuridae Berlese, 1917
Genus: *Trichouropoda* Berlese, 1916
Species: *T. polytricha* Vitzthum, 1923
Figure 2e
Distribution: Holarctic region

Note: The deutonymphs of *T. polytricha* are phoretic to various European bark beetle species, such as *Dryocoetes autographus, Hylastes cunicularius, Hylurgops palliatus, Ips amitinus/I. cembrae/I. sexdentatus/I. typographus,* and *Pityogenes chalcographus* [28,39–44].

Figure 2. Five newly recorded mite species of *I. typographus* in Serbia and their position on the body of the bark beetle. (**a**) *Paraleius leontonychus*, (**b**) *Uroobovella ipidis*, (**c**) *Dendrolaelaps quadrisetus*, (**d**) *Histiostoma piceae*, (**e**) *Trichouropoda polytricha*. (**f**) Mites on the elytra declivity of *I. typographus*, (**g**) mites under the elytra of *I. typographus*, (**h**) mites on the thorax of *I. typographus*.

3.2. Abundance of Newly Found Mite Species at the Sites Studied in National Par Tara

At site No. 1, we collected 1099 adults of *I. typographus*, from which we obtained 322 mites (*P. leontonychus*—78 indv.; *U. ipidis*—2 indv.; *D. quadrisetus*—128 indv.; *H. piceae*—10 indv.; *T. polytricha*—104 indv.). At site No. 2, we collected 412 adults of *I. typographus* and obtained 347 mites (*P. leontonychus*—102 indv.; *U. ipidis*—28 indv.; *D. quadrisetus*—102 indv.; *T. polytricha*—115 indv.). At site No. 3, we collected 442 adults of *I. typographus* with 180 mites (*D. quadrisetus*—78 indv.; *T. polytricha*—102 indv.), At site No. 4, we collected 148 adults of *I. typographus*, of which we found only 12 mites (*T. polytricha*—12 indv.). At site No. 5, we collected 1717 adults of *I. typographus* and obtained 354 mites (*P. leontonychus*—54 indv.; *D. quadrisetus*—12 indv.; *U. ipidis*—83 indv.; *T. polytricha*—205 indv.). At site No. 6, we collected 275 adults of *I. typographus*, from which we obtained 192 mites (*P. leontonychus*—33 indv.; *U. ipidis*—78 indv.; *D. quadrisetus*—54 indv.; *T. polytricha*—27 indv.).

4. Discussion

Our study provides an overview of the current knowledge on mites associated with *I. typographus* in Europe and, for the first time, provides information on their diversity in Serbia and fills the gap in their known distribution. Based on their occurrence in other European countries, all recorded species were expected to be present in Serbia as well. The most frequently found mite species were *T. polytricha* (565 individuals) and *D. quadrisetus* (374 individuals). This result agrees with the results from other European countries, where these two species were considered the most common (see Table S1). However, *P. leontonychus* was also found relatively frequently at sites No. 2, 5, and 6, and *U. ipidis* was also among the most frequently found species at sites No. 4 and 5. The first detection of the fungivorous (or detritivorous) *P. leontonychus* and its higher abundance at three studied sites can be considered very interesting, as this morphologically unique species has so far only been found in northern parts of Europe (see Table S1) [50,56]. The most frequent locations on the body of the bark beetle were elytral declivity and the head. This also corroborates previous observations [45]. Similar to previous studies [28,40,43], our methods for detecting mites on European spruce bark beetles most likely did not cover the full diversity of mites associated with them, so more sensitive methods (e.g., examining mites from tree bark and bark beetle galleries) should be used in further studies. Furthermore, information on the life history of the various mite species is still insufficient, as is the actual diversity of mites compared to the number of species currently recognized [57].

5. Conclusions

Our study of mites associated with *I. typographus* has improved the inadequate knowledge of the regional diversity of this potentially important group and also represents an extension of the known distribution of each species. As our study highlighted, some species play only a phoretic role, using the bark beetles only to transport themselves to different locations, while other species can be a good natural enemy, reducing the population by feeding on the eggs or transporting spores of entomopathogenic fungi. In addition, the knowledge gained about the diversity and biology of mites could serve as a potential tool for future biological control research.

Supplementary Materials: The following supporting information can be downloaded at: https://www.mdpi.com/article/10.3390/f13101586/s1; Table S1: The diversity of mites reported from various European countries.

Author Contributions: Conceptualization, M.M. and M.T.-T.; methodology, M.M., M.T.-T. and M.R.; investigation, M.M. and S.E.; resources, M.M. and M.T.-T.; data curation, M.M.; writing the original draft preparation, M.M. and M.R.; original draft editing, M.M., M.R., M.T.-T. and M.P.; visualization, M.M. and M.R.; funding acquisition, L.R. and A.L. All authors have read and agreed to the published version of the manuscript.

Funding: The work of M.M., M.T.-T., S.E., L.R. and A.L. was supported by the Forestry Directorate of the Ministry of Agriculture, Forestry, and Water Management of the Republic of Serbia within the framework of the project "Forecasting and reporting tasks of the Forestry Directorate in the field of diagnosis of harmful organisms and protection of forest plant health" (Agreement No. 401-00-00026/2020-10), the work of M.R. was supported by the grant "Advanced research to support the adaptation of the forestry and wood processing sector to global change and the fourth industrial revolution", No. CZ.02.1.01/0.0/0.0/16_019/0000803, funded by OP RDE.

Data Availability Statement: Data is contained within the article or Supplementary Materials. The data presented in this study are available in Table S1.

Acknowledgments: The authors would like to thank the public enterprise "Tara National Park" for providing the experimental site and human resources. We also thank Željko Tomanović (University of Belgrade, Faculty of Biology, Belgrade, Serbia) for his support and proofreading and Chris Raper (Natural History Museum London, UK) for his help with language editing.

Conflicts of Interest: The authors declare no conflict of interest. The funders had no role in the design of the study; in the collection, analyses, or interpretation of data; in the writing of the manuscript; or in the decision to publish the results.

References

1. Venäläinen, A.; Lehtonen, I.; Laapas, M.; Ruosteenoja, K.; Tikkanen, O.; Viiri, H.; Ikonen, V.; Peltola, H. Climate Change Induces Multiple Risks to Boreal Forests and Forestry in Finland: A Literature Review. *Glob. Chang. Biol.* **2020**, *26*, 4178–4196. [CrossRef] [PubMed]
2. Senf, C.; Buras, A.; Zang, C.S.; Rammig, A.; Seidl, R. Excess Forest Mortality Is Consistently Linked to Drought across Europe. *Nat. Commun.* **2020**, *11*, 6200. [CrossRef] [PubMed]
3. Bentz, B.J.; Jönsson, A.M.; Schroeder, M.; Weed, A.; Wilcke, R.A.I.; Larsson, K. *Ips typographus* and *Dendroctonus ponderosae* Models Project Thermal Suitability for Intra-and Inter-Continental Establishment in a Changing Climate. *Front. For. Glob. Chang.* **2019**, *2*, 1. [CrossRef]
4. Rosenzweig, C.; Iglesias, A.; Yang, X.B.; Epstein, P.R.; Chivian, E. Climate Change and Extreme Weather Events, Implications for Food Production, Plant Diseases, and Pests. *Glob. Chang. Hum. Health* **2001**, *2*, 90–104. [CrossRef]
5. Radulović, Z.; Karadžić, D.; Milenković, I.; Lučić, A.; Rakonjac, L.; Miletić, Z.; Pižurica, R. Declining of Forests—Biotic and Abiotic Stress. *Glas. Sumar. Fak.* **2014**, 71–88. [CrossRef]
6. Tabaković-Tošić, M. The Condition of Tree Crowns at the Sample Plots of Level I-Reliable or Unreliable Indicators of the Vitality of Main Conifer Species in Serbian Forests. In Proceedings of the 3rd ICP Forests Scientific Conference, Athens, Greece, 26–28 May 2014; p. 27.
7. Tabaković-Tošić, M.; Milenković, I.; Radulović, Z. The Coniferous Anthropogenic and Natural Forests Decline in Serbia Driven by Different Abiotic and Biotic Factors. Sustain. *For. Collect.* **2016**, *73–74*, 49–57. [CrossRef]
8. Tabaković-Tošić, M.; Milosavljević, M. The Correlation between the Changes in Climate, the Intensity of Spruce Decline and the Abundance of Spruce Bark Beetles in "Golija" Nature Park. *Sustain. For. Collect.* **2016**, *73–74*, 59–68. [CrossRef]
9. Češljar, G.; Brašanac-Bosanac, L.; Đorđević, I.; Eremija, S.; Milosavljević, M.; Jovanović, F.; Rakonjac, L.; Simović, S. Unfavorable climatic factors and their impact on the decline of spruce at the Kopaonik national park (Central Serbia). *Fresenius Environ. Bull.* **2022**, *31*, 5204–5215.
10. Češljar, G.; Jovanović, F.; Brašanac-Bosanac, L.; Đorđević, I.; Mitrović, S.; Eremija, S.; Ćirković-Mitrović, T.; Lučić, A. Impact of an Extremely Dry Period on Tree Defoliation and Tree Mortality in Serbia. *Plants* **2022**, *11*, 1286. [CrossRef]
11. Tabaković-Tošić, M.; Nevenić, R.; Češljar, G. Bark Beetle Outbreak in Spruce Communities within a Sample Plot (Level II) in the Mountain Kopaonik in the Period 2010–2013. In Proceedings of the 3rd ICP Forests Scientific Conference, Athens, Greece, 26–28 May 2014; p. 28.
12. Medarević, M. *Šume Tare*; Šumarski fakultet: Belgrade, Serbia, 2005; pp. 1–139.
13. Radović, D.; Stevanović, V.; Marković, D.; Jovanović, S.; Džukić, G.; Radović, I. Implementation of GIS technologies in assessment and protection of natural values of Tara national park. *Arch. Biol. Sci.* **2005**, *57*, 193–204. [CrossRef]
14. Gajić, M. *Flora Nacionalnog parka Tara*; Šumarski fakultet: Beograd, Serbija, 1989; p. 146.
15. Karaklić, D. *Osnova Gazdovanja Šumama za Gazdinsku Jedinicu "TARA" 2021–2030. godina*; JP NP Tara: Bajina Bašta, Srbija, 2021; pp. 30–52.
16. Wermelinger, B. Ecology and management of the spruce bark beetle *Ips typographus*-a review of recent research. *For. Ecol. Manag.* **2004**, *202*, 67–82. [CrossRef]
17. Marković, Č.; Stojanović, A. Differences in bark beetle (*Ips typographus* and *Pityogenes chalcographus*) abundance in a strict spruce reserve and the surrounding spruce forests of Serbia. *Phytoparasitica* **2010**, *38*, 31–37. [CrossRef]
18. Milosavljević, M.D.; Tabaković-Tošić, M.G.; Todorov, I.A.; Mitroiu, M.D.; Georgiev, G.T. New Records of Pteromalid Parasitoids (Hymenoptera: Pteromalidae) Reared from the Spruce Bark Beetle *Ips typographus* (L., 1758) from Serbia. *Act. Zool. Bulgar.* **2021**, *73*, 481–484.
19. Milosavljević, M.; Tabaković-Tošić, M.; Radulović, Z.; Marković, M.; Rindoš, M. Isolation, identification and phylogenetic position of entomopathogenic fungus *Beauveria bassiana* from *Ips typographus* in Serbia. *Fresenius Environ. Bull.* **2021**, *30*, 9443–9448.
20. Hofstetter, R.W.; Dikins-Bookwalter, J.; Davis, T.D.; Klepzig, K.D. Symbiotic associations of bark beetles. In *Bark Beetles: Biology and Ecology of Native and Invasive Species*; Vega, F.E., Hofstetter, R.W., Eds.; Elsevier/Academic Press: London, UK, 2015; pp. 209–245. [CrossRef]
21. Walter, D.E.; Proctor, H.C. *Mites: Ecology, Evolution and Behavior*, 2nd ed.; Springer: Dordrecht, The Netherlands, 2013; p. 471. [CrossRef]
22. Athias-Binche, F.; Morand, S. From phoresy to parasitism: The example of mites and nematodes. *Res. Rev. Parasitol.* **1993**, *53*, 73–79.
23. Moser, J.C.; Konrad, H.; Blomquist, S.R.; Kirisits, T. Do mites phoretic on elm bark beetles contribute to the transmission of Dutch elm disease? *Naturwissenschaften* **2010**, *97*, 219–227. [CrossRef] [PubMed]
24. Wirth, S.F.; Weis, O.; Pernek, M. Comparison of phoretic mites associated with bark beetles *Ips typographus* and *Ips cembrae* from Central Croatia. *Sumar. List.* **2016**, *140*, 549–560. [CrossRef]

25. Burjanadze, M.; Moser, J.C.; Zimmermann, G.; Kleesples, R.G. Antagonists of the spruce bark beetle *Ips typographus* L. (Coleoptera: Scolytidae) of German and Georgian populations. *Insect Pathog. Insect Parasit. Nematodes IOBC/Wprs Bull* **2008**, *31*, 245–250.
26. Tomić, M.; Bezarević, B. Control of Bark Beetle Population at the Tara National Park by Pheromone Traps. In Proceedings of the 7th Congress on Plant Protection, Zlatibor, Serbia, 24–28 November 2014; pp. 217–223.
27. Evans, G.O.; Browning, E. LXXV. Techniques for the preparation of mites for study. *Ann. Mag. Nat. Hist.* **1955**, *8*, 631–635. [CrossRef]
28. Moser, J.C.; Bogenschütz, H. A key to the mites associated with flying *Ips typographus* in South Germany. *Z. Angew. Entomol.* **1984**, *97*, 437–450. [CrossRef]
29. Błoszyk, J. *Geograficzne i Ekologiczne Zróżnicowanie Zgrupowań Roztoczy z Kohorty Uropodina (Acari: Mesostigmata) w Polsce. I. Uropodina Lasów Grądowych (Carpinion betuli)*; Poznań: Kontekst, Poland, 1999; pp. 1–245.
30. Bajerlein, D.; Błoszyk, J.; Gwiazdowicz, D.J.; Ptaszyk, J.; Halliday, B. Community structure and dispersal of mites (Acari, Mesostigmata) in nests of the white stork (*Ciconia ciconia*). *Biologia* **2006**, *61*, 525–530. [CrossRef]
31. Gwiazdowicz, D.J.; Kamczyc, J.; Błoszyk, J. The diversity of phoretic Mesostigmata on *Ips typographus* (Coleoptera: Scolytinae) caught in the Karkonosze forest. *Eur. J. Entomol.* **2011**, *108*, 489–491. [CrossRef]
32. Gwiazdowicz, D.J.; Kamczyc, J.; Teodorowicz, E.; Błoszyk, J. Mite communities (Acari, Mesostigmata) associated with *Ips typographus* (Coleoptera, Scolytidae) in managed and natural Norway spruce stands in Central Europe. *Cent. Eur. J. Biol.* **2012**, *7*, 910–916. [CrossRef]
33. Paraschiv, M.; Isaia, G. Disparity of Phoresy in Mesostigmatid Mites upon Their Specific Carrier *Ips typographus* (Coleoptera: Scolytinae). *Insects* **2020**, *11*, 771. [CrossRef] [PubMed]
34. Saito, Y.; Osakabe, M.; Sakagami, Y.; Yasui, Y. A method for preparing permanent specimens of mites with Canada balsam. *Appl. Entomol. Zool.* **1993**, *28*, 59–597. [CrossRef]
35. Krantz, G.W.; Walter, D.E. *A Manual of Acarology*, 3rd ed.; Texas Tech University Press: Lubbock, TX, USA, 2009; p. 807.
36. Kiełczewski, B.; Wiśniewski, J. Bark beetle acarofauna in different types of forest habitat, Part IV: Oribatei. *Bull. Soc. Amis Sci. Lett. Poz.* **1978**, *18*, 119–133.
37. Pernek, M.; Hrasovec, B.; Matosevic, D.; Pilas, I.; Kirisits, T.; Moser, J.C. Phoretic mites of three bark beetles (*Pityokteines* spp.) on Silver fir. *J. Pest Sci.* **2008**, *81*, 35–42. [CrossRef]
38. Pernek, M.; Wirth, S.; Blomquist, S.R.; Avtzis, D.N.; Moser, J.C. New Associations of Phoretic Mites on *Pityokteines curvidens* (Coleoptera, Curculionidae, Scolytinae). *Cent. Eur. J. Biol.* **2012**, *7*, 63–68. [CrossRef]
39. Fernández, M.; Julio, D.; Moraza, M.L. Acarofauna associated with *Ips sexdentatus* in north-west Spain. *Scand. J. For. Res.* **2013**, *28*, 358–362. [CrossRef]
40. Moser, J.C.; Perry, T.J.; Solheim, H. Ascospores hyperphoretic on mites associated with *Ips typographus*. *Mycol. Res.* **1989**, *93*, 513–517. [CrossRef]
41. Takov, D.; Pilarska, D.; Moser, J. Phoretic mites associated with spruce bark beetle *Ips typographus* L. (Curculionidae: Scolytinae) from Bulgaria. *Act. Zool. Bulgar.* **2009**, *61*, 293–296.
42. Kršlak, B.; Zach, P.; Kulfan, J. The role of *Hylastes cunicularius* Erichson (Coleoptera: Scolytidae) in transferring uropodine mites in a mountain spruce forest. *J. For. Sci.* **2010**, *56*, 258–264. [CrossRef]
43. Vrabec, M.; Kalúz, S.; Ferenčík, J. Foretické roztoče na lykožrútovi smrekovom (*Ips typographus*) na vybraných lokalitách vo Vysokých Tatrách. *Entomofauna carpath.* **2012**, *24*, 1–14.
44. Čejka, M.; Holuša, J. Phoretic mites in uni- and bivoltine populations of *Ips typographus*: A 1-year case study. *Turk. J. Zool.* **2014**, *38*, 569–574. [CrossRef]
45. Khaustov, A.A.; Trach, V.A.; Bobylev, A.N. Mites (Acari) phoretic on six-toothed spruce bark beetle, *Pityogenes chalcographus* Linnaeus (Coleoptera: Curculionidae: Scolytinae), in Western Siberia, Russia. *Acarina* **2016**, *24*, 137–151. [CrossRef]
46. Zach, P.; Kršiak, B.; Kulfan, J.; Parák, M.; Kontschán, J. Mites *Trichouropoda* and *Uroobovella* spp. (Uropodoidea) phoretic on bark beetles (Scolytinae): A comparison from a declining mountain spruce forest in Central Europe. *Int. J. Acarol.* **2016**, *42*, 212–217. [CrossRef]
47. Khaustov, A.A.; Klimov, P.B.; Trach, A.V.; Bobylev, A.N.; Salavatulin, V.M.; Khaustov, A.V.; Tolstikov, A.V. Review of mites (Acari) associated with the European spruce bark beetle, *Ips typographus* (Coleoptera: Curculionidae: Scolytinae) in Asian Russia. *Acarina* **2018**, *26*, 3–79. [CrossRef]
48. Gwiazdowicz, D.J. Mesostigmatid mites (Acari) associated with Scolytidae in Poland. In *Selected Problems of Acarological Research in Forests*; Gwiazdowicz, D.J., Ed.; Wydawnictwo Uniwersytetu Przyrodniczego: Poznan, Poland, 2008; pp. 59–95.
49. Moraza, M.L.; Fernandez, M.; Jurc, M. Phoretic mites of the six-spined engraver beetle, *Ips sexdentatus* (Böerner, 1776) (Coleoptera, Scolytinae), on *Pinus halepensis* in Slovenia. *Int. J. Acarol.* **2013**, *39*, 597–599. [CrossRef]
50. Penttinen, R.; Viiri, H.; Moser, J.C. The Mites (Acari) Associated with Bark Beetles in the Koli National Park in Finland. *Acarologia* **2013**, *53*, 3–15. [CrossRef]
51. Poliță, D.; Manu, M.; Marcu, V.M. Relationship among phoretic mites and Norway spruce bark beetles–*Ips typographus* and *Pityogenes chalcographus*. *Rev. pădurilor.* **2016**, *131*, 57–65.
52. Cilbircioğlu, C.; Kovač, M.; Pernek, M. Associations of Phoretic Mites on Bark Beetles of the Genus *Ips* in the Black Sea Mountains of Turkey. *Forests* **2021**, *12*, 516. [CrossRef]

53. Scheucher, R. Systematik und Ökologie der deutschen Anoetinen. In *Beiträge zur Systematik und Ökologie Mitteleuropäischer Acarina: Tyroglyphidae und Tarsonemini*; Stammer, H.J., Ed.; Akademische Verlagsgesllschaft, Geest & Portig K-G: Leipzig, Germany, 1957; Volume 1, pp. 233–384.
54. Lieutier, F. Les acariens associés à *Ips typographus* et *Ips sexdentatus* (Coleoptera: Scolytidae) en région parisienne et les variations de leurs populations au cours du cycle annuel. *Bull. D'écologie* **1978**, *9*, 307–321.
55. Kiełczewski, B.; Wiśniewski, J. Bark Beetle Acarofauna in Different Types of Forest Habitat. Part III. Tarsonemini, Prostigmata, Acaridiae. *Bull. Soc. Amis Sci. Lett. Poz.* **1980**, *20*, 161–175.
56. Moser, J.C.; Eidmann, H.H.; Regnander, J.R. The mites associated with *Ips typographus* in Sweden. *Ann. Zool. Fenn.* **1989**, *55*, 23–27.
57. Schäffer, S.; Koblmüller, S. Unexpected diversity in the host-generalist oribatid mite *Paraleius leontonychus* (Oribatida, Scheloribatidae) phoretic on Palearctic bark beetles. *PeerJ* **2020**, *8*, e9710. [CrossRef]

forests

MDPI

Article

Effective Reduction in Natural Enemy Catches in Pheromone Traps Intended for Monitoring *Orthotomicus erosus* (Coleoptera, Curculionidae)

Milan Pernek [1,*], Tomislav Milas [2], Marta Kovač [1], Nikola Lacković [3], Milan Koren [4] and Boris Hrašovec [5]

1 Croatian Forest Research Institute, Cvjetno naselje 41, 10450 Jastrebarsko, Croatia; martam@sumins.hr
2 Public Institution for Management of Park-Forest Marjan, Cattanijin Put 2, 21000 Split, Croatia; milas.tomislav@hotmail.com
3 Antuna Mihanovića 3, 10450 Jastrebarsko, Croatia; nikolacko@gmail.com
4 Faculty of Forestry, Technical University in Zvolen, T. G. Masaryka 24, 96001 Zvolen, Slovakia; milan.koren@tuzvo.sk
5 Faculty of Forestry, University of Zagreb, Svetošimunska 25, 10000 Zagreb, Croatia; bhrasovec@sumfak.hr
* Correspondence: milanp@sumins.hr

Abstract: Infestations have persisted following a sudden and intense outbreak of the bark beetle *Orthotomicus erosus* along the Croatian coast, necessitating a continuous battle against this pest. A recommended protective action is the utilization of pheromone traps for population surveillance. Previous monitoring efforts have recorded an exceptionally high capture rate of natural enemies using pheromone traps; these traps inadvertently prevented natural enemies from fulfilling their essential role in controlling bark beetle populations. To address and significantly diminish instances of this unintended capture, our study designed a modification to the Theysohn-type pheromone trap by integrating a metal mesh within the trapping container. An experimental setup was established in Marjan Forest Park, situated on a peninsula bordered by the sea on three sides and partly by the city of Split. For monitoring purposes, unmodified standard pheromone traps were deployed at the onset of a significant *O. erosus* outbreak in Croatia in 2018. Catch data from 2020 to 2022 show a marked decrease in the bark beetle population, indicating a shift toward a latent phase. In 2022, modified traps were integrated into the existing monitoring setup, consisting of 10 pairs, to evaluate whether modifications to the traps could significantly reduce the capture of the bark beetle's natural enemies, specifically *Temnoscheila caerulea*, *Thanasimus formicarius*, and *Aulonium ruficorne*. The objective is to offer recommendations for forestry practices on employing pheromone traps with minimal disturbance to the ecological equilibrium. Our findings indicate that the modifications to the traps markedly decreased the capture of natural predators, particularly *T. caerulea*, which was the predominant predatory insect found in the traps. Simultaneously, captures of the target species, all bark beetles in the trap, were marginally reduced. This decrease in the capture rates of the target bark beetle species, *O. erosus*, is not considered problematic when pheromone traps are utilized primarily for monitoring purposes. The modifications to the traps significantly reduced the capture of common bark beetle predators, thereby facilitating a more balanced strategy in forest protection efforts.

Keywords: bark beetle outbreak; mediterranean pine engraver; modification of traps; *Temnoscheila caerulea*; *Thanasimus formicarius*; *Aulonium ruficorne*; *Tomicus destruens*; *Hylurgus miklitzi*

Citation: Pernek, M.; Milas, T.; Kovač, M.; Lacković, N.; Koren, M.; Hrašovec, B. Effective Reduction in Natural Enemy Catches in Pheromone Traps Intended for Monitoring *Orthotomicus erosus* (Coleoptera, Curculionidae). *Forests* **2024**, *15*, 1298. https://doi.org/10.3390/f15081298

Academic Editors: Katarína Pastirčáková and Rostislav Zemek

Received: 23 June 2024
Revised: 17 July 2024
Accepted: 19 July 2024
Published: 25 July 2024

1. Introduction

Bark beetles are considered one of the most damaging groups of insects to forests worldwide [1]. Out of this large group of insects, there are not many that cause huge losses in forestry. For the past few years, outbreaks of the Mediterranean pine engraver (MPE), *Orthotomicus erosus* Woll. (Coleoptera, Curculionidae, Scolytinae) on Aleppo pines

(*Pinus halepensis* Mill.) have become a regular threat in Croatia [2]. Climate change, extreme drought, and secondary bark beetle attacks eventually led to enlarged bark beetle population levels, which then started attacking healthy trees. This increase in bark beetle populations in conifers is well known and documented in European forestry, most commonly for the spruce bark beetle (*Ips typographus* L.) [3–7]. Similar scenarios where MPEs exhibit aggressive behavior and act as primary pests have been documented in Dalmatia, the Southern Mediterranean part of Croatia [2]. Previously, concerning bark beetle attacks in this specific region, only very local outbreaks of *Tomicus piniperda* L. or *T. destruens* Woll. had been recorded [8,9]. The MPE is a reddish-brown beetle that is 2.7 to 3.5 mm long. It tunnels in the living part of the bark; its larvae are legless, white, and about 2.7 to 3.5 mm long; and their appearance does not change as they grow. The eggs are white, partly transparent, and about 1 mm long. The MPE is naturally distributed in Central Asia, the Middle East, Europe, and China. Although it is widespread throughout Europe, so far it has only caused damage in very warm Mediterranean areas. In France, Morocco, and Turkey, two generations have been identified per year, with three to four in South Africa and Tunisia, and three to five in Israel, where adults are active from March to October. Although previous research identified three [10] or four generations [11], more recent studies have shown that even five generations could be in the outbreak site in Marjan (Split, Croatia) [2]. A reason for this could be climate changes, primarily aridification, and a significant extension of the vegetation period in this region, with an unusually warm November and December (Pernek unpublished data). The MPE is also associated with xylophagous fungi, which probably play a role in the colonization of the host trees [12].

MPE attacks, as described in warmer Mediterranean regions [11,13–16], have never occurred before in Croatia [2]. The behavior of pest insects might change because of climate change, so mortality, reproduction, voltinism, and spatial distribution may be factors that favorably affect the pest insect [2,17,18]. Due to the extension of the growing season, the production of several generations above the usual number per year allowed for an exponential population growth and the development of an outbreak that destroyed 50% of the trees in the Marjan Forest Park after 4 years of heavy attack (Pernek unpublished). Control measures include the prompt cutting of infested trees and the use of bait logs, with pheromones used only for monitoring purposes. A mass trapping experiment was conducted in 2019, but it has not shown good results [18].

In general, pheromone traps of various designs are used for insect attraction, mostly for evaluating population densities, i.e., monitoring. Mass trapping is an exception [19–21]. Concerning the design and placement of pheromone traps, there is a large number of studies in the literature, especially for the most researched species of bark beetle: *I. typographus* [22–25]. In designing optimal lures and pheromone traps for the MPE as a target species, Erosowit® (Witasek, Feldkirchen, Austria) has exhibited significantly higher catch rates compared to Pheroprax® (BASF, Ludwigshafen, Germany) and it has also been significantly more selective, while single Theysohn traps (Theysohn, Salzgitter, Germany) have been shown to be an optimal solution [18]. Although traps are designed to target specific insects, such as other bark beetle species, many other species are also attracted. Mass trapping can unintentionally remove high numbers of predators that use bark beetle pheromones as kairomones [19] (Aukema et al., 2000), and such negative side effects have negative impacts [26]. For example, in the pheromone Pheroprax®, components like ipsenol, ipsdienol, and (S)-cis-verbenol are responsible for attracting the antagonist *Thanasimus dubius* Fab. (Coleoptera, Cleridae) (Aukema et al., 2000). Pheromone traps have been used in Marjan to attract the MPE since 2018. Early on, it was apparent that many specimens of *Temnoscheila caerulea* Olivier (Coleoptera, Trogossitidae), an important predator of bark beetles, were present along with the target organism. This is concerning [19], because four or more generations of MPE have been detected each year [2,27], and a high capture rate of natural enemies should be avoided even if the traps are intended for monitoring or mass trapping. According to Martin et al. [28], the number of catches of *T. caerulea* could be reduced by modifying the pheromone traps.

Pheromone traps have been used for several years to capture MPEs. Throughout the years, the capture of non-target entomofauna, particularly predators, has been evident and identified as a major problem. The predatory species remain trapped and feed on the catches of other insects. Sometimes, in specific periods, the number of predators has been approximately equal to the number of MPEs, which is a major drawback of the pheromone trap. This was particularly evident when a large number of pheromone traps were used in an attempt to reduce the number of MPEs. Therefore, developing a way to reduce the number of non-target species has been necessary, especially concerning predatory and entomofauna captures. The use of protective nets in pheromone traps is not new, and their use has often proven effective for several bark beetle species. However, to our knowledge, there have been no such studies for MPEs. Now that the MPE has established itself in Croatia as a pest with the potential to build populations to levels that cause damage over large areas, this research should serve as a basis for future integrated forest protection. When used against the MPE, pheromone traps baited with an Erosowit® lure catch large numbers of natural enemies; hence, the aims of this work were to (i) monitor the population dynamics of the MPE and (ii) improve trapping protocols in order to decrease unintentional captures.

2. Materials and Methods

2.1. Field Work

This study was carried out between 2020 and 2022 in the public forest (forest park) of Marjan, close to the city of Split, in a 200 ha Allepo pine stand. Marjan is a hill peninsula that was declared a forest park in 1976 and is located just a couple of hundred meters from Diocletian's palace, which is in the center of Split. It is of great importance to the citizens of Split as a recreational area and historical site. The whole protected area of Marjan is ca. 300 ha in size, with forest making up approximately two-thirds of it. Throughout history, it was covered with holm oak and manna ash forest (*Fraxino orni, Quercetum ilicis*) that has gradually degraded. Today, it is mostly dominated by Aleppo pine. Geographically, Marjan is a peninsula, surrounded by the sea and connected to the mainland by a narrow strip on which the city of Split continues. Because the influence of the bark beetles' arrival from the outside is almost negligible, this field experiment could be considered an open-air laboratory experiment.

Since 2017, MPE attacks have occurred regularly in the Marjan Forest Park [2]. Some parts of the forest have a greater proportion of deciduous trees, mostly holm oak (*Quercus ilex* L.), and these parts were avoided when installing the traps. Pheromone traps were set in early spring at the end of March 2020, 2021, and 2022, so the first sampling was always conducted on the last week of March. In total, 10 standard Theysohn traps were set each year, separated from each other by at least 100 m (Figure 1). In 2022, 10 modified traps were added as a pairing to the 10 standard traps (Figure 1). The modification was to avoid the capture of non-targeted insects. For this purpose, a special steel mesh with a 6 mm mesh screen was developed from scratch and installed on the collection container (Figure 2). The trap pairs were placed 20 m apart and at least 20 m from a healthy pine tree.

The catches in the pheromone traps were collected mainly from 7 to 14 days (9 days on average) depending on the weather conditions. The pheromone ampoules Erosowit® were changed five times a year each time the liquid ran low, which was temperature-dependent. The catches were transferred into plastic trays, mixed with ethanol (70%), and labeled. Then, they were transported to the laboratory (Croatian Forest Research Institute, Jastrebarsko, Croatia) and examined under a microscope (Olympus BX2, Tokyo, Japan) after 2 weeks maximum.

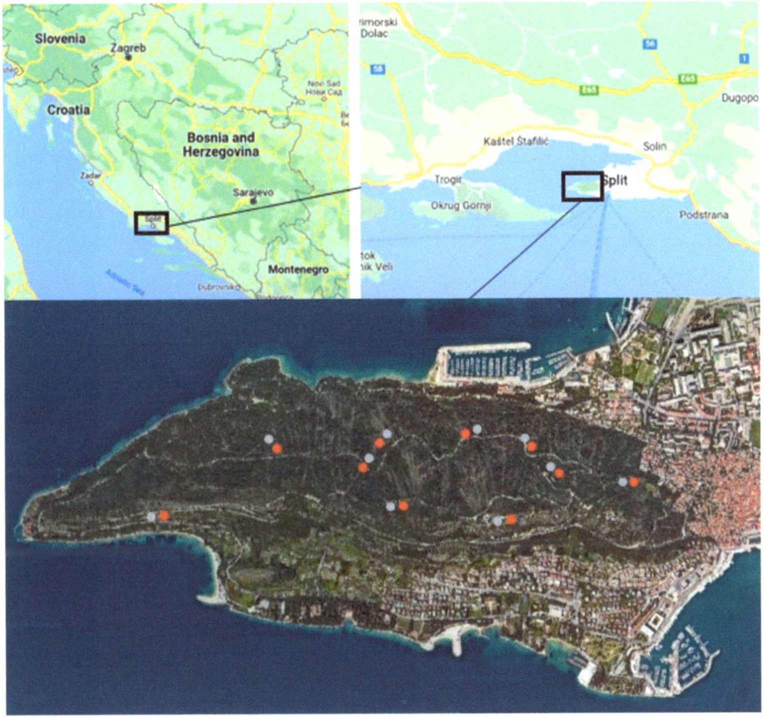

Figure 1. Study area with trap positions (gray dots: standard Theysohn traps installed in 2020; red dots: modified Theysohn traps installed in 2022).

Figure 2. Modification with a steel mesh with a 6 mm screen.

2.2. Laboratory Work

A catch analysis was conducted on the pheromone traps in the laboratory for entomological research at the Croatian Forest Research Institute (Jastrebarsko, Croatia). The laboratory processing involved drying the insects at room temperature and sorting the species under a microscope. The insects were first sorted by taxonomic categories and dried to facilitate counting. Based on the taxonomic categories, the insects were identified using the available morphological keys [29]. The counting of the sorted catch was conducted manually. All larger insects, such as longhorn beetles, beetles with equally sized wings, and natural enemies, were separated and counted by hand in separate Petri dishes.

2.3. Statistical Analysis

A statistical analysis was conducted using R software version 4.2.2 [30]. To assess the statistical significance of the differences in the average number of beetles captured by the standard and modified pheromone traps, we employed the permutation paired *t*-test from the RVAide Memoire package version 0.9-81-2. Permutation methods were chosen due to their numerous benefits over conventional statistical tests. These methods excel in handling multiple comparisons and reducing the Type I errors that arise when numerous hypotheses are tested simultaneously. Unlike traditional tests, permutation tests do not presume any specific population distribution, making them versatile across various data types, including non-normal and non-parametric datasets. They depend on the permutations of the data rather than the actual data values, rendering them more resilient to outliers and other forms of data anomalies.

The effectiveness of the standard and modified traps was gauged by the proportion of captured species. The statistical significance of the differences between the proportions of beetle species captured in the standard and modified traps was assessed using the two-sample test for the equality of proportions, with a continuity correction.

3. Results

Between 2020 and 2022, catches were counted in all the traps to assess the population status of bark beetles. By adding modified traps in 2022, attention was focused on non-selective capture, i.e., the results obtained show to what extent, and to which predatory insects, the trap modification is beneficial. Ultimately, other targeted species of entomofauna were observed in the regular and modified pheromone traps (Table S1).

3.1. Catches of Target Organism MPE

MPEs caught in pheromone traps between 2020 and 2022 were compared with previous years [18]; the catches were constantly decreasing, indicating a population decline. Additionally, the catches between 2020 and 2022 show the collapse of the outbreak very clearly (Figure 3).

3.2. Efficiency of Modified Traps

In 2022, we recorded the number of beetles caught in standard and modified traps (Table 1). We did not use the sample from trap pair #3 on 25 April 2022 for the evaluation, because the standard trap was damaged and the number of beetles could not be determined.

Although the pairs of standard and modified traps were placed close to each other and under similar conditions, the number of beetles captured was different (Figure 4). The average number of beetles captured in the standard traps was statistically greater than the average number of beetles captured in the modified traps (permutation paired *t*-test, $p = 0.03$).

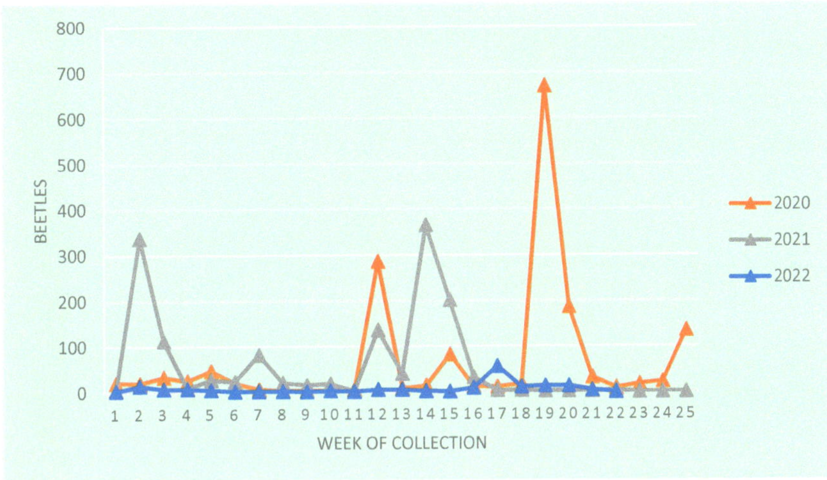

Figure 3. Average catches of *Orthotomicus erosus* in standard pheromone traps 2020–2022 (the first week of each sampling year always began on the last week of March; the sampling was conducted every 9–14 days during the year).

Table 1. Number of beetles caught in standard and modified traps during the entire 2022 season.

Species	Standard Traps				Modified Traps				Total
	Total	Max	Mean	St. Dev.	Total	Max	Mean	St. Dev.	
Orthotomicus erosus	1645	472	164.5	137.4	1017	242	101.7	65.6	2662
Tomicus destruens	176	32	17.6	7.8	135	34	13.5	9.4	311
Hylurgus miklitzi	26,370	5622	2637.0	1543.9	21,051	4119	2105.1	1412.4	47,421
Other Scolytinae	152	67	15.2	19.3	366	157	36.6	46.6	518
Temnoscheila caerulea	874	188	87.4	48.1	486	125	48.6	32.5	1360
Thanasimus formicarius	60	23	6.0	6.8	22	7	2.2	2.0	82
Aulonium ruficorne	111	30	11.1	10.0	70	46	7.0	13.8	181
Monochamus galloprovincialis	30	14	3.0	4.1	1	1	0.1	0.3	31
Buprestidae	446	220	44.6	64.1	33	15	3.3	4.6	479
Other	229	51	22.9	16.7	104	18	10.4	5.0	333
All species	30,093	5970	3000.9	1666.3	23,285	4333	2328.5	1484.7	53,378

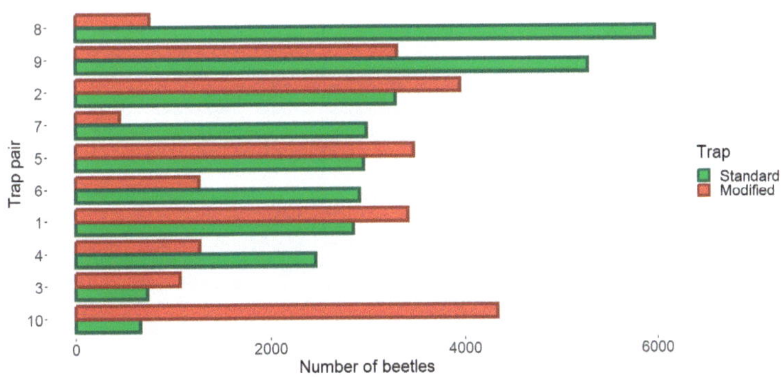

Figure 4. Total number of beetles caught in standard and modified traps.

The largest differences in the number of beetles captured in the standard and modified traps were found in trap pairs #7 and #10. In both cases, several times more beetles were caught in one type of trap than in the other during the summer (Figures 5 and 6).

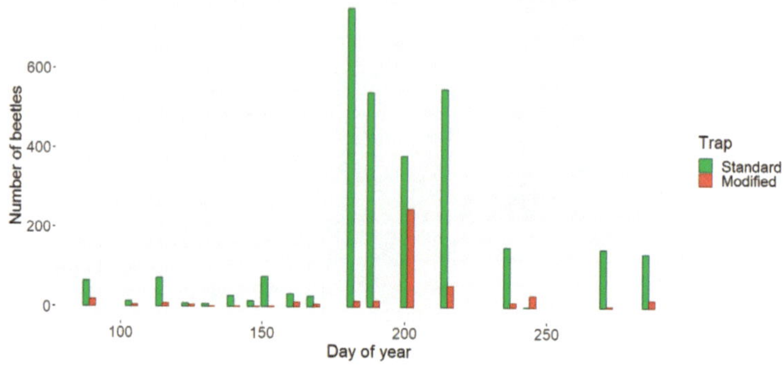

Figure 5. Number of beetles caught in trap pair #7.

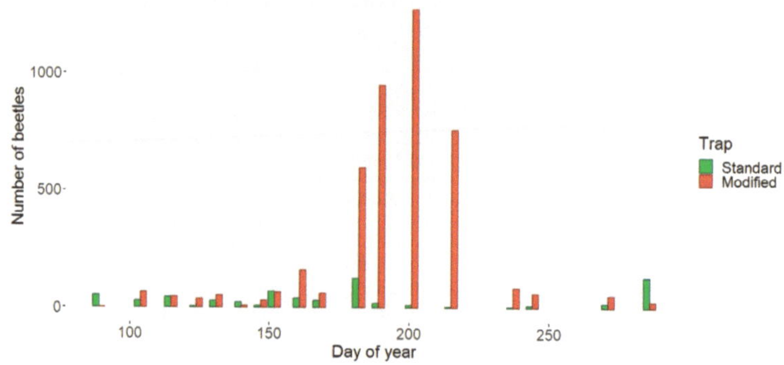

Figure 6. Number of beetles caught in trap pair #10.

According to our observations, the efficiency of the modified traps was about 25% lower than that of the standard traps. We can see that fewer beetles of each species, except for *Scolytinae*, were caught in the modified traps than in the standard traps (Figure 7).

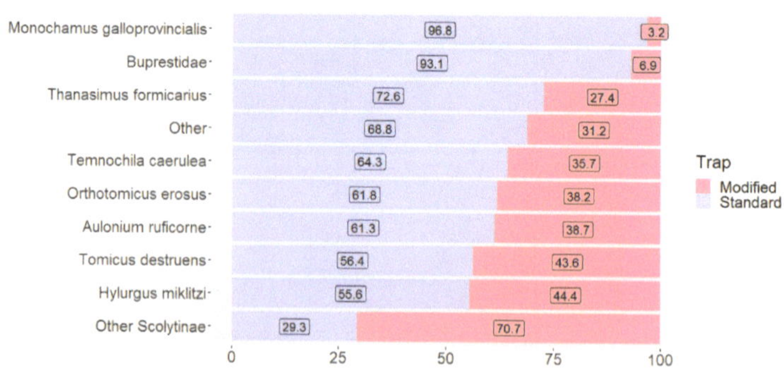

Figure 7. Percentage of beetle species caught in standard and modified traps.

The average number of *O. erosus* captured in the standard traps was statistically higher than the average number captured in the modified traps (Table 1), as determined by the permutation paired *t*-test ($p = 0.004$). The substantial disparity in MPEs captured between the standard and modified traps is primarily attributed to traps #7 and #9 (Figures 8–11). For both pairs of traps, the peak number of beetles captured occurred on the 237th day of the year (25 August 2022).

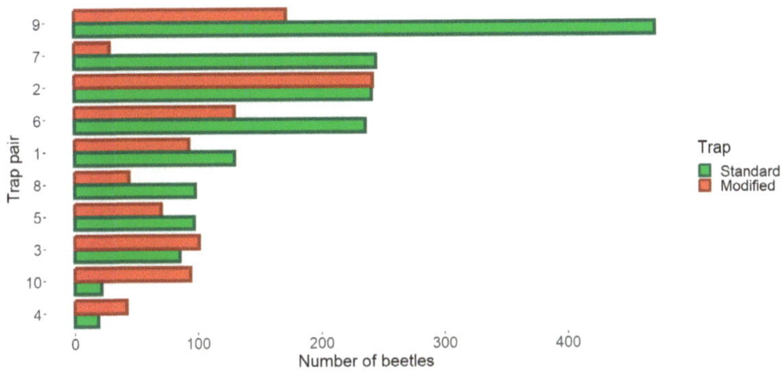

Figure 8. *Orthotomicus erosus* by trap pair.

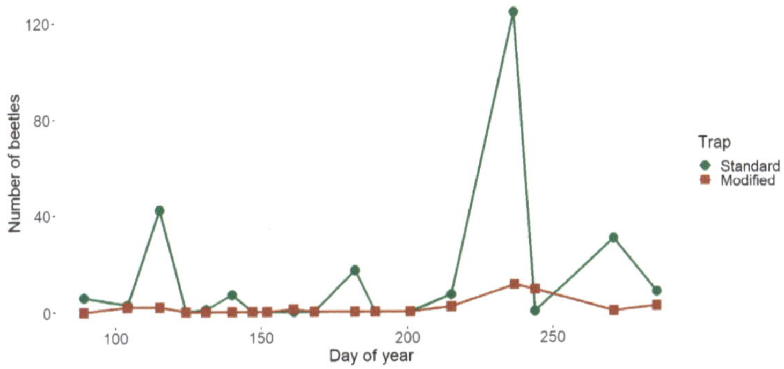

Figure 9. *Orthotomicus erosus* captured in trap #7.

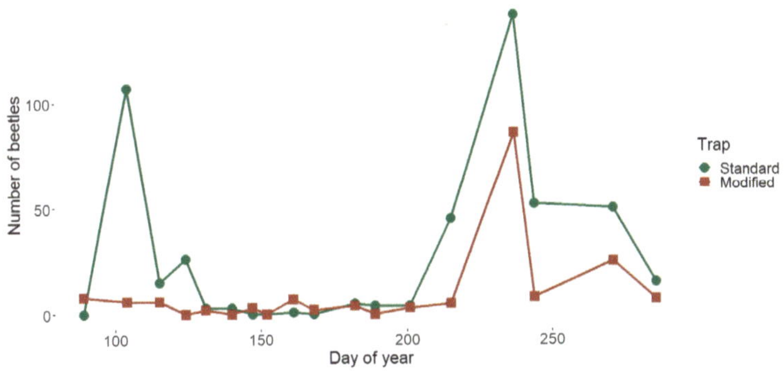

Figure 10. *Orthotomicus erosus* captured in trap #9.

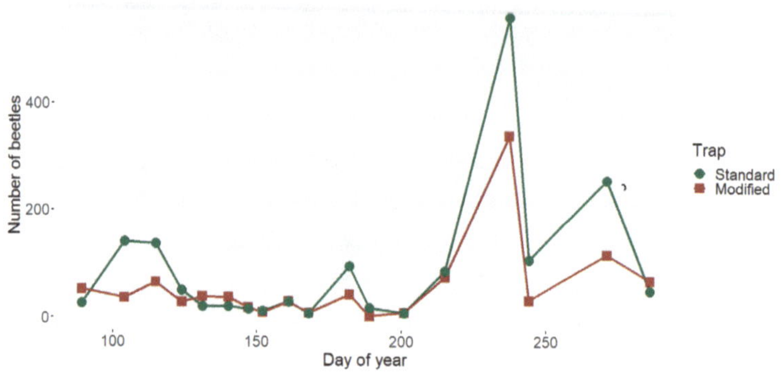

Figure 11. *Orthotomicus erosus* caught in standard and modified traps by day of year.

3.3. Effectiveness of Modified Traps

The main objective of this study was to develop and test modified traps that reduce the risk of trapping natural enemies of bark beetles. Therefore, in the second part of this study, we evaluated the effectiveness of the modified traps in separating bark beetles, their natural enemies, and other non-target species. We expressed the effectiveness of the traps by the proportions of target, non-target, and predatory species captured (Table 2). Consistent with the intent of the modified trap design, the proportion of beetle predators and non-target species captured decreased (Table 2). The proportion of predatory species captured in the modified traps decreased by 1% compared to standard traps. However, the overall efficiency of the modified traps was lower than that of the standard traps. This lower efficiency can be compensated for by placing additional modified traps. For the same anticipated total number of beetles captured, approximately 30% fewer predatory beetles will be captured in the modified traps than in the standard traps. The decrease in the proportion of non-target species was even more significant. For the same projected number of beetles captured, up to four times fewer non-target beetles are expected in the modified traps than in the standard traps. These decreases in the proportions of predator and non-target beetles in the modified traps were statistically significant.

Table 2. Percentage of beetle species captured in standard and modified traps.

	Species	Percentage of Beetle Species in Traps (%)				
		Standard	Modified	L_L	U_L	*p*
Targetspecies	*Orthotomicus erosus*	5.5	4.4	0.73	1.47	$<10^{-8}$
	Tomicus destruens	0.6	0.6	−0.13	1.39	0.98
	Hylurgus miklitzi	87.6	90.4	−3.31	−2.24	$<10^{-15}$
	Other Scolytinae	0.5	1.6	−1.25	−0.88	$<10^{-15}$
	Total	94.2	96.9	−3.10	−2.39	$<10^{-15}$
Predatory species	*Temnoscheila caerulea*	2.9	2.1	0.55	1.08	$<10^{-8}$
	Thanasimus formicarius	0.2	0.1	0.04	0.17	<0.01
	Aulonium ruficorne	0.4	0.3	−0.03	1.70	0.20
	Total	3.5	2.5	0.70	1.28	$<10^{-10}$
Non-target species	*Monochamus galloprovincialis*	0.1	0.0	0.05	0.14	$<10^{-4}$
	Buprestidae	1.5	0.1	1.19	1.49	$<10^{-15}$
	Other	0.8	0.4	0.18	0.45	$<10^{-5}$
	Total	2.3	0.6	1.55	1.95	$<10^{-15}$

L_L—the lower limit of the 95% confidence interval; U_L—the upper limit of the 95% confidence interval; *p*—*p*-value of the proportionality test.

4. Discussion

Although the catch numbers in different types of traps may vary significantly, they still provide information on the population status of bark beetles, indicating that the number of individuals caught is not as important [19]. Generally, pheromone traps do not have a curative function [19,31]. However, the aim is to catch a larger number of target insects, thereby excluding them from the population. However, natural enemies are extremely important in regulating bark beetle population dynamics [32,33]. They contribute to self-sustaining biological control over large areas and long timeframes [34,35]. The selectivity of the pheromone trap is therefore much more important, because by excluding predators from the population, their positive ecological impact is absent [36]. In the extreme case, the number of predators caught and removed from the population is so high that, considering their potential food consumption, we may cause more harm than good. The catch of predators may be so high that the number of bark beetles that would be destroyed in nature becomes significantly smaller than in traps [37]. Furthermore, the removal of predators captured through mass trapping may prolong bark beetle outbreaks [38].

The number of MPE catches in 2022 is negligible compared to catches from previous years, representing the bark beetle outbreak's entry into latency. This is somewhat expected because of the high mortality rate of pine trees caused by the large-scale MPE attacks in previous years, where large numbers of larvae were found under bark. Due to competition within the species (intraspecific competition), the bark beetles do not have the space for normal development anymore [39]. This results in a reduction in their numbers and a decrease in brood productivity. This regulatory mechanism has been shown to be powerful in other species of bark beetles [40]. Additionally, their natural enemies have strengthened and created a strong barrier to the MPE population's resurgence.

Previously used pheromones and traps were used in accordance with the literature [13] or based on research or experience gained in the first years of the outbreak [2]. In this study, the difference in the catches of target insects (MPEs) and non-target insects (predators) was analyzed to provide advice for future monitoring. By placing traps in the same location year after year, we can gain insight into population movements or trends that may help experts implement protection measures correctly and in a timely manner. This study compared pheromones and traps for the capture of MPEs, representing the first such research in Croatia, and the results should be valuable to forestry professionals in the field.

Our results show that the average number of *O. erosus* captured in the standard traps is statistically higher than that in the modified traps (Table 1), as determined by the permutation paired *t*-test ($p = 0.004$). The substantial disparity in MPEs captured between the standard and modified traps is primarily attributed to traps #7 and #9 (Figures 8–11). For both pairs of traps, the peak number of beetles captured occurred on the 237th day of the year (25 August 2022).

The difference in the number of MPEs in the standard trap and modified trap #7 is less extreme. However, the trend in the number of MPEs in modified trap #7 does not match that in standard trap #7. This may be explained by the placement of modified trap #7.

Consistent with the intent of the modified trap design, the proportion of beetle predators and non-target species captured decreased. The 1% reduction in the proportion of predatory species in the standard traps indicates that, given the same total number of beetles caught, the modified traps captured about 30% fewer predatory beetles compared to the standard traps. The catch data from the last three years of research show a clear declining trend for the outbreak. The catches of both target and non-target entomofauna should be viewed in this context. Predator catches were much higher in the previous two years, which could be observed during the emptying of each trap; however, the data were not quantified. However, the predator–bark beetle ratio is likely even more unfavorable than our data suggest.

If we look at the non-target species as a whole, the decrease in proportion was even larger. For the same number of all beetles captured, up to four times fewer beetles of non-target species were captured in the modified traps than in the standard traps. The decreases in the proportion of beetles from predators and non-target species in the modified traps were statistically significant.

The natural enemy *T. caerulea* completely follows the population dynamics of the Mediterranean bark beetle. The data obtained are consistent with the literature [41], but they provide completely new insights into the number of generations and the first occurrences of these species during the year.

When reaching a high population density, MPEs can overwhelm the host defenses, thus causing tree mortality [42]. The MPE can be a serious pest in semi-arid regions, as confirmed by findings on living trees [27], because it is capable of colonizing living trees using the host exhaustion strategy, whereby employing aggregation pheromones attracts other individuals that launch large-scale attacks on the trees [17]. Concerning *T. destruens*, this species is attracted by terpenes emitted from wounds in trees attacked by pioneer bark beetles. This could be the reason why the number of catches of *T. destruens* in our study was relatively low compared to MPE or *H. miklitzi*.

Data on more effective pheromone traps will be valuable for forestry practice and open the possibility of additional innovations for reducing the capture of non-target organisms. Likewise, future research on modifications to pheromone traps should consider other species of harmful bark beetles.

5. Conclusions

In this study, the difference in catches of target insects (MPE) and non-target insects (predators) was analyzed to provide advice for future monitoring. The results showed that the average number of *O. erosus* captured in the standard traps is similar to the ones in the modified traps, while the number of captured beetles of non-target species were up to four times fewer in the modified traps than in the standard traps. The selectivity of the pheromone trap is therefore much more important, because by excluding predators from the population, their positive ecological impact is absent. This study represents the first such research in Croatia, and the results should be valuable to forestry professionals in the field, especially in protected areas where treatment options are limited. Furthermore, it should be kept in mind that the selectivity might change depending on the outbreak phase, and additional research is needed to provide more information about to what extent it is important to modify the pheromone trap.

Supplementary Materials: The following supporting information can be downloaded at: https://www.mdpi.com/article/10.3390/f15081298/s1, Table S1: Total catches of insects by date in 10 traps set in the Marjan Forest Park in 2022.

Author Contributions: Conceptualization and methodology, M.P., M.K. (Marta Kovač) and N.L.; investigation, M.P., N.L. and T.M.; writing—original draft preparation, M.P., T.M. and B.H.; writing—review and editing, M.P., M.K. (Milan Koren), T.M. and B.H.; visualization, M.P. All authors have read and agreed to the published version of the manuscript.

Funding: This research was partly funded by the city of Split, under the research project APSM 2.2, "Population dynamics of the Mediterranean pine engraver".

Data Availability Statement: The data presented in this study are available on request from the corresponding author.

Acknowledgments: We thank Blaženka Ercegovac for her laboratory work in counting beetles and Andrija Jukić and Tomislav Krcivoj for the sampling of the beetle specimens caught in the traps.

Conflicts of Interest: Nikola Lacković is employed by the company Arbofield Ltd., Jastrebarsko. The remaining authors declare that the research was conducted in the absence of any commercial or financial relationships that could be construed as potential conflicts of interest.

References

1. Raffa, K.F.; Gregoire, J.C.; Lindgren, B.S. Natural history and ecology of bark beetles. In *Bark beetles: Biology and Ecology of Native and Invasive Species*; Vega, F.E., Hofstetter, R.W., Eds.; Academic Press: Cambridge, MA, USA, 2015; pp. 1–28.
2. Pernek, M.; Lacković, N.; Lukić, I.; Zorić, N.; Matošević, D. Outbreak of *Orthotomicus erosus* (Coleoptera, Curculionidae) on Aleppo Pine in the Mediterranean Region in Croatia. *South-East Eur. For.* **2019**, *10*, 19–27. [CrossRef]
3. Hlásny, T.; Mátyás, C.; Seidl, R.; Kulla, L.; Merganicová, K.; Trombik, J.; Dobor, L.; Barcza, Z.; Konôpka, B. Climate change increases the drought risk in Central European forests: What are the options for adaptation? *For. J.* **2014**, *60*, 5–18. [CrossRef]

4. Hlávková, D.; Doležal, P. Cambioxylophagous Pests of Scots Pine: Ecological Physiology of European Populations—A Review. *Front. For. Glob. Change* **2022**, *5*, 864651. [CrossRef]
5. Hroššo, B.; Mezei, P.; Potterf, M.; Majdák, A.; Blaženec, M.; Korolyova, N.; Jakuš, R. Drivers of Spruce Bark Beetle (*Ips typographus*) infestations on downed trees after severe windthrow. *Forests* **2020**, *11*, 1290. [CrossRef]
6. Vilardo, G.; Faccoli, M.; Corley, J.C.; Lantschner, V.M. Factors driving historic intercontinental invasions of European pine bark beetles. *Biol. Invasions* **2022**, *24*, 2973–2991. [CrossRef]
7. Wermelinger, B. Development and distribution of predators and parasitoids during two consecutive years of an *Ips typographus* (Col., Scolytidae) infestation. *J. Appl. Entomol.* **2002**, *126*, 521–527. [CrossRef]
8. Pernek, M.; Hrašovec, B.; Županić, M. Beetle fauna captured in traps baited with Tomicus piniperda pheromone blends in a pine stand in central Croatia. In Proceedings of the Ecology, Survey and Management of Forest Insects, Krakow, Poland, 1–5 September 2002; Gen. Tech. Rep. NE-311; McManus, M.L., Liebhold, A.M., Eds.; U.S. Department of Agriculture, Forest Service, Northeastern Research Station: Newtown Square, PA, USA, 2003; pp. 87–89.
9. Hrašovec, B.; Harapin, M.; Pernek, M. Entomological complex of mediterranean forests. In *Forests of the Croatian Mediterranean forests*; Matić, S., Ed.; Academy of Forestry Sciences: Zagreb, Croatia, 2011; pp. 556–572.
10. Lieutier, F.; Ghaioule, D.; Yart, A.; Sauvard, D. Attack behavior of pine bark beetles in Morocco and association with phytopathogenic fungi. *Ann. Rech. For. Maroc.* **2002**, *35*, 96–109.
11. Mendel, Z. Hymenopterous parasitoids of bark beetles [Scolytidae] in Israel: Host relation, host plant, abundance and seasonal history. *Entomophaga* **1986**, *31*, 113–125. [CrossRef]
12. Kovač, M.; Rigling, D.; Pernek, M. Ophiostomatales Associated with Mediterranean Pine Engraver, *Orthotomicus erosus* (Coleoptera, Curculionidae) in Dalmatia, Croatia. *J. Fungi* **2022**, *8*, 788. [CrossRef]
13. Mendel, Z. The relation of bast scale and bark beetle outbreaks to management of pine plantations in Israel. In *XVII International Congress of Entomology, Proceedings of the IUFRO Working Party, Vancouver, BC, Canada, 4 July 1988*; Payne, T.L., Saarenmaa, H., Eds.; Polytechnic Institute and State University: Blacksburg, VA, USA, 1988; pp. 329–335.
14. Sarikaya, O.; Avci, M. Distribution and biology of the Mediterranean pine shoot beetle *Tomicus destruens* (Wollaston, 1985) in the western Mediterranean region of Turkey. *Turk. J. Entomol.* **2010**, *34*, 289–298.
15. Ben Jamaa, M.L.; Lieutier, F.; Yart, A.; Jerraya, A.; Khouja, M.L. The virulence of phytopathogenic fungi associated with the bark beetles *Tomicus piniperda* and *Orthotomicus erosus* in Tunisia. *Forest. Pathol.* **2007**, *37*, 51–63. [CrossRef]
16. Battisti, A.; Larsson, S. Climate change and insect distribution range. In *Climate Change and Insect Pests*; Björkman, C., Niemelä, P., Eds.; CABI: Wallingford, UK, 2016; pp. 1–15.
17. Lieutier, F.; Paine, T.D. Responses of Mediterranean Forest Phytophagous Insects to Climate Change. In *Insects and Diseases of Mediterranean Forest Systems*; Paine, D.T., Lieutier, F., Eds.; Springer: Basel, Switzerland, 2016; pp. 801–858.
18. Pernek, M.; Kovač, M.; Lacković, L. Testing of biological effectiveness of pheromones and traps for catch of Mediterranean bark beetle *Orthotomicus erosus* (Coleoptera, Curculionidae). *Šum. List* **2020**, *144*, 339–350. [CrossRef]
19. Aukema, B.H.; Dahlsten, D.L.; Raffa, K.F. Exploiting behavioral disparities among predators and prey to selectively remove pests: Maximizing the ratio of bark beetles to predators removed during semiochemieally based trap-out. *Environ. Entomol.* **2000**, *29*, 651–660. [CrossRef]
20. Bracalini, M.; Croci, F.; Ciardi, E.; Mannucci, G.; Papucci, E.; Gestri, G.; Tiberi, R.; Panzavolta, T. *Ips sexdentatus* Mass-Trapping: Mitigation of Its Negative Effects on Saproxylic Beetles Larger Than the Target. *Forests* **2021**, *12*, 175. [CrossRef]
21. Faccoli, M.; Stergulc, F. Damage reduction and performance of mass trapping devices for forest protection against the spruce bark beetle, *Ips typographus* (Coleoptera Curculionidae Scolytinae). *Ann. For. Sci.* **2008**, *65*, 309. [CrossRef]
22. Holuša, J.; Hlásny, T.; Modlinger, R.; Lukášová, K.; Kula, E. Felled trap trees as the traditional method for bark beetle control: Can the trapping performance be increased? *For. Ecol. Manag.* **2017**, *404*, 165–173. [CrossRef]
23. Blaženec, M.; Majdák, A.; Jakuš, R. Improvement of Ips typographus catches in pheromone trap barriers by altering of sex assigned pheromone blends. *Folia Oecologica* **2021**, *48*, 25–34. [CrossRef]
24. Heber, T.; Helbig, C.E.; Osmers, S.; Müller, M.G. Evaluation of attractant composition, application rate, and trap type for potential mass trapping of *Ips typographus* (L.). *Forests* **2021**, *12*, 1727. [CrossRef]
25. Lindmark, M.; Wallin, E.A.; Jonsson, B.G. Protecting forest edges using trap logs–Limited effects of associated push-pull strategies targeting *Ips typographus*. *For. Ecol. Manag.* **2022**, *505*, 119886. [CrossRef]
26. Pavlin, R. Problem selektivnosti sintetičnih feromonov. *Zb. Lesar.* **1991**, *38*, 126–160.
27. Moumeni, L.; Gastebois, A.; Gillmann, L.; Papon, N.; Benia, F.; Bouchara, J.-P.; Bounechada, M. Investigating the prevalence of Bark beetles of *Pinus halepensis* in the North East semi-arid region of Algeria. *Biodiversitas* **2021**, *22*, 2755–2762. [CrossRef]
28. Martín, A.; Etxebeste, I.; Pérez, G.; Álvarez, G.; Sánchez, E.; Pajares, J. Modified pheromone traps help reduce bycatch of bark-beetle natural enemies. *Agric. For. Entomol.* **2013**, *15*, 86–97. [CrossRef]
29. Pfeffer, A. *Zentral- und Westpaläarktische Borken- und Kernkäfer*; Naturhistorisches Museum: Basel, Switzerland, 1995.
30. R Core Team. *A Language and Environment for Statistical Computing*; R Foundation for Statistical Computing: Vienna, Austria, 2021.
31. Kuhn, A.; Hautier, L.; San Martin, G. Do pheromone traps help to reduce new attacks of *Ips typographus* at the local scale after a sanitary cut? *PeerJ* **2022**, *10*, e14093. [CrossRef] [PubMed]

32. Raffa, K.F. Differential responses among natural enemies and prey to bark beetle pheromones: Implications of chemical, temporal, and spatial disparities to evolutionary theory and pest management. In *Behavior, Population Dynamics and Control of Forest Insects: Proceedings of a Joint IUFRO Working Party Conference, Maui, Hawaii, 6–11 February 1994*; Hain, F., Salom, S.M., Ravlin, W.F., Payne, T., Raffa, K.F., Eds.; Ohio State University, Ohio Agricultural Research and Development Center: Columbus, OH, USA, 1995; pp. 208–225.
33. Reeve, J.D.; Ayres, M.P.; Lorio, P.L., Jr. Host suitability, predation, and bark beetle population dynamics. In *Population Dynamics: New Approaches and Syntheses*; Cappuccino, N., Price, P.W., Eds.; Academica: San Diego, CA, USA, 1995; pp. 339–357.
34. Dahlsten, D.L.; Whitmore, M.C. The case for and against the biological control of bark beetles. In *Potential for Biological Control of Dendroctonus and Ips Bark Beetles*; Kulhavy, D.L., Miller, M.C., Eds.; School of Forestry, Stephen F. Austin State University: Nacogdoches, TX, USA, 1989; pp. 3–20.
35. Miller, M.C.; Moser, J.C.; McGregor, M.; Gregoire, J.-C.; Baisier, M.; Dahlsten, D.L.; Wrner, R.A. Potential for biological control of native North American *Dendroctonus* beetles (Coleoptera: Scolytidae). *Ann. Entomol. Soc. Am.* **1987**, *80*, 417–428. [CrossRef]
36. Weslien, J.; Regnander, J. The influence of natural enemies on brood production in *Ips typographus* (Col. Scolytidae) with special reference to egg-laying and predation by *Thanasimus formicarius* (Col.: Cleridae). *Entomophaga* **1992**, *37*, 333–342. [CrossRef]
37. Baier, P. Untersuchungen zur abundanzdynamischen Relevaz der Beifänge von *Nemosoma elongatum* L. (Col., Ostomidae) in Chalcoprax beköderten Flugbarierfallen für *Pityogenes chalcographus* L. (Col., Scolytidae). *J. Appl. Ent.* **1994**, *117*, 51–57. [CrossRef]
38. DeMars, C.J.; Dahlsten, D.L.; Sharpnack, N.X.; Rowney, D.L. Tree utilization and density of attacking and emerging populations of the western pine beetle (Coleoptera: Scolytidae) and its natural enemies, Bass Lake, California, 1970-1971. *Can. Entomol.* **1986**, *118*, 881–900. [CrossRef]
39. Grosmann, D.M.; Salon, S.M.; Rowlin, F.W.; Young, R.W. Geographic and gender differences in semiochemicals in emerging adult southern pine beetle (Coleoptera, Scolytidae). *Ann. Ent. Soc Am.* **1986**, *90*, 438–446. [CrossRef]
40. Vanická, H.; Holuša, J.; Resnerová, K.; Ferenčík, J.; Potterf, M.; Véle, A.; Grodzki, W. Interventions have limited effects on the population dynamics of *Ips typographus* and its natural enemies in the Western Carpathians (Central Europe). *For. Ecol. Manag.* **2020**, *470–471*, 118209. [CrossRef]
41. Rando, R.M. Studio de las curvas de vuelo de *Ips sexdentatus* y de sus principales predatores en diversos montes de la provincia de Teruel en 2008 y 2009. Doctoral Dissertation, Universitat Politècnica de València, Valencia, Spain, 2011.
42. Faccoli, M.; Gallego, D.; Branco, M.; Brockerhoff, E.G.; Corley, J.; Coyle, D.R.; Hurley, B.P.; Jactel, H.; Lakatos, F.; Lantschner, V.; et al. A first worldwide multispecies survey of invasive Mediterranean pine bark beetles (Coleoptera: Curculionidae, Scolytinae). *Biol. Invasions* **2020**, *22*, 1785–1799. [CrossRef]

forests

MDPI

Article

Changes in the Parasitism Rate and Parasitoid Community Structure of the Horse Chestnut Leafminer, *Cameraria ohridella* (Lepidoptera: Gracillariidae), in the Czech Republic

Lubomír Volter [1], Eva Prenerová [2], František Weyda [3] and Rostislav Zemek [2,*]

[1] Česká Bříza 216, 33011 Česká Bříza, Czech Republic; lvolter@post.cz
[2] Institute of Entomology, Biology Centre CAS, 37005 České Budějovice, Czech Republic; eva.prenerova@seznam.cz
[3] Department of Medical Biology, Faculty of Science, University of South Bohemia, 37005 České Budějovice, Czech Republic; weydafk@seznam.cz
* Correspondence: rosta@entu.cas.cz

Abstract: The horse chestnut leafminer, *Cameraria ohridella*, Deschka and Dimić, is a moth that has invaded most of Europe since it was first recorded in Macedonia near Lake Ohrid in 1985. It attacks horse chestnut trees and causes aesthetic and vitality problems. The parasitism rate, other mortality rates, and parasitoid structure were studied during a five-year survey at six sites in the Czech Republic. The results showed that the total parasitism rates varied from 1.9% to 20.5%, with an average of 7.2%, similar to other those published studies. The parasitism rate was significantly related to year, the developmental stage of *C. ohridella*, latitude, and greenery maintenance but not to *C. ohridella* population density, altitude, or area size. In contrast, the total other mortality rates varied from 13.7% to 59.5%, with an average of 31%, but overall temporal changes in the values indicated a declining trend. The parasitoid complex was predominantly polyphagous parasitoids of the family Eulophidae, similar to that found previously in south-eastern Europe. The results further revealed that the most abundant parasitoid species, *Minotetrastichus frontalis* (Nees), was gradually replaced by *Pediobius saulius* (Walker). The increasing abundance of *P. saulius* is thus an interesting adaptation of an autochthonous parasitoid to a new host.

Keywords: *Aesculus hippocastanum*; invasive pest; natural enemies; Eulophidae; parasitoids; parasitism rate; mortality rate

Citation: Volter, L.; Prenerová, E.; Weyda, F.; Zemek, R. Changes in the Parasitism Rate and Parasitoid Community Structure of the Horse Chestnut Leafminer, *Cameraria ohridella* (Lepidoptera: Gracillariidae), in the Czech Republic. *Forests* **2022**, *14*, 885. https://doi.org/10.3390/f13060885

Academic Editor: Celia K. Boone

Received: 31 March 2022
Accepted: 6 June 2022
Published: 7 June 2022

1. Introduction

The horse chestnut leafminer, *Cameraria ohridella*, Deschka and Dimić (Lepidoptera: Gracillariidae), is a pest moth that attacks horse chestnut trees (*Aesculus hippocastanum* L. (Sapindales: Sapindaceae)). It was first recorded in 1985 in Macedonia in the area near Lake Ohrid [1], which probably represents the origin of this species [2,3]. Soon after it was first recorded, it started to spread and affect the aesthetics and vitality of the horse chestnut [4], which is a frequently planted ornamental tree species in many parks throughout Europe [5]. Since 2004, horse chestnut leaf miner has also spread in Asia Minor [6–8]. The moth reached the Czech Republic in 1993 [9], and, over the course of five years, it spread across the whole country [6,10].

Although the preferred host tree of *C. ohridella* is *A. hippocastanum*, it lays eggs and forms mines on other *Aesculus* species [11,12] and occasionally attacks sycamore maple (*Acer pseudoplatanus* L. (Sapindales: Sapindaceae)) [13]. Leaf damage inflicted by *C. ohridella* to *A. hippocastanum* in the Czech Republic can exceed 50% in some sites [14] and, together with the horse chestnut leaf blotch caused by *Guignardia aesculi* (Peck) V.B. Stewart (Botryosphaeriales: Botryosphaeriaceae), is the main cause of the decline in horse chestnut trees' ornamental aesthetics [15]. Heavy infestation can cause premature defoliation and negatively affects seed weight [4,16,17].

Despite the extensive efforts of scientists to develop various methods to control or support the control of the populations of *C. ohridella* [18–22], including the application of microbial biocontrol agents [23–31], no other effective and economical solution has yet been found without possible side effects on other organisms, including humans. We therefore need to rely on naturally occurring antagonists.

When invading new regions, leafmining moths often become quickly adopted and controlled by native parasitoids and other natural enemies [32,33]. *Cameraria ohridella*, as an invasive species, was adopted as a host by autochthonous enemies such as predators [34], pathogens [29,30,35], and parasitoids [36–39]. However, the combined impact of natural enemies has not been sufficient for the effective control of *C. ohridella* populations [36,38,40].

Many studies on *C. ohridella* parasitoids to identify natural control factors have been carried out since the first *C. ohridella* invasion in Europe [36–38,40–47]. While parasitism rates vary depending on the calculation method used [48], the reported rates are quite low, ranging between 0.5% and 45% [36,43,47,49–51].

The distribution of parasitoids is affected by geography, which often correlates with the climate and the host distribution [52]. Among geographical factors, altitude [53] and the direction of spread [7] are the most important. *Cameraria ohridella* expanded in the Czech Republic from south to north, so its direction of spread was correlated with latitude [7].

This study aimed to evaluate the effect of geographical variability and temporal changes on mortality rates caused by parasitism, together with other mortality factors during a period of five years in the Czech Republic. Our first objective was to determine whether parasitism and other mortality rates were related to altitude, the population density of *C. ohridella*, and other variables. The second objective was to describe the species within the parasitoid complex of the horse chestnut leafminer.

2. Materials and Methods

2.1. Sampling Sites

Samples of horse chestnut (only *A. hippocastanum* species) leaves infested with *C. ohridella* were collected in six sites representing the main regions of the Czech Republic (Figure 1). The trees sampled were either from city parks, formed alleys, or assemblages of trees near the road with lawn beneath with various intensities of care, e.g., mowing and raking. The altitude of sites ranged from 240 to 463 m above sea level (Table 1). The sampling was carried out over a five-year period from 2006 until 2010.

Table 1. Characteristics of the sampling sites.

Label	City	Local Name	Geographical Coordinates	Altitude(m)	Greenery		
					Type	Area [1]	Care [2]
A	Praha	Obora Hvězda	50°4′57″ N, 14°19′54″ E	369	city park	88	2
B	Plzeň	Lochotínský park	49°45′37″ N, 13°21′49″ E	310	city park	36	1
C	Liberec	Jablonecká street	50°46′5″ N, 15°4′4″ E	382	group of trees	2	1
D	České Budějovice	Sady	48°58′38″ N, 14°28′34″ E	386	city park	3	2
E	Brno	Park Špilberk	49°11′40″ N, 16°36′12″ E	240	city park	21	1
F	Rožnov pod Radhoštěm	Hradisko	49°26′53″ N, 18°7′9 E	463	alley	166	0

[1] Size of the continuous area (in hectares) surrounding the sampled trees covered by lawn, shrubs, and trees.
[2] Intensity of greenery care: 0—no mowing or raking, 1—mowing 1–2 times per season, 2—mowing >2 times per season.

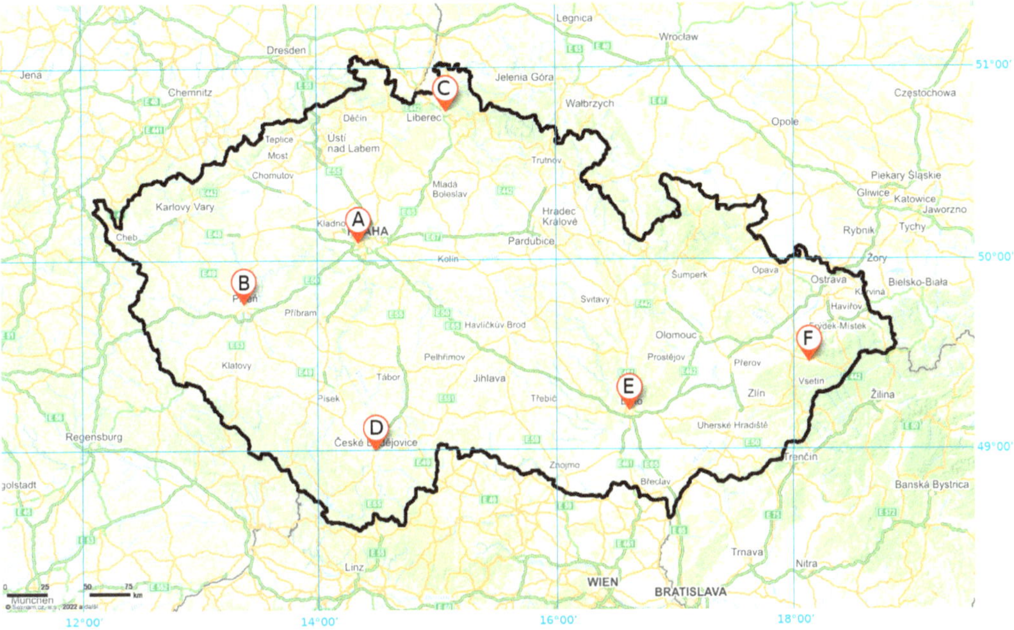

Figure 1. The map of the Czech Republic with marked sites where *Aesculus hippocastanum* leaf samples were collected. Source: Mapy.cz.

2.2. Leaf Sampling

Fifty horse chestnut leaves were randomly picked from the trees (maximum of 5 leaves per tree) up to 2 m above the ground at every site when the first adults of the first *C. ohridella* generation started to emerge. This occurred at approximately the phenological phase of the tree called 'chestnut appearing' [54]. The period of adult emergence was assessed for each site individually. The sampling dates thus differed (Table A1) because leaf miner phenology differed at every site and year. The leaves were put into plastic bags and transported to the laboratory in a cooler, where the leaves were processed immediately or stored in a refrigerator at 4–6 °C for a maximum of 2 days.

2.3. Estimation of Population Density of *C. ohridella*

The population density of *C. ohridella* was assessed as a percentage of the leaf area damaged by the larvae. The visual method proposed by Gilbert and Gregoire [55] was used to estimate leaf damage for each of the 50 compound leaves, and an average was calculated.

2.4. Mine Dissection, Rearing of Parasitoids, and Species Determination

The leaflets were carefully cut from the compound horse chestnut leaves, and 500 mines per sample were dissected under a stereomicroscope with a maximum of 10 mines selected from a single leaflet. The number of living, dead, and parasitized larvae and pupae were recorded. The developmental stage of larvae was determined according to Freise and Heitland [49] and recorded as follows: L1 and L2 = first and second larval instar, L3 = third larval instar, L4 = fourth larval instar (possible L5 instar was included), SP = first and second spinning instar, P = pupa and obviously emerged adults.

Parasitized individuals, including dead pupae, were put singly into plastic Petri dishes. The Petri dishes were stored in boxes at laboratory temperature in relatively high humidity provided by wet cotton wool on the bottom of plastic boxes. The content of Petri dishes was checked for emerged parasitoids daily, and parasitoid individuals were stored in a freezer for future determination.

Determination of parasitoid species was made by the first author according to Grabenweger et al. [56] and compared with specimens of the author's personal collection of parasitoids previously determined by specialists.

2.5. Calculation of Parasitism and Other Mortality Rates

Stage-specific parasitism rates and other mortality rates were calculated according to Volter and Kenis [47]:

$$PN_i = \frac{N_i}{\sum_{j=i}^{5}(N_j + D_j + A_j)} \quad (1)$$

$$PD_i = \frac{D_i}{\sum_{j=1}^{5}(N_j + D_j + A_j)} \quad (2)$$

where i (j) denotes the developmental stage (1: 1st and 2nd instar larva, 2: 3rd instar larva, 3: 4th instar larva, 4: spinning larva, 5: pupa; moths of the same generation were included in the number of living pupae), PN_i and PD_i denote parasitism and other mortality rates, respectively; N_i (N_j), D_i (D_j), and A_i (A_j) denote the number of parasitized, dead, or missing individuals and living individuals, respectively.

Total parasitism rates and total mortality rates were calculated using the following equations:

$$PtN = \sum_{i=1}^{5}\left[PN_i \prod_{j=0}^{i-1}(1 - PN_j - PD_j)\right] \quad (3)$$

$$PtD = \sum_{i=1}^{5}\left[PD_i \prod_{j=0}^{i-1}(1 - PN_j - PD_j)\right] \quad (4)$$

$$PN_0 = PD_0 = 0 \quad (5)$$

where PtN and PtD denote total parasitism and total other mortality rates, respectively. For other symbols, see Equations (1) and (2).

2.6. Statistical Analysis

A generalized linear model with a binomial distribution and logit link was used to analyze parasitism and other mortality rates data. The following predictors were used: year, altitude, latitude, size of continuous greenery area surrounding the sampled trees, developmental stage of *C. ohridella*, population density of *C. ohridella* expressed as leaf damage, and intensity of greenery care (mowing/raking, ranging from 0 to 2). The analysis was performed in SAS® Studio for Linux using the GLM procedure (PROC GENMOD) of SAS/STAT® module [57]. Since the data represented repeated measurements and thus could not be regarded as independent, the Generalized Estimating Equation (GEE) approach [58] was used to account for within-subject correlations, using option statement REPEATED in GENMOD procedure. We also conducted a Spearman correlation analysis [59] to reveal any significant shift in the ratio between two of the most abundant parasitoid species during the studied period. Data were analyzed using the PROC CORR function in SAS/STAT® module [57]. p values <0.05 were considered statistically significant in all tests.

3. Results

3.1. Dates of the Emergence of the First Generation and Population Density of C. ohridella

First adults of the first generation of *C. ohridella* were observed in June and July, depending on the site and year (Table A1). The population density of *C. ohridella* estimated as a percentage of the leaf area damaged by *C. ohridella* larvae varied substantially among sites and years, ranging from 2.0% in České Budějovice in 2007 to 48.7% in Plzeň in 2010 (Figure 2).

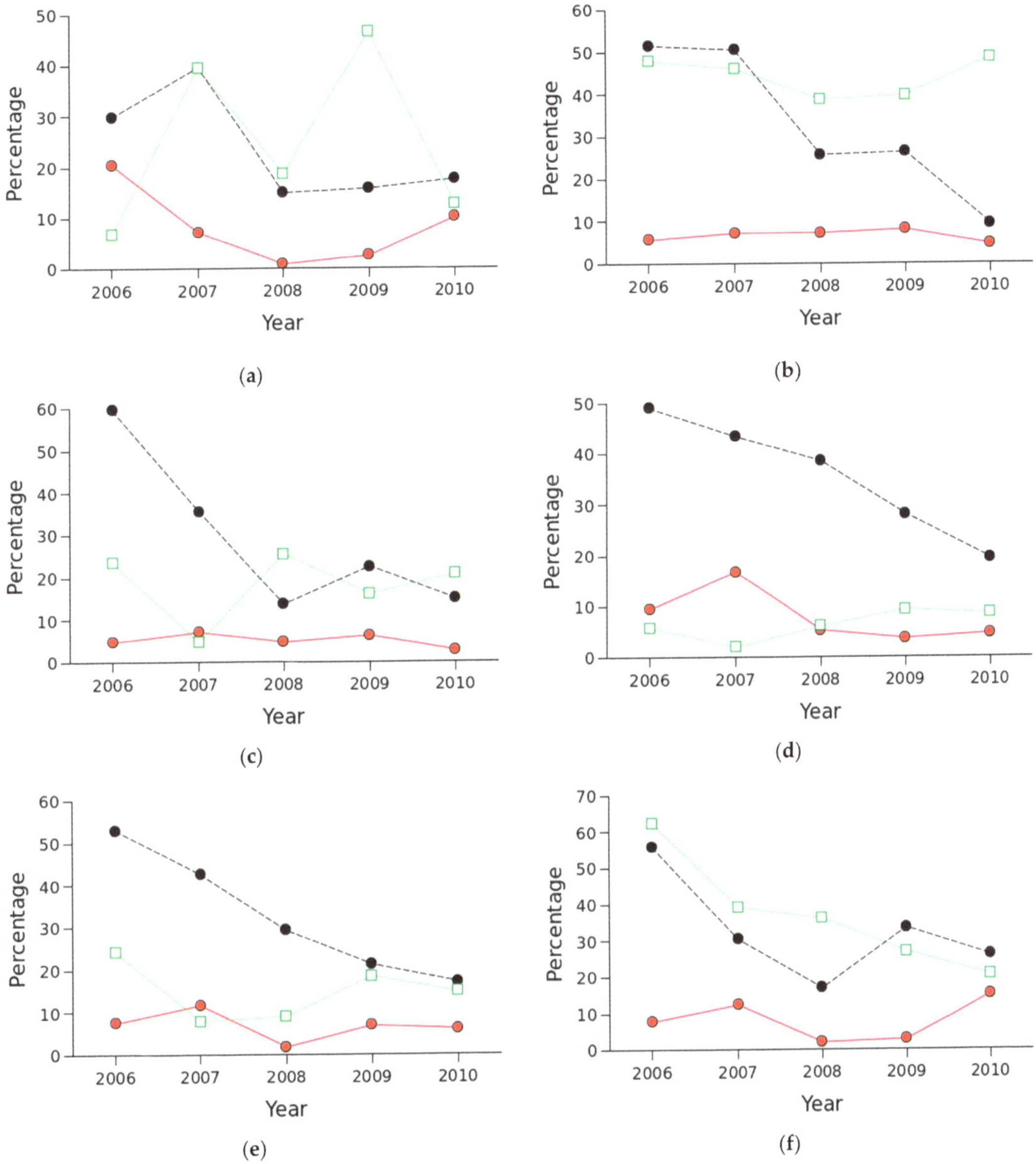

Figure 2. Total parasitism (full red line), total other mortality (dashed black line), and leaf damage inflicted by *C. ohridella* larvae (dotted green line, open squares) at six sites in the Czech Republic from 2006 to 2010. (**a**) Praha; (**b**) Plzeň; (**c**) Liberec; (**d**) České Budějovice; (**e**) Brno; (**f**) Rožnov pod Radhoštěm.

3.2. Parasitism and Other Mortality Rates

Parasitism and other mortality rates of the 15,000 mines dissected showed high variability among developmental stages, years, and sites (Figure 2, Table A1). Parasitism rates of feeding instars reached a maximum of only 1.2%, but were often lower. On the other hand, spinning stages' parasitism was much higher reaching a maximum of 6.4%. The highest parasitism rates were observed in pupae, in which they reached a maximum of 24.1%, with an average value of 7.3%. Besides the positive relationship between the developmental stage of *C. ohridella* and the parasitism rate, the statistical analysis revealed that the parasitism rate was negatively related to year, latitude, and greenery maintenance (Table 2).

Table 2. Results of the analysis of GEE parameter estimates.

Dependent Variable	Parameter	Estimate	Standard Error	Z	*p*
Parasitism rate	Intercept	440.340	136.875	3.22	0.001
	Year	−0.215	0.069	−3.13	0.002
	Altitude	0.001	0.001	1.41	0.157
	Latitude	−0.331	0.052	−6.32	<0.001
	Greenery area	−0.002	0.002	−1.00	0.319
	Developmental stage	1.165	0.091	12.81	<0.001
	C. ohridella population density	−0.009	0.007	−1.15	0.248
	Care (mowing, raking) intensity	−0.157	0.047	−3.31	0.001
Other mortality rate	Intercept	605.940	73.252	8.27	<0.001
	Year	−0.295	0.037	−8.06	<0.001
	Altitude	0.001	0.000	1.65	0.099
	Latitude	−0.315	0.019	−16.80	<0.001
	Greenery area	−0.001	0.001	−2.36	0.019
	Developmental stage	−0.515	0.053	−9.82	<0.001
	C. ohridella population density	0.001	0.002	0.44	0.662
	Care (mowing, raking) intensity	−0.059	0.015	−4.06	<0.001

Total parasitism rates were quite low and varied from 1.9% to 20.5%, with an average value of 7.2%. The highest fluctuation of parasitism rates during the study period was observed in Praha, České Budějovice, and Rožnov pod Radhoštěm (Figure 2).

Mortality caused by other factors was much higher than that caused by parasitism, especially in first/second larval instars and pupae, and varied from 1.4% to 52.9%. Third larval instars up to spinning stages suffered less from other mortality factors; which reached up to 5.8%. Data analysis revealed that in contrast to the parasitism rate, the other mortality rate was also significantly related to the greenery area.

Total other mortality rates varied from 13.7% to 59.5%, with an average value of 31%. The highest rates were found in 2006, followed by an obvious decreasing trend in all sites, except for the Praha site (Figure 2).

3.3. Parasitoids of *C. ohridella*

In total, eight hymenopteran parasitoid species were identified in samples from six sampling sites in the Czech Republic over five years (Table 3). Five Eulophidae species and an undetermined species of the genus *Chrysocharis*, one of the family Eupelmidae, an undetermined species of the genus *Pteromalus* (Pteromalidae), and two ichneumonids were reared in this study. The data show variability in species composition and abundance among sites and hosts (Table 3).

The most abundant parasitoid species was *Pediobius saulius* (Walker) at all sites (Figure A1). Because this species is considered a koinobiont, it is difficult to determine which host stage it attacks; nevertheless, it always emerged from pupae of *C. ohridella*, except for one instance in the spinning stage. The second most abundant parasitoid species was *Minotetrastichus frontalis* (Nees) (Figure A1). However, the comparison of individual years showed that the abundance of these two species changes over time (Figure 3a). The proportion of *M. frontalis* among the total parasitoid complex gradually decreased. On the other hand, it was the only species in this study with more than one larvae observed on one host individual. The multiparasitism was occasionally observed with *M. frontalis*/*P. saulius* and *M. frontalis*/*Closterocetus trifasciatus* (Westwood). *Closterocerus trifasciatus* was the only species reared from the first/second larval instar of *C. ohridella*.

Table 3. Parasitoid species reared from *C. ohridella* and summary data regarding the sampling site and the developmental stage from 2006 to 2010.

Superfamily	Family	Species	Sampling Site [1]						Host Stage [2]				
			A	B	C	D	E	F	L1 and L2	L3	L4	SP	P
Chalcidoidea	Eulophidae	*Cirrospilus vittatus* (Walker)		1									1
		Closterocerus trifasciatus (Westwood)			1	4	2	2	1			2	6
		Chrysocharis sp.	1	2		4	1				1	4	3
		Minotetrastichus frontalis (Nees)	3	6	1	19	10	12		1		22	28
		Pediobius saulius (Walker)	11	28	23	26	19	40				1	146
		Pnigalio agraules (Walker)		7	2		2	1		1	3	5	3
		Pnigalio sp.		3			6			1	2	5	1
	Eupelmidae	*Eupelmus urozonus* (Dalman)					1					1	
	Pteromalidae	*Pteromalus* sp.		3								1	2
Ichneumonoidea	Ichneumonidae	*Itoplectis alternans* (Gravenhorst)		1								1	
		Scambus annulatus (Kiss)			2	1						3	

[1] See Table 1 for site descriptions. [2] Developmental stage of *C. ohridella* attacked: L1 and L2—first and second larval instar, L3—third larval instar, L4—fourth larval instar, SP—first and second spinning instar, P—pupa and obviously emerged adults.

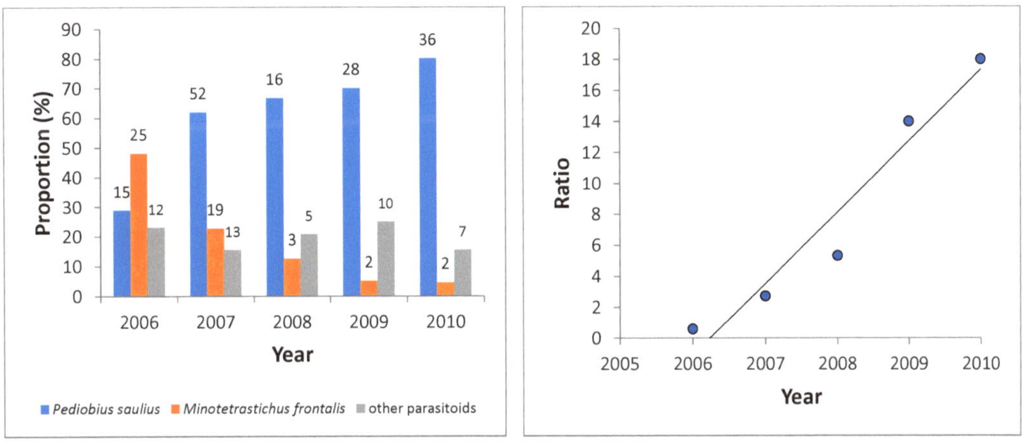

(**a**) (**b**)

Figure 3. Temporal changes in the parasitoid species complex of *C. ohridella* at six sites in the Czech Republic: (**a**) proportion among all parasitoids; numbers above columns represent the number of parasitized host individuals; (**b**) changes in *P. saulis/M. frontalis* ratio fitted by linear trend.

There was an obvious increase in *P. saulius* proportion in the parasitoid complex during the five years of observation, while *M. frontalis* became less abundant (Figure 3b). A Spearman correlation analysis revealed that the *P. saulius/M. frontalis* ratio clearly increased over time ($r_s = 1$; $n = 5$; $p < 0.001$).

4. Discussion

4.1. Parasitism and Other Mortality Rates

The parasitism rates in our study were rather low and did not differ from the results published by other authors [36,43,47,49–51]. Our study showed that parasitoids do not often emerge from the first to fourth larval instars of *C. ohridella*, while higher parasitism is observed in older stages [47]. Pupal parasitism rates ranged from 0 to 12% and from 4 to 28% in Belgrade and Istanbul, respectively [60]. Similar values were observed in this study, where *P. saulius* was responsible for the majority of pupal parasitism. This species can, however, develop as a koinobiont, i.e., it can attack some younger stages of the host and finish its development in the host pupa, in contrast to other parasitoids reared during this study considered to be idiobionts. Because *P. saulius* was also a dominant species, the parasitism rates observed in individual *C. ohridella* stages might thus be biased.

Although different methods for calculating the parasitism rates were used in other studies, the ability of parasitoids to manage *C. ohridella* populations is still too low. Such low parasitism rates probably result from insufficient adaptation of the local parasitoids to the new host [52]. While the composition of the assemblage of parasitoids is latitude-dependent, the higher proportion of *P. saulius* has not yet improved the overall rate of parasitism. Nevertheless, in comparison with Volter and Kenis [47], a stronger pressure of parasitoids on pupae was seen in the present study.

Variability in parasitism rates between sampling sites could be due to different local climatic conditions and factors related to the city's characteristics, such as the structure of the streets and buildings near sampling sites and the prevailing wind direction. Our results confirmed that the parasitism rate was positively related to the developmental stage of the host and negatively related to year, latitude, and greenery care. A study on chalcid wasps that emerged from horse chestnut leaf litter samples collected from 35 sites in the Czech Republic [52] showed that wasp abundance, considered an indicator of parasitism, was also negatively related to latitude, but contrary to our findings, it was positively related to altitude as well as *C. ohridella* abundance. The positive correlation of the number of parasitoids with the number of moths that emerged from leaf litter in spring was confirmed in a study by Kopačka and Zemek [14].

High variability in the percentage of leaf area damaged by *C. ohridella* among sampling sites found in the present study is likely to be geographically related, but variability in leaf damage can be quite high, even within a city area [14]. A key factor seems to be the maintenance of public green areas, particularly raking and removing leaf litter in autumn or early spring, which is a recommended and effective way to regulate the damage caused by *C. ohridella* in the spring [20,61]. However, the removal of fallen horse chestnut leaves also reduces parasitoids that also overwinter in horse chestnut leaf litter [14,52,62]. Our results confirmed a significant negative effect of greenery maintenance intensity, i.e., mowing frequency, on the parasitism rate.

The other mortality rates combined all other mortality factors except parasitism, including the influence of predators such as birds, spiders, or some insects [34]; insect pathogens; and environmental stress. Fungi were observed on some *C. ohridella* cadavers, but it was not possible to determine whether the fungal growth was caused by entomopathogenic or saprophytic fungi on already dead individuals. Although this study did not focus on this, natural infections of *C. ohridella* by entomopathogenic fungi have been previously reported [29,30,35]. The other mortality rates of the first and second larval instars were higher in the present study compared to previous studies [47], which could indicate that the controlling factors have a higher impact on the young larval stages. In comparison to parasitism rates, other mortality rates were much higher and negatively related to the greenery area size as well as greenery care.

There was no obvious difference between the total other mortality rates recorded in the present study and those of Volter and Kenis [47]. The cause of the decline in other mortality rates during the study period observed in all sites is unclear, and further studies should be conducted to elucidate the validity of this trend.

4.2. Parasitoids of *C. ohridella*

The parasitoid species found at all sites in the Czech Republic do not differ from the records in the literature [36–38,40–47,63–66]. The abundance of individual species recorded by us was similar to the data reported from Bulgaria, Greece, Croatia, Macedonia, and Yugoslavia or generally from south-eastern Europe [36,40]. The most abundant parasitoid species in 2006 was *M. frontalis*, but the abundance of *P. saulius* increased in all consecutive years, while the abundance of *M. frontalis* declined. According to published reports, *P. saulius* was not recorded in Bayern (Germany) in 1998, while *M. frontalis* was the most abundant species [50]. The most abundant parasitoid species were *M. frontalis* together with *Pnigalio agraules* (Walker) in Wien and Gerasdorf (Austria) from 1996 to 1999, with no presence or very rare occurrence of *P. saulius* [43,51]. *Minotetrastichus frontalis* and *P. agraules* were also the most abundant species in Bayern in 2000 [49] and leaf litter collected in the Czech Republic in early spring 2002 [52]. The latter study also reported findings of *P. saulius*, contrary to Volter and Kenis [47], who did not detect this species between 2001 and 2003 in the western Czech Republic. It was also found in Ilava (Slovakia), but rarely [47]. In contrast, *P. saulius* dominated in Lipica (Slovenia) [47]. *Pediobius saulius* was not present on *C. ohridella* in 2003–2004 in Olten (Switzerland), Jübeck (Germany), and Brixen (Italy) at all, but it was present and even the most abundant species in Bulgaria in 2003–2004 [42].

These data suggest that the original parasitoid complexes in different countries started to change. There is a clear tendency that *P. saulius*, a parasitoid already known from other hosts in Yugoslavia, Hungary, Czech Republic, Slovakia, Germany, France, Switzerland, and Italy [42,67], is having a gradually increasing impact on *C. ohridella* from the south to north. The mechanism of its sudden appearance or increasing influence on *C. ohridella* is not clear, but it could be attributed to its adaptation to a new host species. It is assumed that when the parasitoid species adapt to *C. ohridella*, then they can follow its geographic dispersal. This was, for example, confirmed for *Pnigalio mediterraneus* Ferriére and Delucchi using molecular markers [68]. While generalist parasitoids may adapt to *C. ohridella* with a shorter time lag, the adjustment of specialist parasitoids probably requires more than a few decades [36]. Despite currently low parasitism rates reported in previous studies (e.g., [39,52]) as well as our findings, the successful biological control of this invasive pest by natural enemies may develop in the future.

5. Conclusions

Our study showed that the parasitism rate remained relatively low and that it was significantly related to year, *C. ohridella* population density and developmental stage, and latitude, but not to altitude or greenery area size and its maintenance. The parasitoid complex is predominantly formed by generalist parasitoids of the family Eulophidae, and it seems that the role of increasing *P. saulius* abundance does not positively influence the parasitism rate, but higher parasitism of *C. ohridella* pupae by *P. saulius* in combination with other mortality factors could improve the control of *C. ohridella* in the future and could be much more effective than *M. frontalis*, which does not exclusively attack pupae. Mortality in the earlier juvenile stages of hosts is caused by many factors, and the pupa, as the last juvenile stage, can be vulnerable to attack by *P. saulius*. If the abundance of *P. saulius* continues to increase, this species might become an important agent for more efficient biological control of *C. ohridella* populations in the future.

Author Contributions: Conceptualization, E.P. and R.Z.; methodology, L.V.; field sampling, L.V., F.W. and R.Z.; sample processing and species identification, L.V.; writing—original draft preparation, L.V.; writing—review and editing, L.V., R.Z., F.W. and E.P.; project administration, E.P.; funding acquisition, E.P. and R.Z. All authors have read and agreed to the published version of the manuscript.

Funding: This research was supported by the Czech Academy of Sciences (RVO: 60077344).

Institutional Review Board Statement: Not applicable.

Informed Consent Statement: Not applicable.

Data Availability Statement: The data presented in this study are available on request from the corresponding author.

Acknowledgments: The authors thank Jitka Pflegerová and Jana Jabůrková for their technical assistance.

Conflicts of Interest: The authors declare no conflict of interest. The funders had no role in the design of the study; in the collection, analyses, or interpretation of data; in the writing of the manuscript, or in the decision to publish the results.

Appendix A

Table A1. Parasitism and other mortality rates (%) on the different developmental stages of *C. ohridella* at six sites in the Czech Republic from 2006 to 2010.

Sampling Site [1]	Year	Sampling Date	Parasitism [2]					Other Mortality [2]				
			L1 and L2	L3	L4	SP	P	L1 and L2	L3	L4	SP	P
	2006	27 July	0	0	0	2.9	24.1	19.6	2.9	0	0	10.3
	2007	20 June	1.0	0	0	0.6	8.7	30.8	2.2	0.4	0	10.7
A	2008	2 July	0	0	0.3	0	0.8	12.2	1.2	0.8	0	1.2
	2009	2 July	0	0.2	0	0	2.6	9.0	0.5	0	0	6.9
	2010	2 July	0.2	0	0	0.9	10.8	13.4	0	0	1.8	3.1
	2006	15 July	0	0.8	0.4	2.2	4.8	26.0	1.6	0.8	1.7	32.7
	2007	17 June	0	0.3	0.3	3.0	7.4	28.2	4.3	0.9	1.5	27.1
B	2008	29 June	0.2	0.5	1.0	6.4	1.8	19.8	3.2	2.7	1.7	0
	2009	27 June	0.2	0	0.7	3.8	6.1	21.0	1.7	0.3	0	5.1
	2010	26 June	0	0	0	0	4.9	6.0	0	0	0	3.7
	2006	26 July	0	0	0	2.9	2.9	13.2	4.1	0	0	52.9
	2007	29 June	0.2	0	0	0.3	9.2	22.0	4.3	0.6	0	13.4
C	2008	10 July	0	0	0	0.3	5.1	8.6	0.4	0.7	0	4.6
	2009	9 July	0.2	0	0.3	0.4	6.6	12.6	1.9	0.3	0	9.4
	2010	13 July	0.2	0	0	0.9	2.1	13.6	0.5	0	0	1.1
	2006	17 July	0	0.4	0.4	3.8	11.1	34.2	3.7	1.2	0	19.4
	2007	20 June	1.2	0.6	0.6	4.5	17.9	25.6	4.2	1.0	0.7	20.8
D	2008	1 July	0.2	0.2	0.6	0.7	5.4	18.2	5.2	3.9	0	17.8
	2009	2 July	0.2	0	0	0.4	4.2	22.2	4.1	0	0	3.6
	2010	2 July	0.4	0	0.7	1.0	3.6	15.6	2.6	0	0	2.2
	2006	10 July	0	0.7	1.1	5.3	3.9	26.6	1.1	0.8	1.6	36.2
	2007	12 June	0.4	0.4	0	1.8	15.6	29.4	5.8	1..9	0	12.2
E	2008	25 June	0	0	0.3	0.4	1.7	17.6	1.5	0.3	1.2	11.7
	2009	25 June	0.2	0.8	0.3	0	6.9	1.4	4.3	0.3	0	6.9
	2010	25 June	0	0.2	0.4	1.5	5.1	12.4	1.7	0	0	3.8
	2006	25 July	0	0	0	0.7	10.3	26.6	1.9	0	0	39
	2007	27 June	0.2	0	0	1.7	13.4	14	2.2	2.2	0	15.6
F	2008	7 July	0	0	0.3	0	2.2	12.6	1.2	0	0	3.9
	2009	8 July	0.2	0	0.4	1.2	2.4	28	3.1	0	0	4.8
	2010	7 July	0.6	0.3	0	0.5	18.1	21	0.6	0	0	5.8

[1] See Table 1 for description of sites. [2] Developmental stages: L1 and L2—first and second larval instar, L3—third larval instar, L4—fourth larval instar, SP—first and second spinning instar, P—pupa and obviously emerged adults.

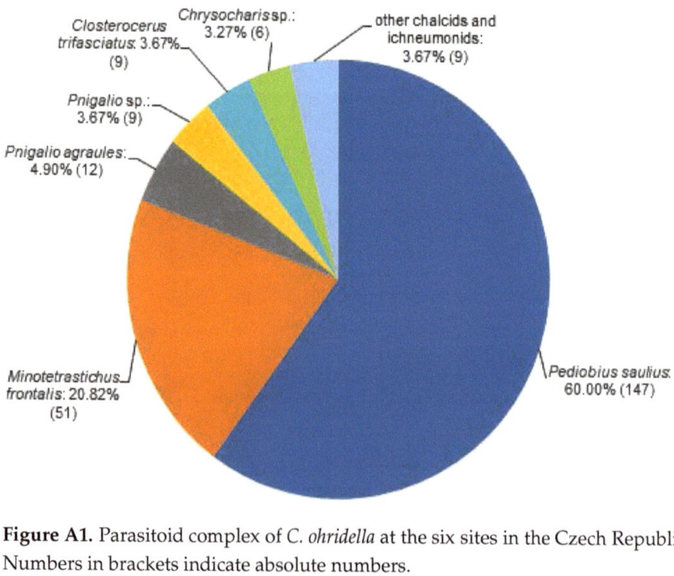

Figure A1. Parasitoid complex of *C. ohridella* at the six sites in the Czech Republic from 2006 to 2010. Numbers in brackets indicate absolute numbers.

References

1. Simova-Tosic, D.; Filev, S. Contribution to the Horse Chestnut Miner. *Zast. Bilja* **1985**, *36*, 235–239.
2. Grabenweger, G.; Grill, R. On the Place of Origin of *Cameraria Ohridella* Deschka & Dimic (Lepidoptera: Gracillariidae). *Beiträge Zur Entomofaunist.* **2000**, *1*, 9–17.
3. Valade, R.; Kenis, M.; Hernandez-Lopez, A.; Augustin, S.; Mari Mena, N.; Magnoux, E.; Rougerie, R.; Lakatos, F.; Roques, A.; Lopez-Vaamonde, C. Mitochondrial and Microsatellite DNA Markers Reveal a Balkan Origin for the Highly Invasive Horse-Chestnut Leaf Miner *Cameraria Ohridella* (Lepidoptera, Gracillariidae). *Mol. Ecol.* **2009**, *18*, 3458–3470. [CrossRef]
4. Thalmann, C.; Freise, J.; Heitland, W.; Bacher, S. Effects of Defoliation by Horse Chestnut Leafminer (*Cameraria Ohridella*) on Reproduction in *Aesculus hippocastanum*. *Trees-Struct. Funct.* **2003**, *17*, 383–388. [CrossRef]
5. Thomas, P.A.; Alhamd, O.; Iszkuło, G.; Dering, M.; Mukassabi, T.A. Biological Flora of the British Isles: *Aesculus hippocastanum*. *J. Ecol.* **2019**, *107*, 992–1030. [CrossRef]
6. Liška, J. *Verbreitung der Rosskastanienminiermotte in Tschechien*; Forstschutz-Aktuell: Wien, Austria, 1997; Volume 21.
7. Šefrová, H.; Laštůvka, Z. Dispersal of the Horse-Chestnut Leafminer, *Cameraria Ohridella* Deschka & Dimic, 1986, in Europe: Its Course, Ways and Causes (Lepidoptera: Gracillariidae). *Entomol. Z.* **2001**, *111*, 194–198.
8. Cebeci, H.; Acer, S. The Occurrence of Some Lepidopterous Species on the Horse Chestnut (*Aesculus hippocastanum* L.) at Istanbul-Belgrad Forest in Turkey. *Acta Agric. Slov.* **2007**, *89*, 95–102. [CrossRef]
9. Laštůvka, Z.; Liška, J.; Vávra, J.; Elsner, V.; Laštůvka, A.; Marek, J.; Dufek, T.; Dvořák, M.; Kopeček, F.; Petrů, M.; et al. Faunistic Records from the Czech Republic. *Klapalekiana* **1994**, *18*, 197–206.
10. Wittenberger, G. Die Roßkastanien-Miniermotte in Nordböhmen nebst einigen Hinweisen zum Vorkommen in Schlesien (Polen). *Nassau. Ver. Für Nat.* **1998**, *98*, 75–78.
11. D'Costa, L.; Koricheva, J.; Straw, N.; Simmonds, M.S.J. Oviposition Patterns and Larval Damage by the Invasive Horse-Chestnut Leaf Miner *Cameraria Ohridella* on Different Species of *Aesculus*: Oviposition Patterns and Larval Damage. *Ecol. Entomol.* **2013**, *38*, 456–462. [CrossRef]
12. Skuhravý, V. Zur Kenntnis der Blattminen-Motte *Cameraria ohridella* Desch. & Dim. (Lep., Lithocolletidae) an *Aesculus hippocastanum* L. in der Tschenchischen Republik. *Anz. Schadl. Pflanzenschutz Umweltschutz* **1998**, *71*, 81–84. [CrossRef]
13. Péré, C.; Augustin, S.; Turlings, T.C.J.; Kenis, M. The Invasive Alien Leaf Miner *Cameraria Ohridella* and the Native Tree *Acer Pseudoplatanus:* A Fatal Attraction? *Agric. For. Entomol.* **2010**, *12*, 151–159. [CrossRef]
14. Kopacka, M.; Zemek, R. Spatial Variability in the Level of Infestation of the Leaves of Horse Chestnut by the Horse Chestnut Leaf Miner, *Cameraria Ohridella* (Lepidoptera: Gracillariidae) and in the Number of Adult Moths and Parasitoids Emerging from Leaf Litter in an Urban Environment. *Eur. J. Entomol.* **2017**, *114*, 42–52. [CrossRef]
15. Kopačka, M.; Nachman, G.; Zemek, R. Seasonal Changes and the Interaction between the Horse Chestnut Leaf Miner *Cameraria ohridella* and Horse Chestnut Leaf Blotch Disease Caused by *Guignardia aesculi*. *Forests* **2021**, *12*, 952. [CrossRef]
16. Nardini, A.; Raimondo, F.; Scimone, M.; Salleo, S. Impact of the Leaf Miner *Cameraria Ohridella* on Whole-Plant Photosynthetic Productivity of *Aesculus hippocastanum*: Insights from a Model. *Trees* **2004**, *18*, 714–721. [CrossRef]
17. Salleo, S.; Nardini, A.; Raimondo, F.; Lo Gullo, M.A.; Pace, F.; Giacomich, P. Effects of Defoliation Caused by the Leaf Miner *Cameraria Ohridella* on Wood Production and Efficiency in *Aesculus hippocastanum* Growing in North-Eastern Italy. *Trees* **2003**, *17*, 367–375. [CrossRef]
18. Freise, J.F.; Heitland, W. The Waipuna Hot Foam System–a Chance for *Cameraria Ohridella* Control? *Nachrichtenbl. Deut. Pflanzenschutz* **2003**, *55*, 242–244.
19. Heitland, W.; Kopelke, J.-P.; Freise, J. Die Roßkastanien-Miniermotte–19 Jahre Forschung Und Noch Keine Lösung in Sicht. *Nat. Und Mus.* **2003**, *133*, 221–231.
20. Kehrli, P.; Bacher, S. Date of Leaf Litter Removal to Prevent Emergence of *Cameraria Ohridella* in the Following Spring. *Entomol. Exp. Appl.* **2003**, *107*, 159–162. [CrossRef]
21. Kehrli, P.; Lehmann, M.; Bacher, S. Mass-Emergence Devices: A Biocontrol Technique for Conservation and Augmentation of Parasitoids. *Biol. Control* **2005**, *32*, 191–199. [CrossRef]
22. Pavela, R.; Bárnet, M. Systemic Applications of Neem in the Control of *Cameraria Ohridella*, a Pest of Horse Chestnut (Aesculus hippocastanum). *Phytoparasitica* **2005**, *33*, 49–56. [CrossRef]
23. Zemek, R.; Prenerová, E.; Volter, L.; Weyda, F.; Skuhravý, V. Perspectives for the Biological Control of *Cameraria ohridella*. *Commun. Agric. Appl. Biol. Sci.* **2007**, *72*, 521–526. [PubMed]
24. Prenerová, E.; Zemek, R.; Volter, L.; Weyda, F. Strain of Entomopathogenic Fungus Isaria Fumosorosea CCM 8367 (CCEFO.011.PFR) and the Method for Controlling Insect and Mite Pests. EPO Patent No. EP2313488, 29 April 2015.
25. Barta, M. In Planta Bioassay on the Effects of Endophytic *Beauveria* Strains against Larvae of Horse-Chestnut Leaf Miner (*Cameraria Ohridella*). *Biol. Control* **2018**, *121*, 88–98. [CrossRef]
26. Richter, D.; Sermann, H.; Jackel, B.; Buttner, C.B. Pathogenicity of Entomopathogenic Fungi on Hibernating Pupae of *Cameraria Ohridella* Deschka & Dimic 1986 (Lepidoptera, Gracillariidae). Part 1: Pathogenicity against the Naked Pupa. *Commun. Agric. Appl. Biol. Sci.* **2007**, *72*, 399–410. [PubMed]
27. Richter, D.; Sermann, H.; Jackel, B.; Buttner, C. Pathogenicity of Entomopathogenic Fungi on Hibernating Pupae of *Cameraria Ohridella* Deschka & Dimic 1986 (Lepidoptera, Gracillariidae). Part 2: Efficacy of Entomopathogenic Fungi against Pupa in Her Pupal Cell. *Commun. Agric. Appl. Biol. Sci.* **2007**, *72*, 411–421.

28. Kalmus, M.; Sermann, H.; Buettner, C. Efficacy of Entomopathogenic Fungi against Eggs and Larvae of the Horse Chestnut Leafminer *Cameraria ohridella*. *J. Plant Dis. Prot.* **2007**, *114*, 92.
29. Nedveckytė, I.; Pečiulytė, D.; Būda, V. Fungi Associated with Horse-Chestnut Leaf Miner Moth *Cameraria Ohridella* Mortality. *Forests* **2021**, *12*, 58. [CrossRef]
30. Robert, S.; Petra, C.; Juraj, M.; Marek, B. Natural Prevalence of Entomopathogenic Fungi in Hibernating Pupae of *Cameraria Ohridella* (Lepidoptera: Gracillariidae) and Virulence of Selected Isolates. *Plant Protect. Sci.* **2016**, *52*, 199–208. [CrossRef]
31. Matuska-Lyzwa, J.; Kaca, W.; Zarnowiec, P. Biological Activity of Wild Isolates of Entomopathogenic Nematodes to Horse-Chestnut Leaf Miner (*Cameraria Ohridella*). *Pol. J. Environ. Stud.* **2015**, *24*, 1181–1184. [CrossRef]
32. Godfray, H.C.J.; Agassiz, D.J.L.; Nash, D.R.; Lawton, J.H. The Recruitment of Parasitoid Species to Two Invading Herbivores. *J. Anim. Ecol.* **1995**, *64*, 393. [CrossRef]
33. Kirichenko, N.; Augustin, S.; Kenis, M. Invasive Leafminers on Woody Plants: A Global Review of Pathways, Impact, and Management. *J. Pest Sci.* **2019**, *92*, 93–106. [CrossRef]
34. Grabenweger, G.; Kehrli, P.; Schlick-Steiner, B.; Steiner, F.; Stolz, M.; Bacher, S. Predator Complex of the Horse Chestnut Leafminer *Cameraria Ohridella*: Identification and Impact Assessment. *J. Appl. Entomol.* **2005**, *129*, 353–362. [CrossRef]
35. Zemek, R.; Prenerova, E.; Weyda, F. The First Record of Entomopathogenic Fungus *Paecilomyces Fumosoroseus* (Deuteromycota: Hyphomycetes) on the Hibernating Pupae of *Cameraria Ohridella* (Lepidoptera: Gracillariidae). *Entomol. Res.* **2007**, *37*, A135–A136.
36. Grabenweger, G.; Kehrli, P.; Zweimüller, I.; Augustin, S.; Avtzis, N.; Bacher, S.; Freise, J.; Girardoz, S.; Guichard, S.; Heitland, W.; et al. Temporal and Spatial Variations in the Parasitoid Complex of the Horse Chestnut Leafminer during Its Invasion of Europe. *Biol. Invasions* **2010**, *12*, 2797–2813. [CrossRef]
37. Lupi, D. A 3 Year Field Survey of the Natural Enemies of the Horse-Chestnut Leaf Miner *Cameraria Ohridella* in Lombardy, Italy. *Biocontrol* **2005**, *50*, 113–126. [CrossRef]
38. Girardoz, S.; Kenis, M.; Quicke, D.L.J. Recruitment of Native Parasitoids by an Exotic Leaf Miner, *Cameraria Ohridella*: Host-Parasitoid Synchronization and Influence of the Environment. *Agric. Forest Entomol.* **2006**, *8*, 49–56. [CrossRef]
39. Tóth, P.; Lukáš, J. Parasitic Ichneumonoidea on the Horse Chestnut Leaf Miner, *Cameraria Ohridella* (Lepidoptera: Gracillariidae) in Slovakia. *J. Pest Sci.* **2005**, *78*, 151–154. [CrossRef]
40. Grabenweger, G.; Avtzis, N.; Girardoz, S.; Hrasovec, B.; Tomov, R.; Kenis, M. Parasitism of *Cameraria Ohridella* (Lepidoptera, Gracillariidae) in Natural and Artificial Horse-Chestnut Stands in the Balkans. *Agric. Forest Entomol.* **2005**, *7*, 291–296. [CrossRef]
41. Freise, J.F.; Heitland, W.; Tosevski, I. Parasitism of the Horse Chestnut Leaf Miner, *Cameraria Ohridella* Deschka and Dimic (Lep., Gracillariidae), in Serbia and Macedonia. *Anz Schadl.* **2002**, *75*, 152–157. [CrossRef]
42. Girardoz, S.; Volter, L.; Tomov, R.; Quicke, D.L.J.; Kenis, M. Variations in Parasitism in Sympatric Populations of Three Invasive Leaf Miners. *J. Appl. Entomol.* **2007**, *131*, 603–612. [CrossRef]
43. Grabenweger, G.; Lethmayer, C. Occurrence and Phenology of Parasitic Chalcidoidea on the Horse Chestnut Leafminer, *Cameraria Ohridella* Deschka & Dimic (Lep., Gracillariidae). *J. Appl. Entomol.* **1999**, *123*, 257–260. [CrossRef]
44. Grabenweger, G. Parasitism of Different Larval Stages of *Cameraria ohridella*. *BioControl* **2003**, *48*, 671–684. [CrossRef]
45. Marchesini, E.; Pasini, M.; Galbero, G. Parassitoidi Della Minatrice Fogliare Degli Ippocastani. *L'informatore Agrar.* **2002**, *27*, 75–77.
46. Stojanović, A.; Marković, C. Parasitoid Complex of *Cameraria Ohridella* (Lepidoptera: Gracillariidae) in Serbia. *Phytoparasitica* **2004**, *32*, 132–140. [CrossRef]
47. Volter, L.; Kenis, M. Parasitoid Complex and Parasitism Rates of the Horse Chestnut Leafminer, *Cameraria Ohridella* (Lepidoptera: Gracillariidae) in the Czech Republic, Slovakia and Slovenia. *Eur. J. Entomol.* **2006**, *103*, 365–370. [CrossRef]
48. Van Driesche, R.G. Meaning of "Percent Parasitism" in Studies of Insect Parasitoids. *Environ. Entomol.* **1983**, *12*, 1611–1622. [CrossRef]
49. Freise, J.F.; Heitland, W. Parasitierungsraten: Ermittlung Und Interpretation Am Fallbeispiel Der Roßkastanien-Miniermotte, *Cameraria Ohridella*, Einer Invasiven Und Faunenfremden Schmetterlingsart in Europa. *Mitt. Dtsch. Ges. Allg. Angew. Ent.* **2004**, *14*, 347–350.
50. Moreth, L.; Baur, H.; Schönitzer, K.; Diller, E. Zum Parasitoid-Komplex Der Roßkastanien-Miniermotte in Bayern (*Cameraria Ohridella*, Gracillariidae, Lithocolletinae). *Mitt. Dtsch. Ges. Allg. Angew. Ent.* **2000**, *12*, 489–492.
51. Stolz, M. Untersuchungen Zur Befallsregulierung Der Kastanienminiermotte Durch Natürliche Gegenspieler. *Der Förderungsdienst* **2000**, *48*, 193–195.
52. Nováková, P.; Holuša, J.; Horák, J. The Role of Geography and Host Abundance in the Distribution of Parasitoids of an Alien Pest. *PeerJ* **2016**, *4*, e1592. [CrossRef]
53. Lomolino, M.V.; Riddle, B.R.; Brown, J.H.; Brown, J.H. *Biogeography*, 3rd ed.; Sinauer Associates: Sunderland, MA, USA, 2006; ISBN 978-0-87893-062-3.
54. Myśkow, E.; Sokołowska, K.; Słupianek, A.; Gryc, V. Description of Intra-Annual Changes in Cambial Activity and Differentiation of Secondary Conductive Tissues of *Aesculus hippocastanum* Trees Affected by the Leaf Miner *Cameraria ohridella*. *Forests* **2021**, *12*, 1537. [CrossRef]
55. Gilbert, M.; Gregoire, J.-C. Visual, Semi-Quantitative Assessments Allow Accurate Estimates of Leafminer Population Densities: An Example Comparing Image Processing and Visual Evaluation of Damage by the Horse Chestnut Leafminer *Cameraria Ohridella* (Lep., Gracillariidae). *J. Appl. Entomol.* **2003**, *127*, 354–359. [CrossRef]

56. Grabenweger, G.; Stolz, M.; Jeziorny, C. *A Key to the Parasitoids of Cameraria Ohridella (Lep., Gracillariidae)*; CD-ROM v. 1.0; Verlag Naturhistorisches Museum: Vienna, Austria, 2003.
57. SAS Institute. *SAS/STAT 14.3: User's Guide*; SAS Institute: Cary, NC, USA, 2017.
58. Liang, K.-Y.; Zeger, S.L. Longitudinal Data Analysis Using Generalized Linear Models. *Biometrika* **1986**, *73*, 13–22. [CrossRef]
59. Siegel, S.; Castellan, N.J. *Nonparametric Statistics for the Behavioral Sciences*, 2nd ed.; McGraw-Hill: Boston, MA, USA, 2003; ISBN 978-0-07-057357-4.
60. Cebeci, H.; Markovic, C.; Grabenweger, G.; Ayberk, H.; Dobrosavljevic, J.; Goltas, M.; Stojanovic, A. Preliminary Notes on Pupal Parasitism Rates of the Horse Chestnut Leafminer, *Cameraria Ohridella* (Lepidoptera: Gracillariidae) in Belgrade and Istanbul. *Fresenius Environ. Bull.* **2018**, *27*, 7122–7124.
61. Baraniak, E.; Walczak, U.; Tryjanowski, P.; Zduniak, P. Effect of Distance between Host Trees and Leaf Litter Removal on Population Density of *Cameraria Ohridella* Deschka & Dimic, 1986 (Lepidoptera, Gracillariidae)-Pest of Chestnut (*Aesculus* Sp.) Trees. *Pol. J. Ecol.* **2004**, *52*, 569–574.
62. Klug, T.; Meyhöfer, R.; Kreye, M.; Hommes, M. Native Parasitoids and Their Potential to Control the Invasive Leafminer, *Cameraria Ohridella* DESCH. & DIM. (Lep.: Gracillariidae). *Bull. Entomol. Res.* **2008**, *98*, 379–387. [CrossRef] [PubMed]
63. Balázs, K.; Thuróczy, C. Über Den Parasitoidkomplex von *Cameraria Ohridella* Deschka et Dimi ü 1986, (Lepidop-Tera, Lithocolletidae). *Entomol. Basil.* **2000**, *22*, 269–277.
64. Balázs, K.; Pál, B. Parasitoidender Roßkastanien-Miniermotte (*Cameraria Ohridella* Deschka & Dimić, 1986) in Nordwest-Ungarn. *Acta Agron. Ovar.* **2001**, *43*, 35–48.
65. Čapek, M. Parasitoids of the Horse-Chestnut Leafminer. In *Horse-Chestnut Leafminer (Cameraria ohridella Deschka & Dimić, 1986)*; Laštůvka, Z., Ed.; Veronica: Brno, Czech Republic, 1999; pp. 1–12.
66. Rämert, B.; Kenis, M.; Kärnestam, E.; Nyström, M.; Rännbäck, L.-M. Host Plant Suitability, Population Dynamics and Parasitoids of the Horse Chestnut Leafminer *Cameraria Ohridella* (Lepidoptera: Gracillariidae) in Southern Sweden. *Acta Agric. Scand. Sect. B-Soil Plant Sci.* **2011**, *61*, 480–486. [CrossRef]
67. Noyes, J. *Interactive Catalogue of World Chalcidoidea. CD-ROM*; Taxapad: Vancouver, BC, Canada, 2002.
68. Gebiola, M.; Lopez-Vaamonde, C.; Nappo, A.G.; Bernardo, U. Did the Parasitoid *Pnigalio Mediterraneus* (Hymenoptera: Eulophidae) Track the Invasion of the Horse Chestnut Leafminer? *Biol. Invasions* **2014**, *16*, 843–857. [CrossRef]

forests

MDPI

Article

Deadwood-Dwelling Beetles (Coleoptera: Eucnemidae) in a Beech Reserve: A Case Study from the Czech Republic

Václav Zumr *, Oto Nakládal and Jiří Remeš

Faculty of Forestry and Wood Sciences, Czech University of Life Sciences Prague, Kamýcká 129, 165 00 Prague, Czech Republic; nakladal@fld.czu.cz (O.N.); remes@fld.czu.cz (J.R.)
* Correspondence: zumr@fld.czu.cz

Abstract: The saproxylic beetles (deadwood-dependent) belong to frequently studied groups of forest insects. Eucnemidae is a rare and poorly studied saproxylic family with a hidden life strictly related to deadwood. We studied the family Eucnemidae in a beech reserve, using 59 window traps placed on standing deadwood (snags) and lying logs. A total of 348 specimens in eight species were recorded in two seasons. The identified species included one critically endangered species (CR): *Hylis cariniceps*; five endangered species (EN): *H. olexai*, *H. foveicollis*, *Isorhipis melasoides*, *Eucnemis capucina*, and *Microrhagus lepidus*; one new species found in Bohemia (a region of the Czech Republic): *Clypeorhagus clypeatus*; and one common species: *Melasis buprestoides*. Most species preferred lying logs, but *E. capucina* and *M. buprestoides* preferred snags. Species richness (q = 0) was higher on lying logs than on snags, and similarly, Shannon diversity (q = 1) was significantly higher on lying logs compared to snags. The species *C. clypeorghagus*, *H. foveicollis*, *H. cariniceps*, and *M. lepides* preferred moist lying logs, while *M. buprestoides* and *E. capucina* preferred drier snags with cavities. The results suggest that in beech forests, lying logs serve as a fundamental habitat for the existence of Eucnemids. This could be due to the more stable microclimatic conditions inside the lying deadwood. From this perspective, our study may help better understand the biology of hidden and understudied rare saproxylic Eucnemids.

Keywords: conservation; forest biodiversity; coleoptera; deadwood moisture; melasidae; deadwood microclimate

Citation: Zumr, V.; Nakládal, O.; Remeš, J. Deadwood-Dwelling Beetles (Coleoptera: Eucnemidae) in a Beech Reserve: A Case Study from the Czech Republic. *Forests* **2024**, *15*, 469. https://doi.org/10.3390/f15030469

Academic Editors: Rostislav Zemek and Katarína Pastirčáková

Received: 30 January 2024
Revised: 24 February 2024
Accepted: 1 March 2024
Published: 2 March 2024

1. Introduction

In recent decades, an increasing interest has been in a more detailed understanding of invertebrate life in forests [1]. The focus of research ranges from broad multitaxon studies, e.g., [2,3], to more detailed investigations of orders, families, and individual genera and taxa, e.g., [4–6]. One of the most studied groups of organisms in forests is saproxylic beetles [7]. Saproxylic beetle species are associated with deadwood, as defined by Speight [8]. During some part of their life cycle, saproxylic invertebrates depend upon the dead or dying wood of moribund or dead trees (standing or fallen), upon wood-inhabiting fungi, or the presence of other obligate saproxylic. Increased volumes and variable deadwood microhabitats are crucial for this group's high richness in forests [9,10]. Targeted removal of deadwood from forests results in a dramatic decline in beetle biodiversity [11], as saproxylic species contribute substantially to forest species richness [5].

The Eucnemidae (Eschscholtz, 1829), also called Melasidae (Leach, 1817), are rare beetles inhabiting deadwood. According to the *Catalogue of Palaearctic Coleoptera*, the valid name of this family is Eucnemidae [12]. Eucnemids are strongly associated with forest environments and are considered a vital indicator group for forest conservation [13,14]. Many species of Eucnemids are listed as primeval forest relicts [15]. The Eucnemids in Europe and the European Union consist of 31 and 29 species, respectively [16]. Currently, in Central Europe, 23 species of the family Eucnemidae have been recorded [12]. In the Czech Republic, 22 species are documented [17]. The 2017 Red List itemizes 19 species, and 95% of them are categorized as threatened species (18 species: 9 in CR, 7 in EN, and 2 in VU) [14]. The following 18 species are included: *Dirrhagofarsus attenuatus* (Mäklin, 1845), *Farsus dubius* (Piller et Mitterpacher, 1783), *Hylis cariniceps* (Reitter, 1902), *Hylis procerulus* (Mannerheim, 1823), *Hylis simonae* (Olexa, 1970), *Microrhagus emyi* (Rouget, 1856), *Microrhagus pyrenaeus* (Bonvouloir, 1872), *Rhacopus sahlbergi* (Mannerheim, 1823), *Xylophilus corticalis* (Paykull, 1800), endangered (EN) species *Eucnemis capucina* (Ahrens, 1812), *Hylis foveicollis* (C. G. Thomson, 1874), *Hylis olexai* (Palm, 1955), *Isorhipis marmottani* (Bonvouloir, 1871), *Isorhipis melasoides* (Laporte, 1835), *Microrhagus lepidus* (Rosenhauer, 1847), *Xylophilus testaceus* (Herbst, 1806), vulnerable (VU) species *Dromaeolus barnabita* (A. Villa et J. B. Villa. 1838), and *Microrhagus pygmaeus* (Fabricius, 1792). Three species were recorded in the Czech Republic for the first time after the valid Red List [18] was published [19–21], but these species are also extremely rare: *Clypeorhagus clypeatus* (Hampe, 1850), *Nematodes filum* (Fabricius, 1801), and *Otho sphondyloides* (Germar, 1818). The only relatively common species of Eucnemids is *Melasis buprestoides* (Linnaeus, 1761). Eucnemids are primarily linked to lowlands up to hilly areas, and their life is mainly associated with deciduous tree species [13]. In beech forests and beech-dominated stands of Europe, up to 15 Eucnemids species have been recorded [22]. Although the bionomics of the Eucnemids is poorly studied, recent advancements in knowledge include insights into the diurnal activity of *E. capucina* [23] and the description of the life requirements of several species in protected lowland oak forest sites [13]. In the species database of the Czech Nature Conservation Agency [24], only 2347 individuals of the family Eucnemidae have been recorded in over fifty years. However, it is likely that many records, including published ones, may not have been incorporated into this database. This family represents an important yet rare group of saproxylic species, and understanding this taxonomic group is crucial for formulating effective conservation strategies. The identification of suitable attributes for the targeted support of Eucnemids in beech forests can be applied in forestry management practices. The aim of the study was to investigate the rare and understudied obligately saproxylic family Eucnemidae in a beech reserve established circa 70 years ago. The research focused on the differences between two types of deadwood, i.e., snags (standing) and lying logs (lying deadwood). Study subjects include: (i) What type of deadwood provides a more important habitat for Eucnemids in terms of species richness and abundance? (ii) What variables of deadwood affected the Eucnemids?

2. Materials and Methods

2.1. Study Site

The National Nature Reserve (NNR) Voděradské bučiny (49°58′ N, 14°48′ E) is located at an altitude ranging from 345 to 501 m [25]. The parent rock is granodiorite, and the prevailing soil type is Cambisols, characterized by a low humus content. The mean annual temperature is 7.8 °C, with an annual precipitation of 623 mm. During the period from April to September, the mean temperature is 14.0 °C, and the precipitation is 415 mm. The vegetation period, defined by a mean temperature above 10 °C, spans over 158 days. The NNR Voděradské bučiny was established in 1955 on a total area of 658 ha to protect old-growth beech stands with a semi-natural stand structure (mainly associations *Luzulo-Fagetum* and *Asperulo-Fagetum*) and scattered geomorphological peri-glacial phenomena. Presently, over 65% of the area is classified as unmanaged beech stands and semi-managed, close-to-nature areas, with potential for transitioning to a non-intervention zone. The NNR Voděradské bučiny currently covers an area of 677 ha. The long-term objective of nature protection is the spontaneous development of most forest stands without direct human interventions [26]. This approach aligns with the focus on biodiversity research, e.g., [6,27].

The old beech stands that currently form the majority of the nature reserve were formerly managed beech stands. The main regeneration method in the past was a shelterwood system, which retained 42 seed trees per hectare. After 1838, a three-phase shelterwood felling approach was implemented, with silver fir being preferentially harvested during the first phase. The whole parent stand was removed within 12–15 years. Subsequently, release felling was followed by secondary felling, and after the next four or five years, the process was finished by final cutting [28]. This very short regeneration period resulted in almost pure and even-aged beech stands [29]. From 1810 to 1850, almost 500 ha of the area (i.e., 76% of Voděradské bučiny) was felled and regenerated. The current beech stands are over 180 years old.

2.2. Deadwood Variables

Eucnemids are strictly related to deadwood microhabitats [13,14]. For this reason, the study focused on the main components of old-growth beech forests and reserves, concentrating on snags and lying logs (minimum diameter of 5 cm) [30]. The deadwood types were analyzed within permanent research plots established in 2000 and 2005 [31]. Due to the irregular distribution of deadwood types, traps were attached to logs or snags within the research plot, i.e., in the same stand with identical structural conditions. Thus, it was possible to assess the influence of the deadwood types studied. The minimum distance between traps was 25 m. Deadwood was selected for trapping where there were no significant pieces of deadwood within the minimum distance (25 m). Several variables were measured for the studied deadwood types (Table 1). Moisture levels were assessed using a point-resistance moisture meter, and measurements were conducted on the same day for all the studied deadwood units (September 2023). This procedure avoids the influence of wet weather on the obtained moisture content of deadwood. The diameter was measured at breast height for snags and mid-length for lying logs using a diameter caliper. The height of the snags was measured with an altimeter, and the length of the logs was measured using a tape measure. The volume of lying logs was calculated using the Huber formula (V), considering the log's length (L) and cross-sectional area at the mid-point (gmid): V = L × gmid. For snags, the formula was based on Brunet and Isacsson [32]: $V = \pi \times d2 \times h/6$ (V = volume, d = DBH, and h = snag height in meters). The formula results in a volume corresponding to 2/3 of a cylinder of a given height and diameter. The canopy openness (canopy) was determined using spherical photographs taken with a fisheye lens above each trap. The analysis was performed using the Gap Light Analyzer software v2.0 [33]. The software converts the open sky to vegetation cover ratio into percentages for each trap site. The number of cavities (birds, base cavity, and dendrotelms) was counted on snags and logs, according to Winter and Möller [34]. Fungi were quantified based on the presence of fruiting bodies, mainly *Fomes fomentarius* and fungal groups (>5 cm in diameter). The proportion of bark loss and bryophytes was recorded as the percentage of the surveyed deadwood surface of the assessed microhabitat, ranging from 0% to 100%. Classification of deadwood into decay classes, according to [35], included the following: Class 1, freshly dead (1–2 years); 2, initiated decomposition (loose bark, tough sapwood); 3, advanced decomposition (soft sapwood and partly tough hardwood); and 4/5, extremely decomposed and moldered. The relationships between the characteristics of deadwood are shown in Figure 1.

Table 1. Values of the variables recorded in deadwood: mean (min–max). Differences were evaluated using one-way ANOVA. * $p < 0.05$ ** $p < 0.01$ *** $p < 0.001$; n.s. = non-significant.

Treatment	Units	Log	Snag	
moisture	%	23.2 (13–59)	14.1 (12–21)	***
diameter	cm	43.9 (9–110)	55.3 (21–131	n.s.
length	m	12 (4–21)	11.1 (3.6–22.2)	n.s.
volume	m^3	2.7 (0.03–12.7)	3.3 (0.6–9.1)	n.s.
canopy	%	8.3 (4.3–21.7)	11.4 (4.7–42.5)	*
cavity	pcs	0.1 (0–3)	0.9 (0–7)	*
fungi	pcs	3.2 (0–20)	5.3 (0–78)	n.s.
bark loss	%	54.3 (0–100)	42.5 (5–100)	n.s.
decay class	class	1st:15; 2nd:6; 3rd:5; 4th:3	1st:9; 2nd:14; 3rd:7	
bryophytes	%	10.9 (0–75)	0.2 (0–5)	**

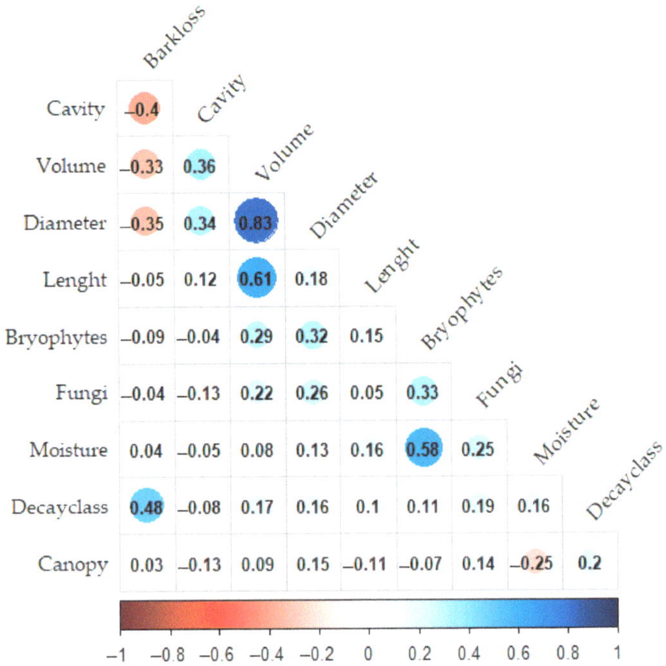

Figure 1. Spearman correlogram showing the relationship between deadwood characteristics (R package Corrplot; Wei and Simko [36]).

2.3. Beetle Sampling and Determination

Data collection was conducted from April to September in both 2022 and 2023. For both years, traps were set for different types of deadwood, consisting of 30 snags (2022) and 29 lying logs (2023). A total of 59 traps were used for the study. Window traps were used to capture beetles for the study objectives. This type of trap is suitable for recording saproxylic beetles [37,38] and thus is the most used type for beetles in many recent entomological studies, e.g., [2,5,6]. Window traps (unbaited) with a preservative solution of propylene glycol, water (1:1.5), and a drop of detergent to remove the surface tension of the solution were used to collect beetles. The catch was retrieved from the traps every two to three weeks. The trap consisted of a roof, a Plexiglass barrier, a funnel, and a collection container. The trap roof was made of a 45 cm diameter plastic bowl. Under the roof, two perpendicular Plexiglass panels created a barrier measuring 40 cm in width and 50 cm in height. Traps on the snags were hung using wire at a height of 1.3 m (southern direction, one trap per snag, Figure 2A). For recording beetles on lying logs, traps were placed directly against the log, with a catch basin on small (5 cm) pegs to prevent the capture of non-target groups (e.g., epigeic beetles) (Figure 2B). To determine the Eucnemids, all specimens (excluding *M. buprestoides*) from this family were first mounted on entomological tags (Figure 2C). Specimens were identified by the specialist Oto Nakládal, following the identification key, e.g., [39] and relying on his extensive experience and expertise in the family Eucnemidae, e.g., [20,23]. The conservation status of the Czech Republic and Germany (Deutschland) (CZ, DE) was also considered [18,40].

Figure 2. Illustration of the traps used in snag (**A**), lying log (**B**) and specimens of Eucnemids prepared for identification (**C**).

2.4. Data Analysis

The following software was used for data analyses: R 4.0.2 [41]. We tested the differences in abundance and number of species of the studied Eucnemidae between logs and snags using a generalized linear effect model (Poisson distribution and log link function) implemented in the package glmmTMB [42]. To assess the significance of preferences for snags/logs in each recorded species, the indicator species values approach was employed using the package "indicspecies" with the "multipatt" function with 9999 permutations [43]. For the species richness curves and gamma diversity assessment among snags/logs, we used the method by Chao et al. [44] based on the Hill numbers q = 0 (species richness) and q = 1 (the exponential of Shannon's entropy index) with the package "iNEXT" [45]. The curves were generated from incidence and abundance data. Subsequently, we used the method of estimating the total number of Eucnemidae species in the study site based on the incidence of each species captured in the study (sample coverage), according to Chao and Jost [46].

The analyses below were performed in Canoco 5 [47]. To determine the preference of each Eucnemidae species, we employed a constrained unimodal correspondence analysis (CCA) with a gradient of 4.3 SD units based on abundance and incidence data. Both input datasets were used without logarithmic transformation. CCA analyses were computed with 4999 unrestricted Monte Carlo permutations with the first constrained axis test. A symbol plot with two axes was used to visualize CCA results. The ordination method, detrended correspondence analysis (DCA), was employed to evaluate and intersperse the species across deadwood characteristics. We first used a comparison between unimodal unconstrained (DCA) vs. unimodal constrained analysis (CCA) to evaluate the highest explained variability in the data. The analysis revealed that 51% of the total variability was explained in the first two canonical axes, with DCA accounting for 47.99% and CCA for 33.60%. The ordination analysis was tested with the Monte Carlo permutation test, applying 4999 unrestricted permutations. A generalized linear model was used to evaluate and determine the significance (p-value) of Eucnemid's response to each deadwood characteristic. We used approaches based on abundance data (quasi-Poisson distribution) and incidence data (binomial distribution). Models were performed based on stepwise selection according to the lowest AIC [48].

3. Results

In total, 348 adults of Eucnemids belonging to eight species were recorded (Table 2). Trap captures between snags and logs did not show significant differences in abundance (p = n.s.) and the number of species (p = n.s.) (Figure 3). Most species preferred lying logs: based on abundance data, pF 6.5, $p < 0.001$ (Figure 4A), and based on incidence data, pF 5.5, $p < 0.001$ (Figure 4B). The significant association with the studied deadwood types among the recorded Eucnemids is shown in Table 3. The cumulative curve (gamma diversity) showed full species richness of old-growth beech stands with no increasing trend in the number of species (coverage-based 100%; eight species). Lying logs showed a higher species richness (q = 0) and, in particular, a significantly higher diversity of Eucnemids (q = 1; Figure 5A). Based on the abundance data, the difference has increased and is statistically significant q = 0 and q = 1 (Figure 5B). The ordination space of DCA illustrated that most Eucnemids species were clustered between higher moisture and advanced decay of deadwood. Deadwood characteristics and species preferences are shown via DCA (Figure 6). Deadwood moisture positively affected the occurrence and abundance of *Clypeorhagus clypeatus* and the abundance of *Hylis foveicollis* (Table 4). The presence of cavities, the thickness and length of deadwood, and canopy openness negatively affected the preference of most Eucnemids species recorded (Figure 6). For other deadwood parameters, there was no clear positive or negative effect on most of the species recorded (Table 4).

Table 2. Species of Eucnemidae and their number of individuals (incidence) recorded in the study.

Species	Abbreviation	Czechia RL	German RL	Log	Snag	Sum
Hylis olexai	*HylsOlex*	EN	NT	30 (17)	149 (10)	179
Melasis buprestoides	*MelsBupr*			24 (13)	87 (23)	111
Hylis foveicollis	*HylsFove*	EN	NT	25 (13)	0	25
Hylis cariniceps	*HylsCarn*	CR	EN	9 (5)	2 (2)	11
Isorhipis melasoides	*IsorMels*	EN	EN	8 (3)	3 (2)	11
Eucnemis capucina	*EucnCapc*	EN	EN	0	5 (5)	5
Clypeorhagus clypeatus	*ClypClyp*	*	-	3 (2)	1 (1)	4
Microrhagus lepidus	*MicrLepd*	EN		2 (2)	0	2

RL—Red List; Czechia: Hejda et al. [18]; Germany: Schmidl et al. [40]; * new species for the CZ Nakládal and Vávra, [20], (-) not included in RL.

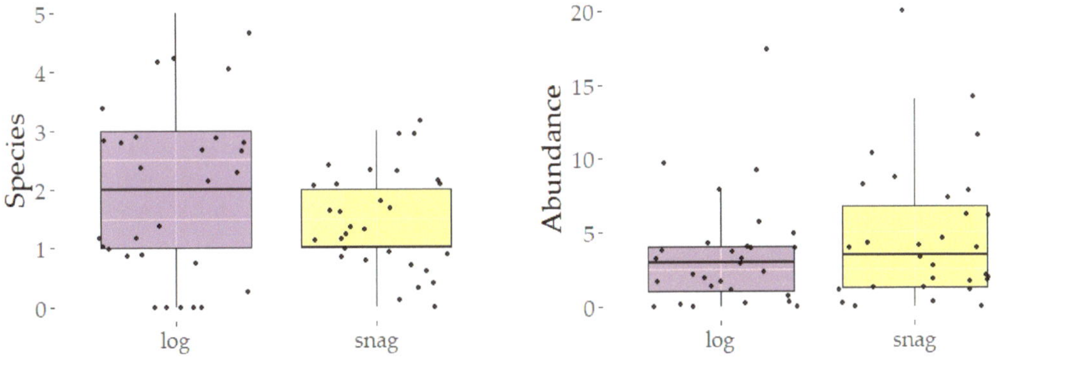

Figure 3. Number of species and number of individuals captured among the logs and snags. Solid lines in boxes represent median values; boxes indicate interquartile range (25%–75%), and whiskers min–max values. One outlier (snag 123 indi.) was removed from the abundance for better readability (right graph).

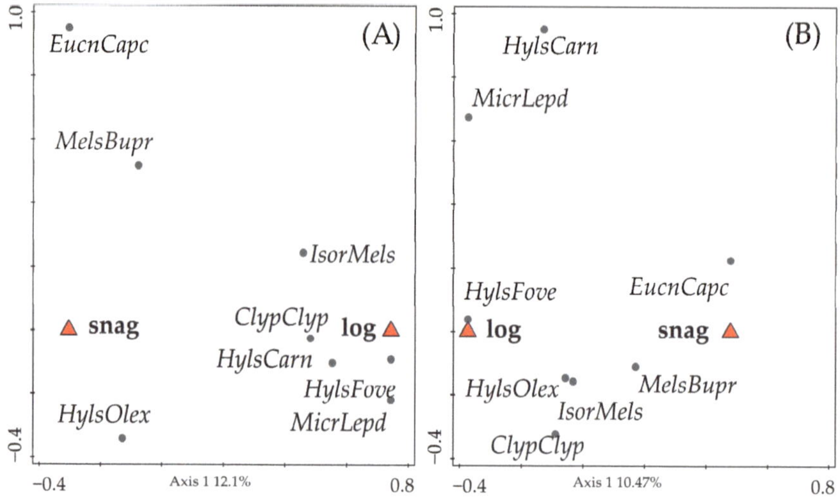

Figure 4. Species symbol plot summarizing the preference of beetle species. CCA's first two axes plotted—abundance data (**A**) and incidence data (**B**).

Table 3. Indicator species values of recorded species of the Eucnemidae and their importance in preferences per snags/logs. * $p < 0.05$, *** $p < 0.001$. Significant values are in bold.

	Logs			
	Stat	***p*-Value**	**Stat**	***p*-Value**
	incidence		abundance	
Hylis foveicollis	0.537	**0.0001** ***	0.438	**0.0001** ***
Hylis olexai	0.254	0.0719		
Microrhagus lepidus	0.189	0.2387	0.189	0.2340
Hylis cariniceps	0.163	0.2559	0.214	0.1201
Clypeorhagus clypeatus	0.081	0.6115	0.112	0.4847
Isorhipis melasoides	0.066	0.6695	0.105	0.5720
	Snags			
Melasis buprestoides	0.326	**0.0189** *	0.439	**0.0004** ***
Eucnemis capucina	0.302	0.0539	0.302	0.0512
Hylis olexai			0.125	0.6061

Table 4. Results of the generalized linear models. Abundance data (quasi-Poisson distribution) and incidence data (binomial distribution) were used. Significant values are in bold ($p < 0.05$). Positive (+: green) or negative (−: red) responses are indicated via *p*-value. *p*-values close to significance ($p < 0.07$) are also shown.

		Moisture	**Diameter**	**Length**	**Volume**	**Canopy**	**Cavity**	**Fungi**	**Bark Loss**	**Decay Class**	**Bryophytes**
Clypeorhagus clypeatus	Abundance	**+0.008**									+0.054
	Incidence	+0.068									
Eucnemis capucina	Abundance								**−0.004**		
	Incidence						+0.062		**−0.018**		

Table 4. *Cont.*

		Moisture	Diameter	Length	Volume	Canopy	Cavity	Fungi	Bark Loss	Decay Class	Bryophytes
Hylis cariniceps	Abundance										
	Incidence										
Hylis foveicollis	Abundance		**−0.047**		−0.058		**−0.022**				
	Incidence	**+0.022**	**−0.020**		**−0.046**	**−0.036**	**−0.016**		**+0.018**		
Hylis olexai	Abundance					**+0.001**					
	Incidence						**−0.008**				
Isorhipis melasoides	Abundance										
	Incidence										
Melasis buprestoides	Abundance	**−0.002**									**−0.025**
	Incidence	−0.064						**−0.013**	**−0.035**		**−0.002**
Microrhagus lepidus	Abundance								**+0.002**		
	Incidence								**+0.023**		

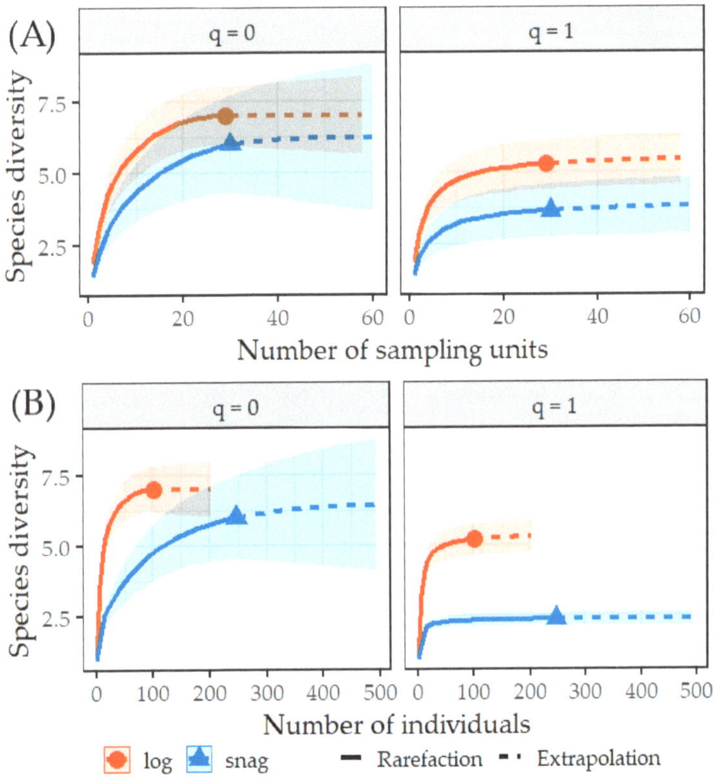

Figure 5. Sample-size-based rarefaction and extrapolation sampling gamma diversity of Eucnemidae—showing the Hill numbers' incidence data (**A**) and abundance data (**B**), q = 0 (species richness) and q = 1 (the exponential of Shannon's entropy index). Colored shaded areas represent the 95% confidence intervals. Solid symbols represent the total number of species and extrapolation (dashed lines) up to double the reference sample size.

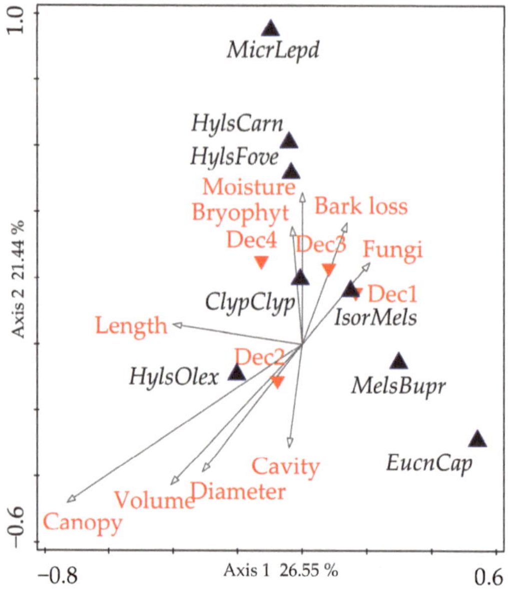

Figure 6. Results of detrented correspodence analysis show the first two canonical axes. The black triangles are Eucnemids species. Red triangles are decay classes, and arrows are environmental continuous characteristics of deadwood.

4. Discussion

We captured 348 beetles representing eight species over a two-year study with 59 traps. With more than a thousand traps distributed throughout Europe, [22] captured 15 species and 1816 individuals in beech-dominated forests. Our study area consists of relatively recent, pure autochthonous beech-protected stands [31] compared to Müller et al. [22], where a wide range of other tree species was represented <50%. This likely led to the higher capture of Eucnemids species, not only those associated with beech stands and beechwood. Another factor affecting the higher species counts in the study by [22] may be the inclusion of sites in Eastern Europe, renowned for its highly conserved forest complexes [49]. Eucnemids have a minimum of at least two years of development [13]. This could affect analyses comparing deadwood host substrates. We sampled the Eucnemids separately each year. We assume that Eucnemids' development is continuous over time and that annual fluctuations are insignificant at the study site. Thus, we also assume that the position of deadwood and its characteristics are important factors, as opposed to different sampling years.

Therefore, our study provides insight into the typical species composition of Eucnemidae in beech forests without further bias. Supporting evidence for the significance of the habitat in the studied beech stands is found compared to similar studies conducted in natural conditions and beech reserves. Procházka and Schlaghamerský [50] identified four species of Eucnemidae with 102 individuals, while Müller et al. [51] found five species with 14 individuals. *Melasis buprestoides*, *Hylis olexai*, and *H. foivecollis* proved to be the most abundant Eucnemids in our study. Our results correspond with [22], who also found the same Eucnemids species to be the most abundant in beech forests a decade earlier. *M. buprestoides* is the most common species in the family and has been observed mainly on snags. In contrast, the other species, especially *H. olexai* and *H. foivecollis*, were more abundantly trapped on lying logs, although most of the 123 individuals of *H. olexai* were captured by one trap on a snag with a large and heavily sunlit cavity. The association of *H. foivecollis* with lower deadwood volumes, as indicated by Procházka and Schlaghamerský 2019 [51], contrasts with our findings, which suggest that the moisture content of the deadwood may play a more significant role in this Eucnemid in beech stand. Results showed that the species *E. capucina* and *M. buprestoides* were associated with snags. These species are cavity-dwelling and thus may be negatively affected by the higher moisture content of fallen logs. Additionally, cavities were practically absent in fallen logs. *E. capucina* has a strong preference for cavity trees [23,52], which is also supported by the results of our study. On the contrary, *M. buprestoides* appears to prefer drier deadwood at the transition between hard wood and soft, partially decomposed wood; therefore, this species favors habitat trees to cavity trees [52]. *Clypeorhagus clypeatus*, a species recently discovered in the Czech Republic [20], was recorded during the study, and our records are the first for all of Bohemia [53]. This species is poorly known, with only a few specimens documented [17]. It is mainly associated with beech forests and beech deadwood in areas without interrupted forest continuity characterized by minimal forest management. The bionomy of this species is not well-described by Mertlik [39].

The preference of most Eucnemids for deadwood position was observed in lying logs. Similar faunistic findings of Eucnemids favoring lying logs have been reported in previous studies, e.g., [21,54] or other beetle species *Tragosoma depsarium* (Cerambycidae) [55]. The position of deadwood influences both its moisture content and the process of wood decomposition [55]. In contrast, [56] found standing oak deadwood generally richer in saproxylic beetles than in lying deadwood. Franc [57] reported differences in favor of lying logs in the case of oak deadwood. In general, oak deadwood is recognized for its richness in saproxylic beetle species [58]. We focused on beech deadwood, which exhibits significantly faster decomposition processes compared to oak [59]. This characteristic may have implications for Eucnemids, as higher moisture content is a very important parameter of lying logs. In addition, natural beech stands are characterized by low heterogeneity due to their darkness. Therefore, in our study, canopy openness was an insignificant variable for most of the recorded Eucnemids species. Consequently, the beech Eucnemids do not seem to support the thesis that sun exposure is a critical element for endangered species [9,60].

Dry and wet periods are likely to influence the preferences of Eucnemids. In current drought periods, deadwood moisture becomes a limiting factor, potentially leading Eucnemids to prefer lying logs. This is probably due to lying logs having a more stable temperature throughout the year and not fluctuating much compared to snags [55]. Conversely, during wet periods, they may exhibit a stronger preference for snags, seeking sunlight as a source of heat and a drier woody substrate. In general, deadwood absorbs water depending on the degree of decomposition and also helps maintain the forest microclimate by evaporation [61]. Likewise, the beech Eucnemids species may be more likely associated with shadier and wetter conditions and to seek out moist pieces of deadwood. We identified the higher moisture levels and higher levels of deadwood decomposition as important variables influencing the presence of *C. clypeatus*, *H. foivecollis*, *H. cariniceps*, and *M. lepidus*. While lying beech deadwood is known to be relatively unstable over time, quickly becoming an unsuitable habitat for beetles [62] and more closely resembling the soil environment [62], standing deadwood increases the persistence of deadwood in stands over several decades [63]. However, our findings indicate that snags exhibit low moisture content, resulting in a lower abundance of Eucnemids. Climate change, particularly alterations in the seasonal distribution of precipitation, may also have an influence on the life of Eucnemids. The overall drier climate in recent times [64] may also negatively impact the Eucnemids. The primary driver of deadwood moisture is rainfall [65]. Nevertheless, their high requirements for deadwood quality and, often, forest continuity appear to be the main limiting factors for their existence [13]. Bark loss was found to be one of the influential variables, especially affecting the species *M. lepidus*. The absence of bark may also affect deadwood moisture. The loss of bark exposes the wood, allowing rainfall to penetrate deeper layers of the wood directly. We observed bark loss on logs with higher moisture content, which also indicates advanced decomposition. However, the microclimate inside deadwood is also affected by forest characteristics [55].

5. Conclusions

The study investigated the habitat requirements of a truly rare saproxylic family in beech stands. Moisture conditions within lying logs play an important role in the successful occurrence and survival of Eucnemidae species. The increasing prevalence of drier and precipitation-free periods in recent times [64,66] may impact the overall abundance and richness of Eucnemids species. Consequently, their preferences are directed primarily to lying logs, which, being in contact with the ground, absorb moisture and provide a favorable environment for the life cycle of beech Eucnemids.

Author Contributions: Conceptualization, V.Z., O.N. and J.R.; methodology, V.Z., O.N. and J.R.; validation, V.Z.; formal analysis, V.Z.; investigation, V.Z. and O.N.; data curation, V.Z.; resources V.Z. and O.N.; writing—original draft preparation, V.Z.; writing—review and editing, V.Z., O.N. and J.R.; visualization, V.Z.; project administration J.R.; funding acquisition, J.R. All authors have read and agreed to the published version of the manuscript.

Funding: The study was funded by grant No. QK23020008, funded by the Ministry of Agriculture of the Czech Republic.

Data Availability Statement: Data available on request.

Acknowledgments: We are also grateful to Markéta Macháčová who proofread the language of this paper. We thank the nature conservation authority for permission to study invertebrates in the reserve. We are also grateful to two anonymous reviewers for valuable comments on improving this scientific paper.

Conflicts of Interest: The authors declare no conflicts of interest.

References

1. Oettel, J.; Lapin, K. Linking Forest management and biodiversity indicators to strengthen sustainable forest management in Europe. *Ecol. Indic.* **2021**, *122*, 107275. [CrossRef]
2. Leidinger, J.; Blaschke, M.; Ehrhardt, M.; Fischer, A.; Gossner, M.M.; Jung, K.; Kienlein, S.; Kózak, J.; Michler, B.; Mosandl, R.; et al. Shifting tree species composition affects biodiversity of multiple taxa in Central European forests. *For. Ecol. Manag.* **2021**, *498*, 119552. [CrossRef]
3. Schauer, B.; Thorn, S.; Blaschke, M.; Kudernatsch, T. Conversion of pure spruce to mixed spruce beech stands: Effects on alpha and beta diversity of multiple taxonomic groups. *For. Ecol. Manag.* **2023**, *545*, 121297. [CrossRef]
4. Kostanjsek, F.; Sebek, P.; Baranova, B.; Seric Jelaska, L.; Riedl, V.; Cizek, L.; Didham, R.; Müller, J. Size matters! Habitat preferences of the wrinkled bark beetle, *Rhysodes sulcatus*, the relict species of European primeval forests. *Insect Conserv. Divers.* **2018**, *11*, 545–553. [CrossRef]
5. Graf, M.; Seibold, S.; Gossner, M.M.; Hagge, J.; Weiß, I.; Bässler, C.; Müller, J. Coverage based diversity estimates of facultative saproxylic species highlight the importance of deadwood for biodiversity. *For. Ecol. Manag.* **2022**, *517*, 120275. [CrossRef]
6. Zumr, V.; Nakládal, O.; Remeš, J.; Brestovanská, T.; Zumr, V. Diversity of click beetles in managed nonnative coniferous and native beech stands: Consequences of changes in the structural and species composition of tree stands in Central Europe. *For. Ecosyst.* **2022**, *9*, 100057. [CrossRef]
7. Seibold, S.; Bässler, C.; Brandl, R.; Gossner, M.M.; Thorn, S.; Ulyshen, M.D.; Müller, J. Experimental studies of dead-wood biodiversity—A review identifying global gaps in knowledge. *Biol. Conserv.* **2015**, *191*, 139–149. [CrossRef]
8. Speight, M.C.D. Saproxylic invertebrates and their conservation. *Nat. Environ. Ser.* **1989**, *42*, 1–79.
9. Seibold, S.; Brandl, R.; Buse, J.; Hothorn, T.; Schmidl, J.; Thorn, S.; Müller, J. Association of extinction risk of saproxylic beetles with ecological degradation of forests in Europe. *Conserv. Biol.* **2015**, *29*, 382–390. [CrossRef]
10. Parajuli, R.; Markwith, S.H. Quantity is foremost but quality matters: A global meta-analysis of correlations of dead wood volume and biodiversity in forest ecosystems. *Biol. Conserv.* **2023**, *283*, 110100. [CrossRef]
11. Thorn, S.; Bässler, C.; Brandl, R.; Burton, P.J.; Cahall, R.; Campbell, J.L.; Castro, J.; Choi, C.-Y.; Cobb, T.; Donato, D.C.; et al. Impacts of salvage logging on biodiversity: A meta-analysis. *J. Appl. Ecol.* **2018**, *55*, 279–289. [CrossRef] [PubMed]
12. Muona, J. Eucnemidae. In *Catalogue of Palaearctic Coleoptera: Elateroidea-Derodontoidea-Bostrichoidea-Lymexyloidea-Cleroidea-Cucujoidea*; Löbl, I., Smetana, A., Eds.; Apollo Books: Stenstrup, Denmark, 2007; Volume 4, pp. 81–87. 935p.
13. Vávra, J.; Škorpík, M. False click beetles (Coleoptera: Eucnemidae) in the Podyjí National Park and surrounding area, with notes to their bionomics. *Thayensia* **2013**, *10*, 53–90.
14. Vávra, J. Eucnemidae. In *Red List of Threatened Species of the Czech Republic*; Hejda, R., Farkač, J., Chobot, K., Eds.; Agentura Ochrany Přírody a Krajiny České Republiky: Příroda, Praha, 2017; p. 355.

15. Eckelt, A.; Müller, J.; Bense, U.; Brustel, H.; Bußler, H.; Chittaro, Y.; Cizek, L.; Frei, A.; Holzer, E.; Kadej, M.; et al. "Primeval Forest Relict Beetles" of Central Europe: A Set of 168 Umbrella Species for The Protection of Primeval Forest Remnants. *J. Insect Conserv.* **2018**, *22*, 15–28. [CrossRef]
16. Nieto, A.; Alexander, K.N.A. *European Red List of Saproxylic Beetles*; Publications Office of the European Union: Luxembourg, 2010; p. 45.
17. Mertlik, J. Faunistic Maps of Species of the Families Cerophytidae, Elateridae, Eucnemidae and Throscidae (Coleoptera: Elateroidea) of the Czech Republic and Slovakia.—Permanent Electronic Publication 2007–2023 (akt. 1.1.2023). 2023. Available online: https://www.elateridae.com/pag_uni.php?idp=19 (accessed on 6 January 2024).
18. Hejda, R.; FarkaČ, J.; Chobot, K. *Red List of Threatened Species of the Czech Republic*; Agentura Ochrany Přírody a Krajiny České Republiky: Praha, Czech Republic, 2017; Volume 36, pp. 1–612, ISBN 978-80-88076-53-7.
19. Foit, J. Faunistic records from the Czech Republic—436. Coleoptera: Eucnemidae. *Klapalekiana* **2018**, *54*, 4.
20. Nakládal, O.; Vávra, C.J. Faunistic records from the Czech Republic—454. Coleoptera: Eucnemidae. *Klapalekiana* **2018**, *54*, 290.
21. Bednařík, M. Otho Sphondyloides (Germar, 1818) (Coleoptera, Eucnemidae), a New Species for the Fauna of the Czech Republic. 2021. Available online: https://www.elateridae.com/elateridarium/page.php?idcl=335 (accessed on 3 January 2024).
22. Müller, J.; Brunet, J.; Brin, A.; Bouget, C.; Brustel, H.; Bussler, H.; Foerster, B.; Isacsson, G.; Koehler, F.; Lachat, T.; et al. Implications from large-scale spatial diversity patterns of saproxylic beetles for the conservation of European Beech forests. *Insect Conserv. Divers.* **2013**, *6*, 162–169. [CrossRef]
23. Nakládal, O.; Synek, J.; Zumr, V. Diurnal and Sex Ratio Flight Activity of Rare Cavity-Dweller *Eucnemis capucina* Ahrens, 1812 (Coleoptera: Eucnemidae) in Lowland Deciduous Forest. *Forests* **2023**, *14*, 720. [CrossRef]
24. NCA CR. *Species Occurrence Database*; Nature Conservation Agency of the Czech Republic, NCA CR: Prague, Czech Republic, 2024. Available online: https://portal.nature.cz/publik_syst/ctihtmlpage.php?what=3&nabidka=hlavni (accessed on 19 January 2024).
25. *Care Plan for NPR Voděradské Bučiny 2021–2029*; Agency for Nature Conservation and Landscape Protection of the Czech Republic: Prague, Czech Republic, 2023.
26. Bílek, L.; Remeš, J.; Podrázský, V.; Rozenbergar, D.; Diaci, J.; Zahradník, D. Gap Regeneration in Near-Natural European Beech Forest Stands in Central Bohemia—The Role of Heterogeneity and Micro-Habitat Factors. *Dendrobiology* **2013**, *71*, 59–71. [CrossRef]
27. Zumr, V.; Nakládal, O.; Bílek, L.; Remeš, J. The diameter of beech snags is an important factor for saproxylic beetle richness: Implications for forest management and conservation. *For. Ecosyst.* **2023**, *10*, 100143. [CrossRef]
28. Šrámek, O. SPR Voděradské bučiny I. a II. *Památky A Příroda* **1983**, *166–171*, 241–248.
29. Pokorný, J. Vývoj a druhová skladba černokosteleckých lesů do konce 18. století, Část II. Vývoj a obraz druhové skladby revíru Bohumile, Voděrady a Oplany. *Sborník Lesn. Fak. Vysok. Školy Zemědělské V Praze* **1963**, *6*, 31–107.
30. Christensen, M.; Hahn, K.; Mountford, E.P.; Ódor, P.; Standovár, T.; Rozenbergar, D.; Diaci, J.; Wijdeven, S.; Meyer, P.; Winter, S.; et al. Dead wood in European beech (*Fagus sylvatica*) forest reserves. *For. Ecol. Manag.* **2005**, *210*, 267–282. [CrossRef]
31. Bilek, L.; Remes, J.; Zahradnik, D. Managed vs. unmanaged. Structure of beech forest stands (*Fagus sylvatica* L.) after 50 years of development, Central Bohemia. *For. Syst.* **2011**, *20*, 122–138. [CrossRef]
32. Brunet, J.; Isacsson, G. Influence of snag characteristics on saproxylic beetle assemblages in a south Swedish beech forest. *J. Insect Conserv.* **2009**, *13*, 515–528. [CrossRef]
33. Frazer, G.W.; Canham, C.D.; Lertzman, K.P. *Gap Light Analyzer (GLA), Version 2.0: Imaging Software to Extract Canopy Structure and Gap Light Transmission Indices from True-Colour Fisheye Photographs, User's Manual and Program Documentation*; Simon Fraser University/Institute of Ecosystem Studies: Burnaby/Millbrook, UK, 1999.
34. Winter, S.; Möller, G.C. Microhabitats in lowland beech forests as monitoring tool for nature conservation. *For. Ecol. Manag.* **2008**, *255*, 1251–1261. [CrossRef]
35. Gossner, M.; Lachat, T.; Brunet, J.; Isacsson, G.; Bouget, C.; Brustel, H.; Brandl, R.; Weisser, W.W.; Müller, J. Current near-to-nature forest management effects on functional trait composition of saproxylic beetles in beech forests. *Conserv. Biol.* **2013**, *27*, 605–614. [CrossRef]
36. Wei, T.; Simko, V. R Package 'corrplot': Visualization of a Correlation Matrix (Version 0.92). 2021. Available online: https://github.com/taiyun/corrplot (accessed on 5 January 2024).
37. Wikars, L.-O.; Sahlin, E.; Ranius, T. A Comparison of Three Methods to Estimate Species Richness of Saproxylic Beetles (Coleoptera) in Logs and High Stumps of Norway Spruce. *Can. Entomol.* **2005**, *137*, 304–324. [CrossRef]
38. Alinvi, O.; Ball, J.P.; Danell, K.; Hjältén, J.; Pettersson, R.B. Sampling Saproxylic Beetle Assemblages in Dead Wood Logs: Comparing Window and Eclector Traps to Traditional Bark Sieving and A Refinement. *J. Insect Conserv.* **2007**, *11*, 99–112. [CrossRef]
39. Mertlik, J. The species of the family Melasidae (Coleoptera: Elateroidea) Czech and Slovak Republics. *Elateridarium* **2008**, *2*, 69–137. Available online: https://www.elateridae.com/elateridarium/page.php?idcl=88 (accessed on 1 September 2023).
40. Schmidl, J.; Sausage, C.; Bussler, H. Red list and complete list of varieties of the "Diversicornia" (Coleoptera) of Germany. In *Red List of Endangered Animals, Plants and Fungi of Germany, Volume 5: Invertebrates (Part 3)*; Nature Protection and Biodiversity; Ries, M., Balzer, S., Gruttke, H., Haupt, H., Hofbauer, N., Ludwig, G., Matzke-Hajek, G., Eds.; Agriculture Publisher: Münster, Germany, 2021; Volume 70, pp. 99–124.

41. R Core Team. *R: A Language and Environment for Statistical Computing*; R Foundation for Statistical Computing: Vienna, Austria, 2020. Available online: https://www.R-project.org/ (accessed on 5 June 2023).
42. Brooks, M.E.; Kristensen, K.; van Benthem, K.J.; Magnusson, A.; Berg, C.W.; Nielsen, A.; Skaug, H.J.; Mächler, M.; Bolker, B.M. Glmmtmb Balances Speed And Flexibility Among Packages For Zero-Inflated Generalized Linear Mixed Modeling. *R J.* **2017**, *9*, 378–400. [CrossRef]
43. De Cáceres, M.; Legendre, P. Associations between species and groups of sites: Indices and statistical inference. *Ecology* **2009**, *90*, 3566–3574. [CrossRef]
44. Chao, A.; Gotelli, N.J.; Hsieh, T.C.; Sander, E.L.; Ma, K.H.; Colwell, R.K.; Ellison, A.M. Rarefaction and extrapolation with Hill numbers: A framework for sampling and estimation in species diversity studies. *Ecol. Monogr.* **2014**, *84*, 45–67. [CrossRef]
45. Hsieh, T.C.; Ma, K.H.; Chao, A. iNEXT: iNterpolation and EXTrapolation for Species Diversity. R Package Version 2022, 3.0.0. Available online: http://chao.stat.nthu.edu.tw/wordpress/software-download/ (accessed on 20 January 2024).
46. Chao, A.; Jost, L. Coverage-based rarefaction and extrapolation: Standardizing samples by completeness rather than size. *Ecology* **2012**, *93*, 2533–2547. [CrossRef]
47. Šmilauer, P.; Lepš, J. *Multivariate Analysis of Ecological Data Using CANOCO 5*; Cambridge University Press: Cambridge, UK, 2014. [CrossRef]
48. Akaike, H. A Bayesian analysis of the minimum AIC procedure. *Ann. Inst. Stat. Math.* **1978**, *30*, 9–14. [CrossRef]
49. Sabatini, F.M.; Burrascano, S.; Keeton, W.S.; Levers, C.; Lindner, M.; Pötzschner, F.; Verkerk, P.J.; Bauhus, J.; Buchwald, E.; Chaskovsky, O.; et al. Where Are Europe's Last Primary Forests? *Divers. Distrib.* **2018**, *24*, 1426–1439. [CrossRef]
50. Procházka, J.; Schlaghamerský, J. Does dead wood volume affect saproxylic beetles in montane beech-fir forests of Central Europe? *J. Insect Conserv.* **2019**, *23*, 157–173. [CrossRef]
51. Müller, J.; Bußler, H.; Kneib, T. Saproxylic Beetle Assemblages Related to Silvicultural Management Intensity And Stand Structures In A Beech Forest In Southern Germany. *J. Insect Conserv.* **2008**, *12*, 107–124. [CrossRef]
52. Müller, J.; Jarzabek-Müller, A.; Bussler, H.; Gossner, M.M. Hollow beech trees identified as keystone structures for saproxylic beetles by analyses of functional and phylogenetic diversity. *Anim. Conserv.* **2014**, *17*, 154–162. [CrossRef]
53. Nakládal, O.; Zumr, V. Faunistic Records from the Czech Republic—526: Coleoptera: Eucnemidae. *Klapalekiana* **2022**, *28*, 112.
54. Mertlik, J.; Pelikán, J. New data about *Hylis olexai* (Coleoptera: Eucnemidae) for the area of the Czech Republic and Slovakia. *Elateridarium* **2013**, *7*, 45–54. Available online: https://www.elateridae.com/elateridarium/page.php?idcl=207 (accessed on 2 January 2024).
55. Lindman, L.; Öckinger, E.; Ranius, T. Microclimatic conditions mediate the effect of deadwood and forest characteristics on a threatened beetle species, *Tragosoma depsarium*. *Oecologia* **2022**, *199*, 737–752. [CrossRef] [PubMed]
56. Bouget, C.; Nusillard, B.; Pineau, X.; Ricou, C. Effect of deadwood position on saproxylic beetles in temperate forests and conservation interest of oak snags. *Insect Conserv. Divers.* **2012**, *5*, 264–278. [CrossRef]
57. Franc, N. Standing or downed dead trees—Does it matter for saproxylic beetles in temperate oak-rich forest? *Can. J. For. Res.* **2007**, *37*, 2494–2507. [CrossRef]
58. Kappes, H.; Topp, W. Emergence of Coleoptera from deadwood in a managed broadleaved forest in central Europe. *Biodivers. Conserv.* **2004**, *13*, 1905–1924. [CrossRef]
59. Hararuk, O.; Kurz, W.A.; Didion, M. Dynamics of dead wood decay in Swiss forests. *For. Ecosyst.* **2020**, *7*, 36. [CrossRef]
60. Jonsell, M.; Weslien, J.; Ehnström, B. Substrate requirements of red-listed saproxylic invertebrates in Sweden. *Biodivers. Conserv.* **1998**, *7*, 749–764. [CrossRef]
61. Floriancic, M.G.; Allen, S.T.; Meier, R.; Truniger, L.; Kirchner, J.W.; Molnar, P. Potential for Significant Precipitation Cycling by Forest-Floor Litter and Deadwood. *Ecohydrology* **2023**, *16*, e2493. [CrossRef]
62. Bartlett, B. *The Afterlife of a Tree*; WWF Poland: Warszawa-Hajnówka, Poland, 2005; ISBN 83-920712-1-2.
63. Oettel, J.; Zolles, A.; Gschwantner, T.; Lapin, K.; Kindermann, G.; Schweinzer, K.-M.; Gossner, M.M.; Essl, F. Dynamics of standing deadwood in Austrian forests under varying forest management and climatic conditions. *J. Appl. Ecol.* **2023**, *60*, 696–713. [CrossRef]
64. Allen, C.D.; Macalady, A.K.; Chenchouni, H.; Bachelet, D.; McDowell, N.; Vennetier, M.; Kitzberger, T.; Rigling, A.; Breshears, D.D.; Hogg, E.H.; et al. A global overview of drought and heat-induced tree mortality reveals emerging climate change risks for forests. *For. Ecol. Manag.* **2010**, *259*, 660–684. [CrossRef]
65. Green, M.B.; Fraver, S.; Lutz, D.A.; Woodall, C.W.; D'Amato, A.W.; Evans, D.M. Does Deadwood Moisture Vary Jointly with Surface Soil Water Content? *Soil Sci. Soc. Am. J.* **2022**, *86*, 1113–1121. [CrossRef]
66. Park Williams, A.; Allen, C.D.; Macalady, A.K.; Griffin, D.; Woodhouse, C.A.; Meko, D.M.; Swetnam, T.W.; Rauscher, S.A.; Seager, R.; Grissino-Mayer, H.D.; et al. Temperature as a potent driver of regional forest drought stress and tree mortality. *Nat. Clim. Chang.* **2013**, *3*, 292–297. [CrossRef]

forests MDPI

Article

Hidden Potential of the Subdominant Ant *Formica lemani* Bondroit (Hymenoptera: Formicidae): The Formation of Large Nest Complexes and Restructuring Behavioural Stereotypes

Tatiana Novgorodova * and Dmitry Taranenko

Institute of Systematics and Ecology of Animals SB RAS, Novosibirsk 630091, Russia
* Correspondence: tanovg@yandex.ru

Abstract: The potential of subdominant ants of the *Formica fusca* group and their role in forests are still underestimated. Since ant behaviour is dependent on colony size, studying the functional organisation of nest complexes (NC) is most promising for a more accurate assessment of species capabilities. The study focused on the main ecological and ethological issues of the life activity of *Formica lemani* Bondroit within large NC (>150 nests) and beyond. After preliminary mapping of the *F. lemani* NC (main nests, trails, foraging trees), off-nest activity, aggressiveness, and trophobiotic relationships with aphids in and outside the NC territory were studied. Within the NC, the dynamic density, the intensity of movement on trails, and aggressiveness of *F. lemani* were significantly higher than beyond; the range of symbiont aphids was twice as small, with aphids on birches playing a key role in carbohydrate nutrition of *F. lemani*. The latter ensures accelerated restoration of trophobiotic interactions in spring and stability of the food supply until autumn. Combined with the lack of pressure from *F. rufa* group ants, this allowed *F. lemani* to maintain high population densities, and significantly increased its competitiveness, and role in plant protection against phytophages.

Keywords: aspen–birch forest; *Formica fusca* group; ant assemblages; hierarchical status; dynamic density; aggressiveness; trophobiosis; ant–aphid interactions

Citation: Novgorodova, T.; Taranenko, D. Hidden Potential of the Subdominant Ant *Formica lemani* Bondroit (Hymenoptera: Formicidae): The Formation of Large Nest Complexes and Restructuring Behavioural Stereotypes. *Forests* **2024**, *15*, 1322. https://doi.org/10.3390/f15081322

Academic Editors: Rostislav Zemek and Katarína Pastirčáková

Received: 27 June 2024
Revised: 25 July 2024
Accepted: 28 July 2024
Published: 30 July 2024

1. Introduction

Ants of the *Formica rufa* group with complex social and territorial behaviour, forming large colonies and supercolonies, play one of the leading roles in terrestrial ecosystems, including forests, and are the major structuring force of ant assemblages [1–9]. Due to the complex organisation of the protected foraging area, specialisation of worker ants [10,11], and task partitioning among foragers collecting honeydew [12,13], obligate dominants of the *Formica rufa* group actively protect their resources from competitors, including natural enemies [14,15]. This allows the ants to utilise available resources more efficiently, maintain a stable resource base, and, at the same time, protect plants from leaf-eating insects.

Representatives of the *Formica fusca* group and the *Formica rufibarbis* group, previously referred to the subgenus *Serviformica* (hereinafter—the Serviformica group), have the status of subdominants in multispecies ant assemblages. In these species, foraging territories are usually unprotected, there is no specialisation among foragers, and solitary foraging is observed [8,16]. At the same time, the results of some studies indicate the presence of hidden potential in ants of this group. It was found experimentally that behaviour of subdominant ants of the genus *Formica*, in particular their social and territorial organisation, depend, to a large extent, on colony size and dynamic density of workers [10,11,16–19]. For example, when colony size of the *Formica cunicularia* Latreille and *F. picea* Nylander exceeds one thousand workers, the complexity of both the social structure of the ant colony and the territorial organisation of these ant species increases: colonies of both species begin to build mound nests and, at least partly, to protect their foraging territories [16,20]. Furthermore, besides the simple models of the organisation of honeydew collection, the use of more complex models with protection of aphid colonies, was noted [13]. In the presence of natural and/or artificial objects that are able to accumulate heat (stone, stump, metal sheet, etc.), the construction of mound nests was also observed in *F. fusca* Linnaeus [10].

However, it should be noted that the dynamic density and colony size of subdominant ants from the Serviformica group are usually controlled and strictly regulated by obligate dominants of the *Formica* genus both within their foraging areas and during the occupation of new territories [7,11,21–23]. Nevertheless, the formation of stable nest complexes was noted in some species of the group—*F. cunicularia, F. cinerea* Mayr, *F. imitans* Ruzsky [10,18,19,24,25]. All this suggests that the hidden capabilities of the subdominant ants from the Serviformica group may be much broader than previously thought, and that the extent of their potential capabilities is still being underestimated.

One of the promising representatives of the group, in terms of studying hidden potential, is the northern black ant, *Formica lemani* Bondroit. It is a transpalaearctic boreomontane species that plays a subdominant role in forest ant assemblages [8,26]. In the forest–steppe and steppe zones, nests of *F. lemani* are confined to forest elements (aspen–birch stands, forest plantations, etc.). Like other species of the group, *F. lemani* primarily builds sectional nests in soil and often settles in dead wood. However, under certain conditions, such as scarcity of suitable habitats, it may form mound nests. The size of monogynous colonies of *F. lemani* does not exceed 1000 individuals, while polygynous colonies may reach from four to six thousand individuals. The foraging territory of *F. lemani* is usually not protected [8], and neither are its carbohydrate food sources, including aphid colonies (Novgorodova, unpublished data). This species usually uses tunnels to communicate between sections in the nest and to forage for food [8]. Formation of nest complexes in *F. lemani* has not been noted until recently. However, in 2020, a large nest complex of *F. lemani* was discovered for the first time in one of the aspen–birch stands in the south of Western Siberia (Novosibirsk Region, Russia) [27]. The nest complex immediately attracted attention due to the powerful ant trails on the trees, which were completely atypical for this species, and clearly visible even from afar. Studying the structural and functional organisation of such single-species nest complexes of subdominants of the genus *Formica* is highly important for assessing the potential functional capabilities of the species, as well as for predicting the effects of forced structural rearrangements of multispecies ant assemblages.

The main aim of the study was to explore the main ecological and ethological issues of the life activity in *F. lemani* from large settlements, utilising the example of the nest complex identified in the Novosibirsk Region (Russia) in 2020. The objectives of this research were: (i) to estimate the size and identify the main infrastructural elements of the territory of the large *F. lemani* complex nest; (ii) to identify behavioural patterns of the *F. lemani* foragers, including off-nest activity, aggressiveness, and organisation of honeydew collection, in and outside the nest complex; and (iii) to identify the features of trophobiotic relationships between *F. lemani* foragers from the large settlement and aphids.

2. Materials and Methods

The study was conducted in August–September 2020 and 2022 in the south of the Novosibirsk Region (Karasuk District, vicinity of Sheinfeld Village; Russia) in the territory of the *F. lemani* nest complex (NC) in an aspen–birch stand, as well as outside the nest complex—in the same or neighbouring aspen–birch stand.

The species identity of the *F. lemani* individuals from the studied NC was additionally confirmed by the results of the molecular genetic analysis (sample F24-1): by COI—genebank number OM722036 (https://www.ncbi.nlm.nih.gov/nuccore/OM722036.1 accessed on 25 May 2023), ITS1—OM728505 (https://www.ncbi.nlm.nih.gov/nuccore/OM728505.1 accessed on 25 May 2023), D2 28S—OM722055 (https://www.ncbi.nlm.nih.gov/nuccore/OM722055.1 accessed on 25 May 2023) [28].

2.1. The Main Characteristics of the F. lemani Nest Complex

To approximate the size and structure of the *F. lemani* nest complex, determine its boundaries and occupied territory, as well as identify the main infrastructural elements of the NC area, a schematic mapping of the *F. lemani* NC was carried out, indicating the main nests and trails, as well as foraging trees (hereafter—mapping). The mapping was carried out in two stages. At the first stage, in late September 2020 (at the time of discovery), the main part of the *F. lemani* NC was mapped. During that period, mainly the largest nests were mapped. Their location was identified by the tree foraging ant trails, strong enough to persist until the end of the season. The mapping was continued in the first half of August 2022 with confirming the location of the large nests, collecting extra information on the other nests, and mapping the system of foraging trails.

In the explored territory, the nests of the *Formica rufa* group were also mapped. The main characteristics of each anthill were noted. The diameter and height of each anthill mound were measured both without and including the soil base—embankment (d/D and h/H, respectively). For dead (degraded) ant nests, the diameter of the crater was measured.

2.2. Off-Nest Activity and Aggressiveness of the F. lemani Foragers

To identify the behavioural characteristics of the *F. lemani* foragers, we took into account the off-nest activity of the species studied (dynamic density of foragers and intensity of movement of foragers on foraging trees), as well as assessed the aggressiveness of foragers on tree trunks and in aphid colonies. The studies were carried out both on and off the area occupied by the *F. lemani* nest complex in the same stand. The investigations outside the NC were carried out at a distance of more than 20 m from the border of NC, which was clearly visible by green foliage and trails of *F. lemani* on birch trunks.

2.2.1. Dynamic Density

Dynamic ant density was assessed in August 2022 using standard methodology [8]. Ants were counted in three replicates in a 10 dm^2 (31.6×31.6 cm^2) plot bounded by a wire frame on legs. The number of ants that visited the site within 5 min was recorded, after which the average dynamic density (individuals/$dm^2 \cdot$min) was calculated. A total of 23 sites were surveyed: 13—in the territory of the NC; 10—beyond. In the territory of the *F. lemani* NC, the study plots were located at least 2 m from the nearest foraging trail of *F. lemani*.

2.2.2. Intensity of Movement of Foragers on Trees

The intensity of foraging movement on the trunks of main forage trees was assessed on birch (*Betula pendula* Roth) in early August 2022. A count of foragers crossing a conventional line in both directions at a height of about 1.3 m was carried out for 5 min (in triplicate) with subsequent averaging and recalculation for 1 min (individuals/min). When traffic of foragers was of high intensity, we used video recording and counted foragers in the laboratory. Testing was carried out on 29 selected trees located within and off the territory of the *F. lemani* nest complex (19 and 10 trees, respectively). The trunk diameter of the birches selected for testing ranged from 9 to 28 cm (at the site of forager counting), and the ratio of the trees of different size categories within and off the *F. lemani* NC was similar. The proportions of selected birches of three size categories in the territory of the NC and beyond were 9–15 cm—31.6% and 30.0%, respectively; 16–20 cm—42.1% and 40.0%; 21–28 cm—26.3% and 30.0%.

2.2.3. Ant Aggressiveness

We compared the aggressiveness of *F. lemani* foragers in the territory of the NC with the aggressiveness data of foragers tested outside it. Aggressiveness was assessed on birch trunks (diameter from 16 to 20 cm) and in aphid colonies of *Chaitophorus populeti* (Panzer) on aspen by evaluating the responses of ants to a simple artificial irritant (a dissecting needle) using the universal scale developed earlier and based on reactions of ant foragers to various stimuli [13,29]. Aggressiveness of foragers was assessed on a 9-point scale: (0) avoidance—dropping down or running away; (1) tolerance—neutral reaction (ants do not react); (2) antennation—investigation of the irritant using antennae; (3) alert pose—standing still with mandibles slightly open and antennae slightly extended towards the irritant; (4) aggressive pose—the stance adopted by ants before an attack (stilt-legged posture; mandibles widely open, antennae directed towards the irritant or slightly upwards; usually with gaster extended forwards in order to spray acid); (5) threatening lunges—usually repeated rapid lunges towards the irritant with open mandibles, but without contacting it; (6) hit-and-run attack—sudden attack on the irritant ($\leq$1 s); (7) biting—short bites (less than 5 s); (8) death grip—a prolonged biting (ant seizes the irritant and does not loosen its grip for more than 5 s). If rapid change in ant reactions from one to another (increasing aggression to the irritant) was observed during testing, only the more aggressive reaction was used in the analysis. For each individual, testing was carried out at least 3 times. A total of 190 *F. lemani* foragers were tested, of which 145 foragers were tested within the NC studied (80 foragers on trunks, 65 honeydew collectors in 15 aphid colonies) and 45 foragers were tested beyond (30 individuals on trunks and 15 honeydew collectors in 15 aphid colonies).

2.3. Trophobiotic Relationships of F. lemani with Aphids

To identify trophobiotic relationships between aphids and representatives of *F. lemani* colonies from the nest complex, the above-ground parts of plants up to a height of about 2–3 m were inspected in the area occupied by the NC. The root part was investigated only in the presence of soil disruptions (result of ant activities) at the base of the plant. Similar studies were carried out outside the territory of the NC in the same and neighbouring aspen–birch stand, where only small single nests of *F. lemani* were recorded. Insects were fixed in 70% ethanol. A total of 362 samples were collected. Analysis of the material and further identification of aphids was carried out using Stemi 2000-C and Zeiss Axiostar Plus microscopes (Carl Zeiss MicroImaging GmbH, Göttingen, Germany). The aphids were studied on microscope slides prepared using Faure–Berlese fluid. When identifying aphids, an online identification and information guide was used, which includes current information on aphids [30]. The material is deposited at the Institute of Systematics and Ecology of Animals, Siberian Branch of the Russian Academy of Sciences (Novosibirsk, Russia).

2.4. Data Analysis

A comparative analysis of the parameters of off-nest activity (dynamic density and intensity of foraging movement on trunks of birch trees), as well as the aggressiveness of *F. lemani* foragers in the territory of the NC and beyond, was carried out using the Mann–Whitney test. To compare the degree of aggressiveness of aggressive and tolerant foragers of *F. lemani* in the territory of the NC and beyond, the Mann–Whitney U Test with Bonferroni correction ($p < 0.017$) was used. To assess the relationship between intensity of foraging movement and trunk diameter, the Pearson correlation coefficient was calculated.

3. Results

3.1. The Main Characteristics of the F. lemani Nest Complex

In 2022, the surveyed *F. lemani* NC covered an area of about 4800 m^2 and included more than 150 large nests (Figure 1). Most nests were located at the bases of birch trees and inside logs or large branches of old fallen trees. The ants used old overgrown (and already hollow inside) large branches of birch trees as tunnels. In addition, a pronounced network of surface trails was observed in the territory of the *F. lemani* nest complex. Exchange trails (for the exchange of individuals and brood) were noted between nests located at the bases of trees at a distance of 4 to 6 m from each other. The main trails led to forage trees—birches and aspens. Visible even from a distance, trails on trunks consisted not only of honeydew collectors, but also of hunters who descended down the trunk with prey.

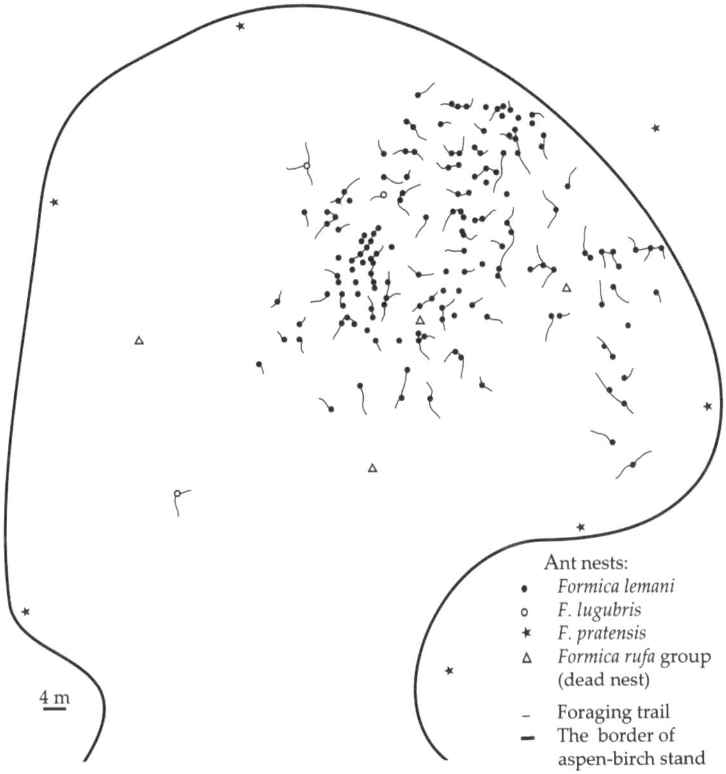

Figure 1. Schematic map of the nest complex of *Formica lemani* in the Karasuk District of the Novosibirsk Region in the vicinity of Sheinfeld Village (Russia).

Only three nests of *Formica rufa* group ants (*Formica lugubris* Zetterstedt) were recorded in the area of the studied nest complex. One *F. lugubris* nest (d/D = 55/90 cm, h/H = 20/40 cm; d/D = 60/105 cm, h/H = 30/45 cm) was located at a distance of about 30–40 m from the boundary of the *F. lemani* NC, 1 nest (d/D = 60/100 cm, h/H = 25/45 cm)—about 10–15 m from the NC. Another small nest with a flat dome (d/D = 45/50 cm; h/H = 5/25 cm) was found on the periphery of the nest complex (Figure 1). Nests of obligate dominants of the *Formica rufa* group with mound diameters (d) greater than 60 cm were absent in the area of the surveyed *F. lemani* NC.

Four craters from long-degraded nests of red wood ants (D = 110–120 cm) were recorded in this part of the aspen–birch stand, two of which were located directly in the territory of the *F. lemani* NC, and two were found outside.

Colonies of *Formica pratensis* Retzius were registered along the boundary of the aspen–birch stand only (Figure 1). The buffer zone between the foraging areas of *F. pratensis* colonies and the territory of *F. lemani* nest complex was about 15–20 m.

3.2. Off-Nest Activity and Aggressiveness of the F. lemani Foragers

3.2.1. Dynamic Density

The dynamic density of *F. lemani* in the territory of the NC was almost 3.9 times higher than beyond (Figure 2). The dynamic density of *F. lugubris* in the vicinity of the small, weakened nest of this species within the studied NC was 7.5 times as low as the *F. lemani* (0.06 and 0.45, respectively). Single individuals of *Camponotus herculeanus* (L.) and *Leptothorax acervorum* (Fabricius) were recorded only on the periphery of the studied NC.

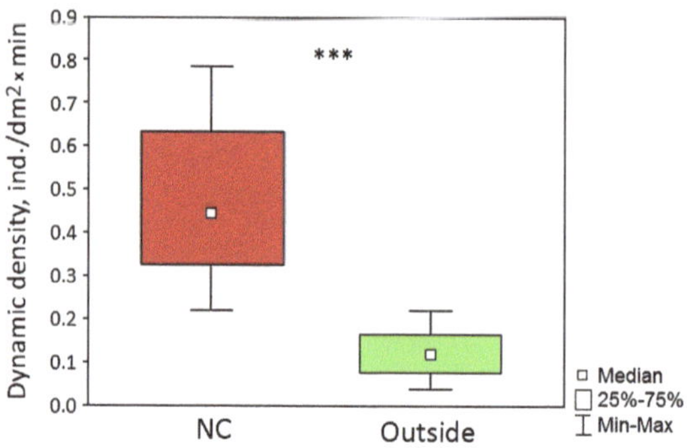

Figure 2. Dynamic density of *Formica lemani* within the territory of the nest complex (NC) and beyond (Outside). Mann–Whitney U Test: ***—$p < 0.0001$.

3.2.2. Intensity of Movement of Foragers on Trees

A significant positive correlation between the intensity of movement of foragers and the diameter of trunks of selected birch trees at the survey site was found only for the territory of the *F. lemani* nest complex (r = 0.577, $p = 0.01$). However, outside the NC, the correlation was insignificant (r = 0.223, $p = 0.537$), so further comparison was made for total data, without taking into account the influence of the trunk diameter. The intensity of movement of *F. lemani* foragers on the birch trunks within the territory of the NC was significantly higher than beyond in the same stand, and reached 123.6 individuals/min (Figure 3). Outside the studied NC, the movement intensity of *F. lemani* on birch trunks did not exceed 5.2 individuals/min (Figure 3).

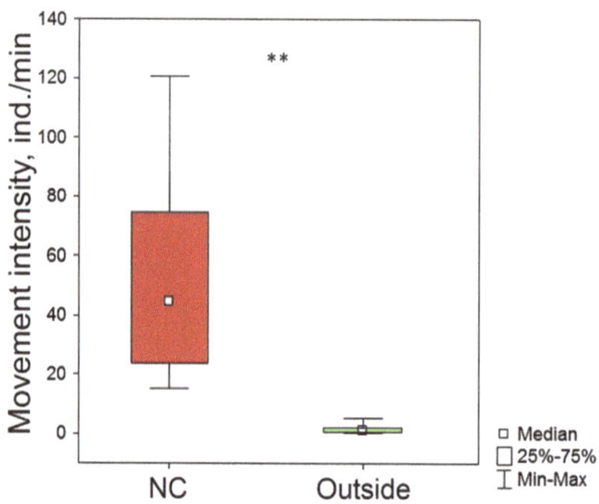

Figure 3. Movement intensity of *Formica lemani* foragers on trunks of birch trees within the territory of the nest complex (NC) and beyond (Outside). Mann–Whitney U Test: **—$p < 0.001$.

3.2.3. Aggressiveness

The aggressiveness of *F. lemani* foragers in the territory of the studied nest complex was generally significantly higher than outside it both on birch trunks (Mann–Whitney criterion, U = 351.5, $p < 0.001$) and in aphid colonies (U = 168.0, $p < 0.001$) (Figure 4).

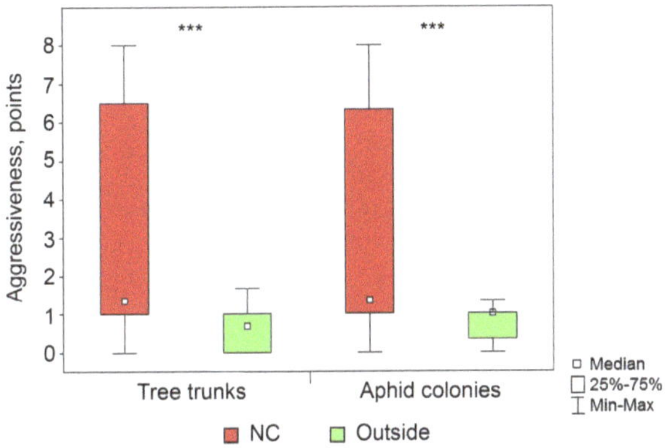

Figure 4. Aggressiveness of *Formica lemani* foragers on the trunks of birch trees and in the aphid colonies of *Chaitophorus populeti* on aspen trees within the territory of the large nest complex (NC) and beyond (Outside). Mann–Whitney U Test: ***—$p < 0.001$.

Outside the NC, both on the trunks of forage trees and in aphid colonies, *F. lemani* foragers showed only non-aggressive reactions (0 to 2 points) in response to stimulus presentation: average aggressiveness indices of most individuals fell within the range (0; 1] points (Figure 5).

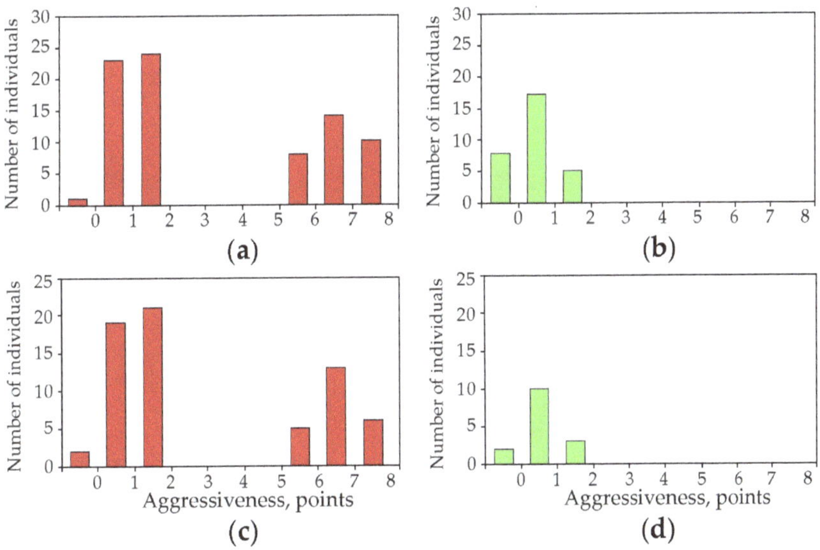

Figure 5. Differentiation of the *Formica lemani* foragers by the degree of aggressiveness in the territory of the large nest complex (red) compared to the foragers beyond the NC (green): (**a**,**b**)—the results of testing on the trunks of forage trees (*Betula* spp.); (**c**,**d**)—the results of testing in the aphid colonies. Aggressiveness scale: 0—avoidance; 1—tolerance; 2—antennation; 3—alert pose; 4—aggressive pose before an attack; 5—threatening lunges; 6—hit-and-run attack; 7—short bites (<5 s); 8—death grip (a prolonged biting for more than 5 s).

In the territory of the nest complex, two groups of foragers with different degrees of aggression—tolerant vs. aggressive—were clearly distinguished both on the trunks of forage trees and among the honeydew collectors tested in aphid colonies. The spectrum of responses of tolerant individuals was similar to that observed outside the NC (0–2 points, very rarely 3 points): the average aggressiveness of most individuals fell within the ranges of (0; 1] and (1; 2] points (Figure 5). Aggressive foragers responded to stimulus presentation by predominantly demonstrating hit-and-run attacks, short bites, and death grip, while threatening lunges were recorded in isolated cases (6–8 points respectively, rarely 5 points): average aggression scores of most tested individuals fell within the range (6; 7] points (Figure 5).

The degree of aggressiveness of honeydew collectors of *F. lemani* beyond the nest complex and tolerant foragers within the territory of the NC (both on the selected trees and in aphid colonies) was similar and significantly lower than that of aggressive foragers in the NC territory (Figure 6).

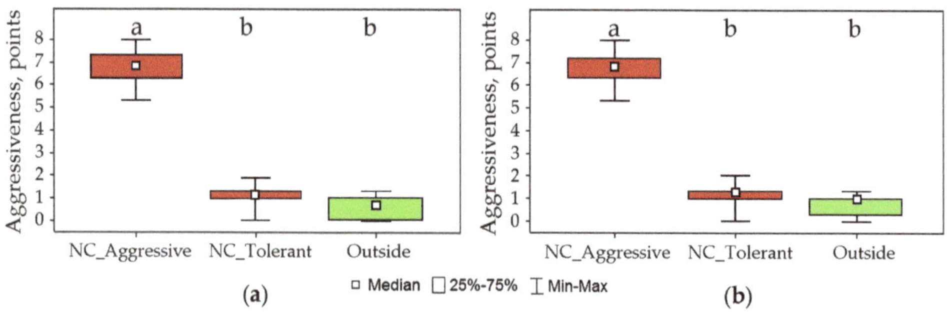

Figure 6. The degree of aggressiveness of aggressive and tolerant foragers of *Formica lemani* in the territory of the studied nest complex (NC) compared to the *F. lemani* foragers beyond the NC territory (Outside): (**a**) the results of testing on the trunks of foraging birch trees; (**b**) the results of testing in the aphid colonies (*Chaitophorus populeti*) on aspen trees. Different letters above the data (a,b) indicate significant differences (Mann–Whitney U Test with Bonferroni correction, $p < 0.0001$); identical letters (b) indicate non-significant differences ($p > 0.017$).

3.3. Trophobiotic Relationships between F. lemani and Aphids

The structure of trophobiotic relationships between *F. lemani* and aphids in the territory of the nest complex and beyond had some differences (Table 1). In the NC territory, *F. lemani* visited aphid colonies of only seven species from seven genera of four subfamilies: Aphidinae—three species, Calaphidinae—two, Thelaxinae and Chaitophorinae—one species each (Table 1). Dendrobiont species predominated and accounted for 57.1% of the total number of symbiont species. Their colonies were located on birches (*Symydobius oblongus* (von Heyden), *Glyphina betulae* (L.), *Callipterinella tuberculata* (von Heyden), and aspens (*Chaitophorus populeti* (Panzer)). On herbaceous plants, only single colonies of aphids *Metopeurum fuscoviride* Stroyan (on *Tanacetum vulgare* L.), *Aphis fabae* Scopoli (on *Sonchus arvensis* L.), and *Semiaphis horvathi* (Kaltenbach) (on *Silaum silaus* (L.) Schinz & Thell.) were noted.

The main source of carbohydrate food for *F. lemani* from the nest complex was honeydew of aphids on woody plants, mainly on birch trees. Powerful trails of *F. lemani* foragers were observed on the majority of birch trees (90%) located in the territory of the studied NC. Outside the boundary of the nest complex (in the areas of similar size both in the same and neighbouring aspen–birch stands), 1.7–1.9 times more aphid species (12 and 13, respectively) associated with *F. lemani* were identified (Table 1). The proportion of aphid species whose colonies were located on herbaceous plants in the same aspen–birch stand was 66.7% (out of 12 species), and in the neighbouring one—69.2% (out of 13 species). In total, 15 species of aphid symbionts of *F. lemani* were identified beyond the NC studied (Table 1). These species belonged to nine genera of four subfamilies.

Table 1. Trophobiotic relationships between *F. lemani* and aphids within the territory of the studied nest complex (NC) and beyond—in the same (Outside_1) and in the neighbouring (Outside_2) aspen–birch stands (+—single cases; ++—common; +++—numerous; – none).

No	Aphids (Subfamily, Genus, Species)	Plants	NC	Outside_1	Outside_2
	Thelaxinae				
1	*Glyphina betulae* (L.)	*Betula pendula* Roth, *Betula* sp.	+++	+	+
	Calaphidinae				
2	*Callipterinella tuberculata* (von Heyden)	*Betula pendula, Betula* sp.	+++	+	+
3	*Symydobius oblongus* (von Heyden)	*Betula pendula, Betula* sp.	+++	+	+
	Chaitophorinae				
4	*Chaitophorus populeti* (Panzer)	*Populus tremula* L.	++	+	+
5	*Sipha (Rungsia) elegans* Del Guercio	*Elytrigia repens* (L.) Desv. ex Nevski., Poaceae	–	+	+
6	*S. (Rungsia) maydis* Passerini	*Elytrigia repens, Festuca pratensis* Hudson	–	+	–
	Aphidinae				
7	*Aphis craccae* L.	*Vicia cracca* L.	–	–	+
8	*A. craccivora* Koch	*Vicia* sp.	–	+	–
9	*A. fabae* Scopoli	*Sonchus arvensis* L.	+	+	+
10	*A. franzi* Holman	*Seseli libanotis* (L.) C. Koch	–	+	+
11	*A. hieracii* Schrank	*Hieracium umbellatum* L.	–	–	+
12	*A. silaumi* Bozhko	*Silaum silaus* (L.)	–	+	+
13	*Rhopalosiphum oxyacanthae* (Schrank)	*Dactylis glomerata* L.	–	–	+
14	*Metopeurum fuscoviride* Stroyan	*Tanacetum vulgare* L.	++	+	+
15	*Semiaphis anthrisci* (Kaltenbach)	*Anthriscus silvestris* L.	–	+	+
16	*Semiaphis horvathi* Szelegiewicz	*Silaum silaus* (L.) Schinz & Thell.	+	–	–
	Total number of species:		7	12	13
	On woody plants		4	4	4
	On herbaceous plants		3	8	9

In aphid colonies beyond the studied nest complex, only solitary unspecialised foragers of *F. lemani* were observed, which did not show any aggressive reactions to the irritant (Figures 5 and 6). These foragers themselves collected and carried honeydew to a nest, leaving the aphid colonies unattended.

In the territory of the NC, *F. lemani* used more complex behavioural models with the clear division of honeydew collection and protection of symbionts between foragers with low and high levels of aggressiveness, respectively (Figures 5 and 6). Therefore, tolerant individuals performed the functions of collecting and transporting honeydew to their nests and were significantly less aggressive (Figures 5 and 6).

Among the honeydew collectors tested in aphid colonies, usually about 20%–30% of foragers (rarely 50% of individuals) demonstrated aggressive behaviour and performed the functions of guards. Aggressive ants almost did not collect honeydew, these individuals quickly responded to the appearance of foreign objects and protected their symbionts from various competitors, including aphidophages and other ant species.

4. Discussion

In the forest–steppe and steppe zones in the south of Western Siberia, as in other parts of the *Formica lemani* range, the northern black ant is confined to forest communities, where it usually plays the role of subdominant [8,26]. In the studied area, an atypical ant assemblage, rather unique for the forest–steppe landscapes of the south of Western Siberia, with the absolute dominance of *F. lemani*, has formed. At first glance, the area of the *F. lemani* nest complex does not appear to be too large (about 0.005 km^2), at least compared to the areas of red wood ant supercolonies that can exceed 2.5 km^2 [9,31]. However, the significant reorganisation of various ecological and ethological aspects of the life of this species is impressive.

According to our long-term research in the south of the Novosibirsk Region (2000–2022), in this territory, *F. lemani* usually lives in sectional nests in the soil, as well as inside old stumps, logs, and large branches of fallen trees, etc. This ant typically uses tunnels to communicate between sections of the nest and for foraging, which is consistent with the results of other studies [8].

In the territory of the studied nest complex, in addition to the tunnels that are quite common for the species, a pronounced network of foraging trails was discovered, which was similar to a functional structure of the trail system of red wood ants. The trail network in the NC territory included exchange trails between nests at the bases of birches, as well as foraging trails leading to trees (birch, aspen), where workers not only actively collected honeydew, but also hunted. Among the foragers descending down a trunk, a large number of honeydew collectors with swollen abdomens, as well as hunters with prey (caterpillars, various dipterans, aphids), were noted. Some of noted exchange trails can be classified as exchange–foraging trails, since they were used by *F. lemani* for both purposes—exchange of individuals and brood, and organisation of foraging (being part of the foraging network). According to our preliminary observations, the presence of exchanges between most of nests in the NC territory, with regular transitions of individuals, may indicate that the studied nest complex of *F. lemani* is a single supercolony, which apparently consists of several polycalic systems (an ant colony simultaneously inhabiting several nests with different functions). However, additional detailed studies are needed to obtain more precise information on this issue.

Searching for food on trees is not something special for *F. lemani*. A striking difference from the usual lifestyle of the northern black ant in the territory of the studied nest complex is the very high off-nest activity (dynamic density, intensity of movement of foragers along the trails). This is manifested in the presence of trunk trails with high intensity of *F. lemani* movement, which were easy to see on birch trees with white bark even from a distance. According to our data, the intensity of *F. lemani* movement on the trunks of foraging trees in the territory of the nest complex was significantly (about 32 times) higher than outside, and the dynamic density of *F. lemani* within the NC territory was almost 4 times as high as beyond.

Outside the studied nest complex of *F. lemani*, small colonies of this species were not detected to have protected foraging areas. The collection of aphid honeydew was carried out by single unspecialised foragers, who independently collected and carried honeydew to their nests, continuously leaving colonies of symbiont aphids unattended. This is the simplest type of honeydew collection in ants [13] that usually corresponds to the lowest level of symbiont protection by ants against aphidophages [14].

In the territory of the NC, the behaviour of *F. lemani* differed significantly from that typical of small colonies of this species. The aggressiveness of foragers on the trunks of forage trees and in aphid colonies was 4–5 times as high. In addition, increased territorial organisation was observed. Due to the highly increased dynamic density and aggressiveness of foragers, the NC territory was almost completely controlled by *F. lemani*. Single individuals of other species (*Camponotus herculeanus* (L.), *Leptothorax acervorum* (Fabricius)) were noted only in the peripheral part of the *F. lemani* nest complex.

The only small degrading nest of *F. lugubris* with an almost flat mound, the height (h) of which was only 5 cm and the diameter (d) of about 50 cm, was also located on the periphery of the nest complex; the dynamic density of *F. lugubris* in the area of this anthill was 7.5 times lower than the dynamic density *of F. lemani*. The only trail of foragers from the *F. lugubris* nest led to the tree adjacent to this anthill. The weak construction, size, and shape of the anthill, combined with low off-nest activity, indicated the grave condition of the colony [8].

As for three other nests of *F. lugubris* found in the same stand but outside the NC, the mound diameter (d) of these anthills did not exceed 60 cm, corresponding to the minimal parameters of permanent residential nests of red wood ants aged 1–2 years [8]. The absence of large colonies of red wood ants in the study area allowed *F. lemani* to avoid interspecific social control typical of ant assemblages that include obligate dominants of *Formica rufa* group [7,21,22].

At the same time, despite the absence of red wood ants, fairly high efficiency in protecting woody plants from leaf-eating insects was noted in the territory of the studied *F. lemani* NC during the years of the field research. Hence, during outbreaks of gypsy moth (*Lymantria dispar* (L.)) in 2020 and 2022, the area occupied by the *F. lemani* nest complex was clearly distinguished from the background of neighbouring areas affected by the pest, mainly due to its preserved green foliage. A special assessment of the degree of damage to the leaves was not carried out; however, the border of the *F. lemani* NC was quite well noticeable (Figure 7).

In aphid colonies within the territory of the *F. lemani* NC, the use of more complex behavioural patterns was observed with the partial division of collecting honeydew and protecting symbionts by more aggressive foragers. Both in aphid colonies and on trunks of forage trees, a clear division of the two groups of tested ant individuals—aggressive and tolerant—was observed according to their responses to the irritant (stimulus). Observations of ant behaviour before testing showed that tolerant workers with a low degree of aggressiveness were engaged in collecting honeydew and transferring the collected honeydew to other individuals. Aggressive ants exhibited the full range of aggressive reactions to any external intervention (the approach of aphidophages, other insects and the researcher, the presentation of an artificial stimulus, etc.). On tree trunks, the lowest rates of aggressiveness were seen in foragers with swollen abdomens, which were engaged in transporting honeydew to their nests. This is consistent with the research data on the aggressiveness of *Formica rufa* group ants, which perform different functions—protecting aphids, and collecting and transporting aphid honeydew [13].

Our work makes clear the presence of specialisation in groups of foragers visiting individual aphid colonies, with the division of functions of collecting honeydew and protecting symbionts corresponding to at least a medium level of specialisation [13]. To more accurately determine the depth of specialisation of honeydew collectors, and to answer the question how exactly the collected honeydew is transported (whether there are specialists in delivering honeydew to the ant nests or not), additional research is required.

In the territory of the nest complex of *F. lemani*, the range of aphid symbionts of the northern black ant turned out to be twice as small as beyond, with aphids on birches playing a key role in carbohydrate nutrition of *F. lemani*. It was birch trees that served as key elements in the formation of the NC structure and its trail network. Large nests of *F. lemani* were located at the bases of birches with a trunk diameter of 20–25 cm (Figure 7), the main foraging trails also led to birch trees. Foragers of *F. lemani* usually reached rare colonies of aphids on herbaceous plants through tunnels, using old hollow large branches of birch trees in the forest litter.

Figure 7. The territory of the studied nest complex (NC) of *Formica lemani* in the aspen–birch stand in the south of the Novosibirsk Region in the vicinity of Sheinfeld Village (Russia): (**a**) the central part of the *F. lemani* NC; (**b**)—the border of the studied NC; (**c**)—view of the nest complex (green foliage) from the adjacent territory; (**d**)—the adjacent territory (the photo is taken from the border of the *F. lemani* NC); (**e–g**)—large nests of *F. lemani* in the territory of the studied NC; (**h–j**)—foraging trails of *F. lemani* on birch trees in the NC territory.

In ants of the *F. rufa* group, the structural and functional organisation of polycalic systems and supercolonies has similar basic principles, including dendrobiont aphids as a main source of carbohydrates, and trees (host plants of myrmecophilous aphids) as a core element of trail system and territorial structure [1,10,11,32,33]. In addition to aphid honeydew, ants may use other sources of carbohydrates, including extrafloral nectaries [1,10,33,34]. Aspen leaves may also have small glands at the base of the leaf that secrete nectar attracting ants [35]. We also noted similar extrafloral nectaries in aspen (*Populus tremula* L.) in the south of Western Siberia. However, according to our observations, ants usually visit foliar nectaries of *Populus tremula* L. only in spring while aphids are absent or limited. During our investigation of the *F. lemani* NC in August–September (2020, 2022), such sort of ant–plant association was not noted both on and off the area occupied by the studied nest complex. To further understand the role of various resources, including foliar nectaries on aspen, in the life of the studied nest complex of *F. lemani*, detailed investigation of this issue throughout the season (starting in April–May) is required.

Beyond the studied nest complex, aphids on herbaceous plants prevailed among *F. lemani* symbiont aphids. Only single foragers of *F. lemani* were found on trees, which, when possible, collected honeydew in colonies of aphids visited by red wood ants. This stealing behaviour is quite common among subdominant ants of the genus *Formica* [13]. The obtained results are consistent with the data that in the presence of red wood ants, the strong ecological pressure from these dominants forces co-occurring species to adapt by changing their activity in space or time, and behaviour, including foraging strategy (e.g., choosing smaller food pieces or using less rewarding sources) [7,36–41]. At the same time, our data clearly demonstrate the great potential of subdominant ant species.

5. Conclusions

Overall, the formation of the nest complex with a high density of nests and numbers in *F. lemani* is apparently caused by the lack of pressure from the obligate dominants of *Formica rufa* group, which allowed *F. lemani* to use the honeydew of obligate myrmecophilous aphids on trees (birch, aspen) as the main resource in carbohydrate nutrition. Close interaction with aphids on wood plants gives ants a number of significant advantages: it promotes accelerated restoration of trophobiotic interactions with aphids in spring, and ensures relative stability of the food supply and the ability to regularly obtain sufficiently large volumes of carbohydrate food throughout the season. Combined with the lack of abundance control by red wood ants, this allowed *F. lemani* not only to increase, but maintain high population densities, as well as to demonstrate more complex behaviours. Altogether, this significantly increased the *F. lemani* competitiveness, as well as the role of this species in the plant protection against phytophages.

Author Contributions: Conceptualisation, T.N.; methodology, T.N.; formal analysis, T.N.; investigation, T.N. and D.T.; data curation, T.N.; writing—original draft preparation, T.N.; writing—review and editing, T.N.; visualisation, T.N.; supervision, T.N. All authors have read and agreed to the published version of the manuscript.

Funding: This research was funded by The Federal Fundamental Scientific Research Programme of the Russian Academy of Sciences, grant number FWGS-2021-0002.

Data Availability Statement: The data presented in this study are available on request from the corresponding author.

Acknowledgments: The authors are grateful to E.B. Fedoseeva (Zoological Museum of Moscow State University, Moscow) for valuable comments when discussing an early version of the manuscript, A.V. Stekolshchikov (Zoological Institute RAS, St. Petersburg, Russia) for verification of the aphid identifications, N. Murrey for checking and improving the English of this manuscript, and also anonymous reviewers for their helpful discussion of the results and valuable comments on the manuscript.

Conflicts of Interest: The authors declare no conflicts of interest. The funders had no role in the design of the study; in the collection, analyses, or interpretation of data; in the writing of the manuscript; or in the decision to publish the results.

References

1. Dlussky, G.M. *Ants of the Genus Formica (Hymenoptera: Formicidae g. Formica)*; Nauka: Moscow, Russia, 1967; p. 236.
2. Pokarzhevskij, A.D. The distribution and accumulation of nutrients in nests of ant *Formica polyctena* (Hymenoptera, Formicidae). *Pedobiologia* **1981**, *21*, 117–124. [CrossRef]
3. Whittaker, J.B. Effects of ants on temperate woodlands. In *Ant-Plant Interactions*; Huxley, C., Cutler, D., Eds.; Oxford University Press: Oxford, UK, 1991; pp. 67–69.
4. Frouz, J.; Santruckova, H.; Kalcik, J. The effect of wood ants (*Formica polyctena* Foerst.) on the transformation of phosphorus in a spruce plantation. *Pedobiologia* **1997**, *41*, 437–447. [CrossRef]
5. Folgarait, P. Ant biodiversity and its relationship to ecosystem functioning: A review. *Biodivers. Conserv.* **1998**, *7*, 1221–1244. [CrossRef]
6. Laakso, J.; Setälä, H. Impacts of wood ants (*Formica aquilonia* Yarr.) on the invertebrate food web of the boreal forest floor. *Ann. Zool. Fenn.* **2000**, *37*, 93–100.
7. Savolainen, R.; Vepsäläinen, K.; Wuorenrinne, H. Ant assemblages in the taiga biome: Testing the role of territorial wood ants. *Oecologia* **1989**, *81*, 481–486. [CrossRef] [PubMed]
8. Zakharov, A.A.; Dlussky, G.M.; Goryunov, D.N.; Gilev, A.V.; Zryanin, V.A.; Fedoseeva, E.B.; Gorokhovskaya, E.A.; Radchenko, A.G. *Monitoring Formica Ants*; KMK Scientific Press: Moscow, Russia, 2013; p. 99.
9. Stockan, J.A.; Robinson, E.J.H. *Wood Ant Ecology and Conservation. Ecology, Biodiversity and Conservation*; Cambridge University Press: Cambridge, UK, 2016; p. 304. [CrossRef]
10. Zakharov, A.A. *Ants of Forest Communities, Their Life and Role in the Forest*; KMK Scientific Press: Moscow, Russia, 2015; p. 404.
11. Zakharov, A.A. *Ant, Colony, Supercolony*; Fiton XXI: Moscow, Russia, 2021; p. 192.
12. Reznikova, Z.I.; Novgorodova, T.A. Division of labour and exchange of information within ant settlement. *Usp. Sovrem. Boil.* **1998**, *118*, 345–357.
13. Novgorodova, T.A. Organization of honeydew collection by foragers in different ant species (Hymenoptera, Formicidae): Effect of colony size and species specificity. *Eur. J. Entomol.* **2015**, *112*, 688–697. [CrossRef]
14. Novgorodova, T.A.; Gavrilyuk, A.V. The degree of protection different ants (Hymenoptera: Formicidae) provide aphids (Hemiptera: Aphididae) against aphidophages. *Eur. J. Entomol.* **2012**, *109*, 187–196. [CrossRef]
15. Novgorodova, T.A.; Kryukov, V.Y. Quarantining behaviour of ants towards infected aphids as an antifungal mechanism in ant-aphid interactions. *Entomol. Exp. Appl.* **2017**, *162*, 293–301. [CrossRef]
16. Reznikova, Z.I. *Interspecific Relationships in Ants*; Nauka: Novosibirsk, Russia, 1983; p. 208.
17. Reznikova, Z. Ants–Individual and Social Cognition. In *Field and Laboratory Methods in Animal Cognition: A Comparative Guide*; Bueno-Guerra, N., Amici, F., Eds.; Cambridge University Press: Cambridge, UK, 2018; pp. 8–30. [CrossRef]
18. Zakharov, A.A. Classification of social structures in ants. *Zool. Zh.* **2005**, *84*, 1272–1288.
19. Burgov, E.V. Spatial-functional structures in ants Serviformica (Hymenoptera: Formicidae) *Vestnik MGPU. Seria. Estestv. Nauki* **2016**, *4*, 19–27.
20. Reznikova, Z.I.; Shillerova, O.A.; Stebaev, I.V. Organization of the foraging area and features of individual behaviour in *Formica cunicularia glauca* Ruzsk. (Hymenoptera, Formicidae). In *Ecological Issues: Species, Populations, Community*; Stebaev, I.V., Ed.; Novosibirsk State University Press: Novosibirsk, Russia, 1978; pp. 157–170.
21. Reznikova, Z.I. Ethological Mechanisms of Population Density Control in Coadaptive Complexes of Ants. *Russ. J. Ecol.* **1999**, *30*, 210–215.
22. Reznikova, Z.I. A new form of interspecific relationships in ants: The hypothesis of interspecific social control. *Zool. Zh.* **2003**, *82*, 816–824.
23. Burgov, E.V. The multispecies formicary association: The transition from an incomplete to a complete association. *Entomol. Rev.* **2015**, *95*, 1010–1020. [CrossRef]
24. Lindström, K.; Berglind, A.; Pamilo, R. Variation of colony types in the ant *Formica cinerea*. *Insectes Sociaux* **1996**, *43*, 329–332. [CrossRef]
25. Grimalsky, V.I.; Entin, L.I. Gray sand ant (*Formica cinerea*) in Belarus. *Zool. Zh.* **1980**, *59*, 1159–1167.
26. Radchenko, A.G. *Ants (Hymenoptera, Formicidae) of Ukraine*; Institute of Zoology by I.I. Shmalgausen NAS of Ukraine: Kiev, Ukraine, 2016; pp. 1–480.
27. Novgorodova, T.A. The ability of *Formica lemani* to form large settlements. In Proceedings of the XVI All-Russian Myrmecological Symposium «Ants and forest protection», Moscow, Russia, 27–31 August 2022; pp. 61–65.
28. Chesnokova, S.V.; Vaulin, O.V.; Zhigulskaya, Z.A.; Novgorodova, T.A. *Formica gagatoides* Ruzsky, 1904, and Siberian *F. kozlovi* Dlussky, 1965 (Hymenoptera: Formicidae); Two or One Species? *Diversity* **2023**, *15*, 686. [CrossRef]
29. Novgorodova, T.A. Testing Ant Aggression. In Proceedings of the XIII All-Russian Myrmecological Symposium «Ants and forest protection», Nizhny Novgorod, Russia, 26–30 August 2009; pp. 274–275.
30. Aphids on the World's Plants. An Online Identification and Information Guide. Available online: http://www.aphidsonworldsplants.info (accessed on 15 February 2023).
31. Higashi, S.; Yamauchi, K. Influence of a supercolonial ant *Formica (Formica) yessensis* Forel on the distribution of other ants in Ishikari Coast (Japan). *Jpn. J. Ecol.* **1979**, *29*, 257–264.

32. Rosengren, R.; Sundström, L. The interaction between red wood ants *Cinara* aphids, and pines. A ghost of mutualism past. In *Ant–Plant Interactions*; Huxley, C.R., Culter, D.F., Eds.; Oxford Scientific Publications: Oxford, UK, 1991; pp. 80–91.
33. Domisch, T.; Risch, A.C.; Robinson, E.J.H. Wood ant foraging and mutualism with aphids. In *Wood Ant Ecology and Conservation. Ecology, Biodiversity and Conservation*; Stockan, J.A., Robinson, E.J.H., Eds.; Cambridge University Press: Cambridge, UK, 2016; pp. 145–176.
34. Nixon, G.E.J. *The Association of Ants with Aphids and Coccids*; Commonwealth Institute of Entomology: London, UK, 1951; p. 36.
35. Doak, P.; Wagner, D.; Watson, A. Variable extrafloral nectary expression and its consequences in quaking aspen. *Can. J. Bot.* **2007**, *85*, 1–9. [CrossRef]
36. Savolainen, R.; Vepsäläinen, K. Niche differentiation of ant species within territories of the wood ant *Formica polyctena*. *Oikos* **1989**, *56*, 3–16. [CrossRef]
37. Gallé, L. Assembly of sand-dune forest ant communities. *Memorab. Zool.* **1990**, *44*, 1–6.
38. Savolainen, R. Interference by wood ant influences size selection and retrieval rate of prey by *Formica fusca*. *Behav. Ecol. Sociobiol.* **1991**, *28*, 1–7. [CrossRef]
39. Punttila, P.; Haila, Y.; Tukia, H. Ant communities in taiga clearcuts: Habitat effects and species interactions. *Ecography* **1996**, *19*, 16–28. [CrossRef]
40. Seima, F.A. *Population Structure of Taiga Ants*; Perm State University Press: Perm, Russia, 2008; p. 166.
41. Czechowski, W.; Markó, B.; Radchenko, A.; Ślipiński, P. Long-term partitioning of space between two territorial species of ants (Hymenoptera: Formicidae) and their effect on subordinate species. *Eur. J. Entomol.* **2013**, *110*, 327–337. [CrossRef]

forests

MDPI

Article

Non-Native Plants Influence Forest Vegetative Structure and the Activity of Eastern Temperate Insectivorous Bats

Morgan Mark [1], Evan Drake [2], Kathleen Kerwin [1] and Brooke Maslo [1,*]

[1] Department of Ecology, Evolution, and Natural Resources, Rutgers, The State University of New Jersey, New Brunswick, NJ 08901, USA; morgan.mark@rutgers.edu (M.M.)
[2] Department of Natural Sciences, Felician University, Rutherford, NJ 07070, USA
* Correspondence: brooke.maslo@rutgers.edu; Tel.: +1-848-932-5572

Abstract: Temperate insectivorous bats value high prey abundance and appropriate vegetative structure when selecting foraging habitats. Forests, particularly in the eastern United States, provide prime foraging habitats for bats but can be heavily impacted by non-native plants, which may alter arthropod diversity and abundance, as well as vegetative structure. To investigate the associations between non-native plants and insect abundance, vegetative structure, and, consequently, bat activity, we performed vegetation surveys, insect trapping, and acoustic monitoring at 23 forested plots in northern New Jersey, USA. We predicted that non-native vegetation would either positively influence bat activity by increasing structural openness (thus, facilitating flight) or negatively influence bat activity by lowering the abundance of putative prey. We also hypothesized that vegetative characteristics, and therefore non-native vegetation, impact bats differently depending on their foraging habitat preferences. The percent of non-native cover of the ground and midstory vegetative layers of our study plots ranged from 0 to 92.92% ($\bar{x}$ = 46.94 ± 5.77 SE) and was significantly correlated with structural vegetative characteristics, such as midstory clutter (β = 0.01 ± 0.006 SE), but not putative prey abundance (β = −0.81 ± 2.57 SE). Generalized linear models with only vegetative characteristics best predicted overall bat activity and foraging, which were greatest in areas with a high percent non-native vegetation and low midstory clutter. Although percent non-native vegetation and midstory clutter were also significant effects for bats that prefer to forage in open areas, neither vegetative characteristics nor prey abundance were significant effects for clutter-loving bats. Such findings suggest that vegetative structure is more important than prey availability for predicting overall insectivorous bat activity, but other factors, such as foraging strategy and life history traits, can impact how bat guilds respond to non-native vegetation. Therefore, more research is required to reveal additional mechanisms by which non-native plants impact bats.

Keywords: acoustic monitoring; bat conservation; Chiroptera; non-native plants; prey availability; vegetation structure; wildlife–habitat relationships

Citation: Mark, M.; Drake, E.; Kerwin, K.; Maslo, B. Non-Native Plants Influence Forest Vegetative Structure and the Activity of Eastern Temperate Insectivorous Bats. *Forests* **2024**, *15*, 711. https://doi.org/10.3390/f15040711

Academic Editors: Katarína Pastirčáková and Rostislav Zemek

Received: 15 February 2024
Revised: 2 April 2024
Accepted: 14 April 2024
Published: 17 April 2024

1. Introduction

Non-native plants can profoundly impact ecological communities by reducing native biodiversity and altering ecosystem function [1–3]. By modifying the biotic (e.g., prey availability) and abiotic characteristics (e.g., structure, light availability, and soil chemistry) of a native species' habitat, non-native plants may affect individual fitness and ultimately population or species persistence [4–6]. Because impacts on individual species can radiate through food webs, non-native plants can also alter the availability, distribution, and quality of resources for consumers at other trophic levels [7–9]. For example, non-native plants compete with and suppress the growth of native plants, altering the pre-existing vegetative community [10,11]. Such changes can directly and indirectly reduce the survival and reproductive success of some native herbivorous insects, which subsequently impact insectivores and their predators [12]. By potentially reducing native biodiversity, non-native plants can also diminish the ecosystem functions that native species perform, such as pollination and tree regeneration, which may consequently alter the succession dynamics of a landscape [13–16].

Forests, particularly those in the eastern temperate zone, are highly susceptible to invasions of non-native plants due to such human-induced stressors as forest fragmentation and soil disturbance [17,18]. Non-native shrubs typically increase understory clutter [17], which could affect how forest-dwelling wildlife use the landscape. Although non-native plants in some cases have been shown to have little impact on native fauna [19], non-native plants often force animals to adapt or suffer negative fitness consequences [16]. Indeed, native eastern North American butterflies, birds, and snakes have been shown to alter their habitat use in response to non-native plants, often resulting in increased mortality. The West Virginia white (*Pieris virginiensis*, Edwards), an imperiled native butterfly, preferentially oviposits on non-native garlic mustard (*Alliaria petiolate*, Cavara & Grande), despite the plant being toxic to its caterpillars [20,21]. Neotropical migrant birds, such as Kentucky Warblers (*Geothlypis Formosa*, Wilson) and Hooded Warblers (*Setophaga citrina*, Boddaert), will not nest in areas dominated by non-native plants [22], leading to decreased reproductive success. Further, nonmigratory birds that do build nests in non-native shrubs experience higher daily mortality and nest failure due to predation than birds using native shrubs [22,23]. Ectotherms also avoid non-native vegetation at multiple spatial scales, suggesting that non-native plants do not provide the required habitat components [24]. Consequently, non-native plants can negatively impact native fauna by increasing mortality, reducing reproductive success, and diminishing habitat quality. Thus, understanding how and to what extent non-native plants impact native fauna requires studying multiple animal taxa.

Despite the growing body of research on the effects of non-native plants on native wildlife, temperate insectivorous bats are understudied in the invasion literature [25]. This is problematic because North American bats rely heavily on forested landscapes and are experiencing significant population declines resulting from multiple threats (e.g., disease, wind energy production, etc.) [26,27]. All bat species native to eastern North America are insectivorous and require forests for roosting, reproduction, and/or foraging [28–30]. When selecting foraging habitats in forested landscapes, bats value both the high abundance of arthropod prey as well as the appropriate vegetative structure. However, studies show that neither of these factors alone can reliably predict bat presence or activity [31,32]. Non-native plants further obfuscate these dynamics by influencing both arthropod abundance and vegetative structure [5,12]. Although most of the literature suggests that non-native vegetation negatively affects habitat structure for bats by acting as clutter that obstructs flight and impedes echolocation, non-native plants often suppress native tree and shrub seedlings and create forest structures with open midstories and canopies. Such open habitats may facilitate flight and predator avoidance [25,33], ultimately increasing their use by bats.

Because many bats, including all temperate North American species, echolocate [34], passive acoustic monitoring has become a valuable tool for evaluating how bats are affected by factors such as non-native plants [35,36]. Bats produce echolocation pulses at a consistent rate to navigate and hunt [37,38]; thus, the number of pulses recorded can be a proxy for bat activity [39]. Moreover, when a bat detects a prey item, it deviates from its consistent echolocation pattern and produces a terminal buzz, whereby the bat directs pulses toward the target prey at an increasingly rapid rate until it is found and consumed [40]. Therefore, the number of terminal buzzes recorded can be a proxy for foraging activity [41]. Acoustic monitoring studies have implicated low prey abundance caused by non-native plants in reduced bat activity [42]. Acoustic monitoring has also revealed that non-native structural clutter impacts open-space foraging bats but not clutter-loving foragers [43]. Bat guilds or species may be impacted by non-native plants differently due to variations in foraging strategy, physiology, and prey preferences, which are all often correlated [37,40]. Insectivorous bats can be characterized into two foraging guilds based on their echolocation pulses. Species that produce low-frequency pulses (~16–33 kHz) typically forage in open habitats, such as above the forest canopy or in openings and corridors. Such low-frequency echolocation is ideal for long-range prey detection [44]. Oppositely, species that forage in cluttered habitats, such as in or below the forest canopy, generally produce high-frequency pulses (~34–50 kHz) that facilitate short-range object detection [44,45]. Thus, it is possible that non-native plants that mediate changes in overall vegetative structure, either by creating clutter or more open landscapes, may have opposite effects on bats based on their preferences in foraging habitats.

Here, we explored the relationship between non-native plants and habitat use by bats in eastern temperate forests. Using bat activity as an indicator of habitat suitability [46,47], we combined acoustic monitoring, conventional insect trapping, and vegetation surveys to investigate if bat activity is correlated with plot non-native percentage, which we defined as the average non-native cover of the ground and midstory vegetative layers. We predicted that survey plots with a high non-native percentage would also have open midstories and canopies, suggesting that non-native plants positively influence bat habitat by facilitating flight. Despite increased levels of overall bat activity, however, we also predicted that plots with high non-native percentages would exhibit reduced bat foraging activity relative to plots with low non-native percentages due to lower arthropod abundance and diversity. However, we hypothesized that non-native plants would affect bats differently depending on their foraging strategy: bats that prefer to forage in open areas (which produce low-frequency pulses) would likely be more active in survey plots with a high percent non-native, which we predicted would be more structurally open, while the opposite would be true for clutter-loving bats (which produce high-frequency pulses).

2. Materials and Methods

2.1. Study Area and Survey Locations

We conducted our study within the Morristown National Historical Park–Jockey Hollow Unit (*hereafter*, Jockey Hollow) in Morristown, New Jersey, USA, between June and August 2020. Jockey Hollow consists of 567 hectares of contiguous eastern temperate forest that is heavily influenced by past land uses [48]. Many of the lower-elevation areas in Jockey Hollow are dominated by early-to-mid successional species, such as tulip poplar (*Liriodendron tulipifera*, L.), that form an open midstory and partially open canopy. These areas often experience severe deer browse, which limits the regeneration of native vegetation and facilitates the invasion of non-native plants that deer often selectively avoid, such as multiflora rose (*Rosa multiflora*, Thunb.), Japanese barberry (*Berberis thunbergii*, DC.), and Japanese stilt grass (*Microstegium vimineum*, A. Camus) [49]. Such non-native plants form a dense thicket and are locally abundant in the shrub layer, outcompeting resident native shrubs such as northern spicebush (*Lindera benzoin*, Blume) and American witch hazel (*Hamamelis virginiana*, L.) [48]. Previous biodiversity surveys of Jockey Hollow confirmed the presence of four of the nine bat species in New Jersey, including big brown bat (*Eptesicus fuscus*, Palisot de Beauvois), eastern red bat (*Lasiurus borealis*, Müller), little brown bat (*Myotis lucifugus*, Le Conte), and northern long-eared bat (*Myotis septentrionalis*, Trouessart) [50].

Within the study area, we selected 23 circular plots, which we broadly classified into three habitat types, as follows: open habitats ($N = 6$) included locations within canopy gaps; forested habitats ($N = 13$) included locations with closed canopies; and stream/corridors ($N = 4$) included locations containing former access roads or water bodies, which are known to influence bat activity [51] (Figure 1). We delineated survey plots at a diameter of 30 m, corresponding to the presumed range of the acoustic detectors within forest habitats. We located survey plots at least 200 m apart [52] to ensure that acoustic detectors in adjacent plots were not sampling the same bat individuals. Because we wanted to explore the relationship between non-native vegetation and bat activity, we selected survey plots that ranged in percent non-native cover and other habitat characteristics that could influence bat locomotion and foraging, such as canopy cover, tree density, and shrub density.

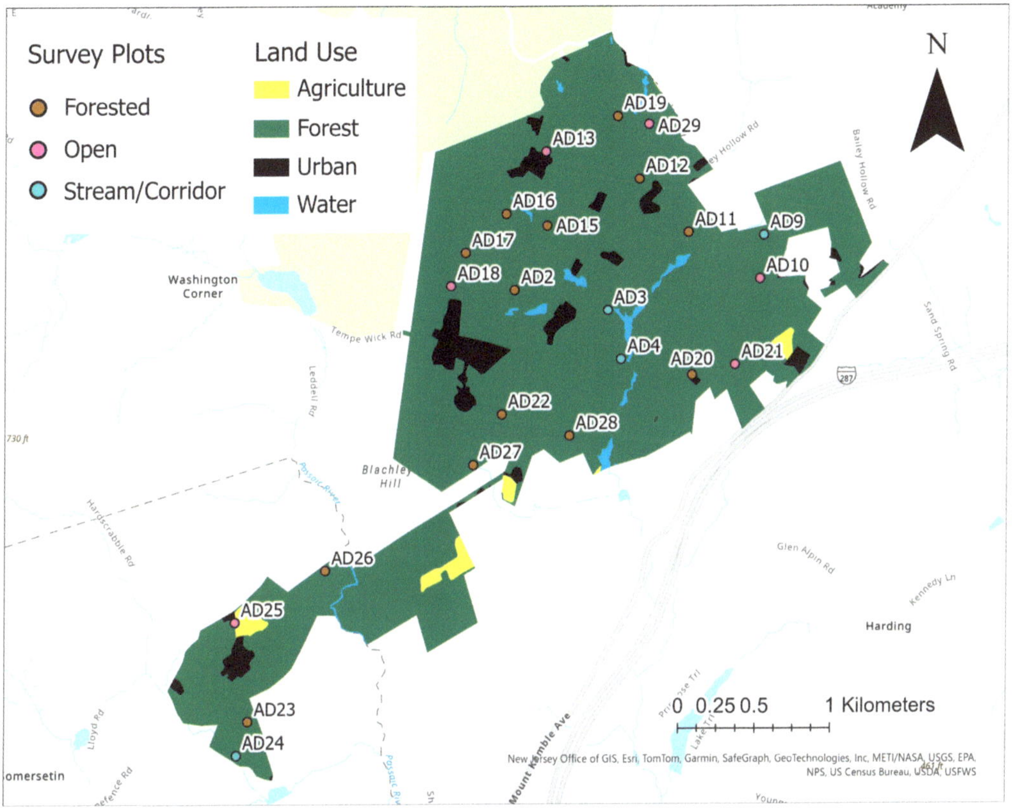

Figure 1. We conducted acoustic monitoring, conventional insect trapping, and vegetation surveys in 23 survey plots (30 m in diameter) in the Morristown National Historical Park–Jockey Hollow Unit, NJ, USA, in summer 2020. Brown markers indicate forested sites, pink markers indicate open sites, and blue markers indicate stream/corridor sites.

2.2. Vegetation Surveys

To determine how plant structure and nativity (i.e., native or non-native) influence bat activity, we conducted a vegetation survey at each site within a 15 m radius of the bat detector. Because structurally cluttered environments can hinder flight and reduce bat activity [43], we quantified measures of clutter at multiple vertical strata. We measured midstory clutter by counting the number of midstory trees and shrubs, defined as woody species 2–10 m in height, in the entire plot. We measured ground layer clutter by averaging the percent ground cover of four 1 m^2 quadrats placed 7.5 m from the detector in each of the four cardinal directions. We also documented what percentage of midstory trees and shrubs in the plot and the ground layer in the four quadrats were non-native.

To test the hypothesis that a closed canopy facilitates bat activity by reducing predation risk, we also measured canopy cover by averaging five spherical crown densiometer measurements recorded in each plot, one from the center of the plot and one from a random point in each cardinal direction. Finally, we calculated the overall plot percent non-native to investigate a possible relationship between plant composition and bat activity. Because every site had a fully native canopy (i.e., no non-native canopy trees), we only considered the ground and midstory layers in our calculation. Therefore, we averaged the percent midstory non-native and four non-native ground cover percentages, one from each quadrat, to obtain the plot percent non-native.

2.3. Acoustic Surveys and Bat Echolocation Analysis

From June to August 2020, we monitored survey plots for bats following the North American Bat (NABat) Monitoring Program protocols for stationary acoustic monitoring [39]. At the center of each survey plot, we placed a Pettersson D500X (Pettersson Elektronik AB, Uppsala, Sweden) acoustic detector equipped with a full-spectrum omnidirectional ultrasonic microphone. We mounted the microphone approximately 3 m off the ground and angled it approximately 45° upward into the airspace [41]. At each survey plot, we recorded bat echolocation pulses for two consecutive nights that were forecasted to have little to no precipitation and wind speeds below 10 km/h [41]. The detectors were active from 30 min before sunset to 30 min after sunrise with a 500 kHz sampling frequency and medium trigger sensitivity to reduce nontarget noise recordings. Once triggered, detectors recorded for three seconds and stored recordings as .WAV files. Because we assumed that each recording was a succession of pulses, or a pass, produced by one individual as it flew near the detector, we quantified general bat activity and foraging activity by counting the total number of recorded passes and terminal buzzes, respectively, from a survey location over the two-night monitoring period.

We analyzed all recordings using the SonoBat 4.4.5 software and the northeastern North America regional library [53]. We first used the file-scrubbing function in SonoBat to eliminate recordings that did not meet the default medium-quality threshold. We then programmed the Batch Classification in SonoBat according to NABat guidelines to classify the remaining recordings that surpassed the acceptable quality threshold of 0.80. After this filtering process, recordings that were below the automatic classification threshold of 0.90, and therefore unidentifiable to species level, were labeled as high-frequency or low-frequency unknowns in SonoBat. Recordings that surpassed the automatic classification threshold of 0.90 and contained at most 16 consecutive echolocation pulses were then automatically classified by SonoBat as one of the nine bat species found in New Jersey. Three bat species in New Jersey produce low-frequency pulses, including big brown bat, hoary bat (*Lasiurus cinereus*, Palisot de Beauvois), and silver-haired bat (*Lasiurus noctivagans*, Le Conte) [54]. We summed the passes produced by these bats and the low-frequency unknown classification to calculate the total number of low-frequency passes. Six species in New Jersey produce high-frequency pulses, including little brown bat, Indiana bat, eastern red bat, northern long-eared bat, evening bat (*Nycticeius humeralis*, Rafinesque), and tricolored bat (*Perimyotis subflavus*, Cuvier). We summed the passes produced by these bats and the high-frequency unknown classifications to calculate the total number of high-frequency passes.

We used the sonogram viewing window in SonoBat to manually verify automatic classifications and to detect terminal buzzes. To minimize false-positive detections, we manually vetted every pass and corrected erroneous automatic classifications made by SonoBat. Because of their often indistinguishable pulse characteristics, we combined the little brown bat (*Myotis LUcifugus*) and Indiana bat (*Myotis SOdalis*, Miller & Allen) classifications into a single LUSO category, which we subsequently treated as a single sonotype [41,55]. A terminal buzz can be heard as a distinct pitch change when played at 10× reduced speed and visualized as a short burst of accelerating pulses in the sonogram viewing window [41]. Although terminal buzzes can be identified using the sonograms produced by SonoBat, they often cannot be reliably classified into species because they are accompanied by changes in typical pulse characteristics. Accordingly, we manually classified all files containing a terminal buzz as either low- or high-frequency unknowns.

2.4. Arthropod Surveys

Concurrent with each acoustic survey, we deployed a blacklight insect bucket trap (Leptraps, LLC, Georgetown, KY, USA) within 10 m of each survey plot. This distance is sufficiently close to the survey plot to appropriately characterize the arthropod community without influencing bat activity [56,57]. Traps consisted of 15 W T8 blacklight bulbs (PestWest, Sarasota, FL, USA), drawing arthropods into a catch bucket lined with Vaportape

II insecticidal strips (Hercon, Emigsville, PA, USA) raised ~0.3 m above the ground. On the morning immediately following each two-day bat survey, we transferred the contents of the trap into an airtight plastic bag and stored them in a freezer at −20 °C until processing. We then soaked all arthropods in 95% ethanol before counting and sorting them by order. We calculated the Shannon diversity index of all arthropod orders at each site using the vegan package in R [58]. We also calculated the order richness of all arthropods at each site. We calculated the abundance of potential arthropod prey at each site by taking the sum of the orders typically eaten by insectivorous bats in New Jersey, i.e., Coleoptera, Diptera, Hemiptera, Hymenoptera, Lepidoptera, and Trichoptera [59].

2.5. Statistical Analysis

We performed all statistical analyses using R v3.3.0+ [60]. We manipulated data using the dplyr package [61] and created plots using the ggplot2 package [62]. To ensure that all measured variables exerted similar influences on the models, we scaled and centered all predictor variables to have a mean of 0 and a standard deviation of 1. To obtain standard error values, we calculated bootstrapped means using 1000 replicates in the boot package [63]. To test the hypothesis that non-native plants affect arthropod prey, we performed linear mixed-effects models (LMERs) in the lme4 package [64] using non-native vegetation percentage as the predictor and either prey abundance or total arthropod order diversity as the response. We included the month as a random effect to account for potential declines in arthropod abundance and diversity as the year progresses from spring to fall [65]. We also conducted a generalized linear model (GLM) in the stats package [60] using percent non-native as the predictor and total arthropod order richness as the response. We used ANOVA in the stats package to confirm that prey abundance and total arthropod order diversity differed significantly by survey month (June, July, August).

To elucidate the relationships between non-native plants and habitat structure, we performed LMERs using non-native percentage as the predictor and either canopy cover or ground cover as the response. We included habitat type as a random effect to account for differing vegetative communities and structures in open, forested, and stream/corridor habitats. We also conducted a GLM using a quasipoisson family to determine if percent non-native was correlated with midstory clutter. We used percent non-native as the predictor and midstory trees and shrubs as the response for this GLM.

We performed GLMs in the MASS package [66] using a negative binomial family to determine the relative effects of vegetative structure, vegetative nativity, and prey availability on bat activity and foraging. We generated fifteen candidate models to test biologically relevant hypotheses about bat behavior (Table 1). To determine which predictors were most strongly associated with bat activity, we used the MuMIn package [67] to conditionally average all candidate models within 2 ΔAICc scores of the top model [68]. This yielded model-averaged coefficients for each of the top predictors, calculated by averaging over the models where the predictor appeared, which we then compared to assess their magnitude and direction on bat activity. To determine the relative effects of the predictors on bats that use differing foraging strategies, we performed two GLMs as described above, one for total low-frequency bat passes and one for total high-frequency bat passes. We again averaged all candidate models within 2 ΔAICc scores of the top model.

Table 1. Candidate model set to test competing hypotheses about predictors of bat activity in northern New Jersey.

Bat Activity Depends On:	**Predictors**				
	Canopy Cover	**Midstory Trees and Shrubs**	**Ground Cover**	**Percent Non-Native**	**Prey Abundance**
All predictors (global model)	X	X	X	X	X
Habitat structure and composition	X	X	X	X	
Habitat structure	X	X	X		
Flight clearance (impacted by non-native plants) and predator avoidance	X	X		X	
Flight clearance, prey availability, and predator avoidance	X	X			X
Flight clearance and predator avoidance	X	X			
Midstory structure		X		X	
Ground structure			X	X	
Prey availability (impacted by non-native plants)				X	X
Canopy structure	X				
Flight clearance (dictated by midstory clutter)		X			
Ground cover			X		
Non-native percentage				X	
Prey availability					X
No predictors (null model)					

3. Results

3.1. Vegetation and Arthropod Surveys

The percent of non-native vegetation in each plot ranged from 0 to 92.9% ($\bar{x}$ = 46.94 ± 5.77 SE) and had a significant positive relationship with both midstory trees and shrubs (β = 0.01 ± 0.006 SE) and ground cover (β = 0.36 ± 0.14 SE; Figure 2).

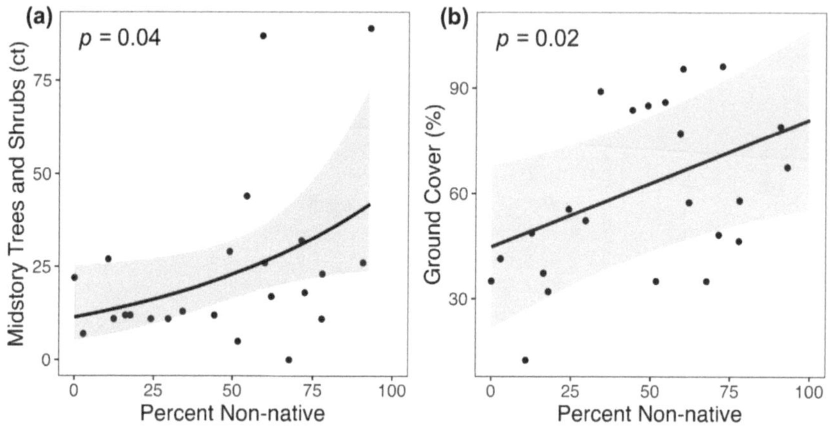

Figure 2. The percentage of a plot (average of the ground and midstory layers occupied by non-native vegetation) was significantly and positively correlated with the (**a**) number of midstory trees and shrubs (β = 0.01 ± 0.006 SE) in the entire plot and (**b**) percentage of the ground covered by vegetation (β = 0.36 ± 0.14 SE) in four quadrats within the plot in northern New Jersey, USA, in summer 2020 (number of plots = 23). Points depict the original data, solid lines depict the fitted model, and shaded areas depict 95% confidence intervals.

Of the 20,474 total arthropods captured, we considered 19,825 (96.8%) to be potential prey for insectivorous bats (Appendix A). Putative prey abundance varied across survey plots, ranging between 96 and 1462 individuals ($\bar{x}$ = 861 ± 85.5 SE). The arthropod community was dominated by lepidopterans (*N* = 8399), coleopterans (*N* = 4496), and trichopterans (*N* = 4318), but hemipterans (*N* = 734) and hymenopterans (*N* = 615) were also well represented at our plots. Coleopterans, dipterans, hymenopterans, and lepidopterans occurred at every survey location. Prey abundance (β = −0.81 ± 2.57 SE), total arthropod order diversity (β = −0.001 ± 0.001 SE), and total arthropod order richness (β = −0.001 ± 0.001 SE) were not significantly correlated with non-native percentage.

3.2. Acoustic Surveys

Over 46 recording nights, we collected 3614 usable bat passes, 303 of which contained terminal buzzes (Appendix B). Bat activity and foraging varied widely by survey plot, ranging from 5 to 684 total passes ($\bar{x}$ = 157 ± 34.28 SE) and from 0 to 80 terminal buzzes ($\bar{x}$ = 13 ± 3.75 SE) recorded over the two-night monitoring period. Because we treated LUSO as a single sonotype, we recorded all "eight" bat species found in New Jersey. Big brown bats were recorded most frequently (51.7%), followed by low-frequency unknowns (18.2%) and hoary bats (11.4%). We recorded 3074 total low-frequency passes (85.0%) and 540 total high-frequency passes (14.9%). We recorded only three northern long-eared bat passes and two tricolored bat passes, both of which produce high-frequency pulses.

3.3. Predictors of Bat Activity

The top models for total bat passes and terminal buzzes were midstory structure (number of midstory trees and shrubs and percent non-native) and non-native percentage, respectively. Percent non-native had a significant positive effect on both total passes (β = 0.92 ± 0.38 SE; Figure 3a) and terminal buzzes (β = 1.57 ± 0.46 SE; Figure 3b). Midstory trees and shrubs had a significant negative effect on total passes (β = −0.58 ± 0.27 SE), but not terminal buzzes (β = −0.37 ± 0.34 SE). Although canopy cover was a covariate in two of the top-ranked models for total passes, it was not a significant predictor. Prey abundance and ground cover were not included in any top-ranked models and were therefore not significant predictors of overall bat activity or foraging.

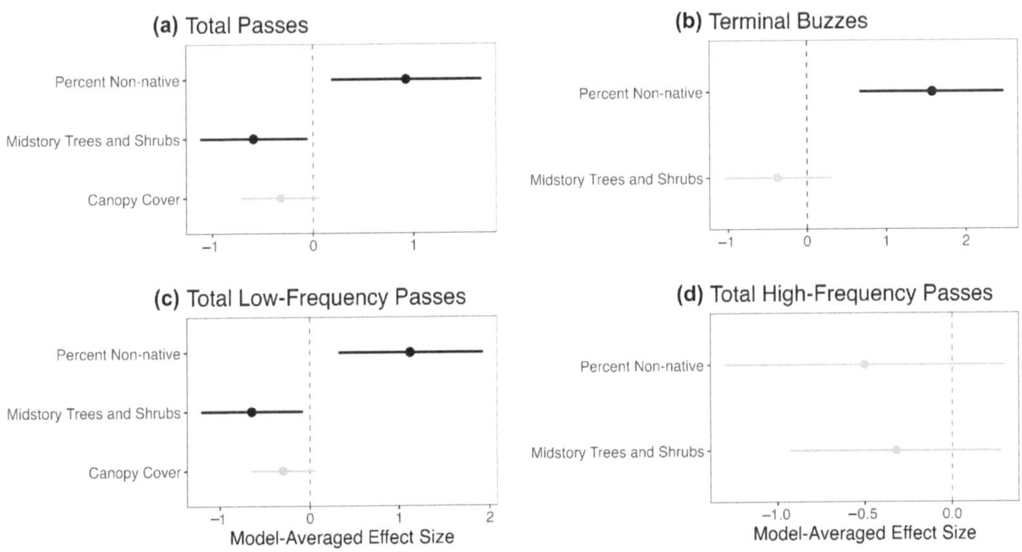

Figure 3. The percentage of a plot (average of the ground and midstory layers) occupied by non-native

vegetation had a significant positive effect on the number of (**a**) total bat passes; (**b**) terminal buzzes; and (**c**) total low-frequency passes recorded in northern New Jersey, USA, in summer 2020, as revealed by model-averaging the effect sizes of predictors from top-ranked generalized linear models (<2 ΔAICc scores from the top model). No measured predictor had a significant effect on (**d**) total high-frequency passes. Lines depict 95% confidence intervals. Black points and black lines indicate significant effects (the model-averaged effect size did not cross 0). Gray points and gray lines indicate effects that were not significant (the model-averaged effect size crossed 0).

The trends for low-frequency passes were nearly identical to those for total passes. The top model for total low-frequency passes included the number of midstory trees and shrubs ($\beta = -0.64 \pm 0.29$ SE) and non-native percentage ($\beta = 1.12 \pm 0.41$ SE), both of which had a significant effect on low-frequency passes (Figure 3c). There were no significant predictors of total high-frequency bat passes (Figure 3d). Detailed GLM results are in Appendix C.

4. Discussion and Conclusions

Overall, our research suggests that non-native vegetation does not have a completely negative impact on forest habitat use by bats. This conclusion is driven by two important findings. First, the percentage of non-native vegetation at a plot had the largest and most positive effect on both total passes, a proxy for overall bat activity, as well as terminal buzzes, a proxy for foraging activity, indicating that bats were more active in plots with higher percentages of non-native plants. Second, non-native percentage was not significantly correlated with putative prey abundance, which itself did not have a significant effect on overall bat activity or foraging activity. Together, such results suggest that bats value habitat structure more than prey availability when using foraging habitats and that non-native plants may benefit bats by creating habitat structures more conducive to foraging.

Contrary to our predictions and the findings of previous studies [69,70], putative prey abundance and total arthropod order diversity and richness were not correlated with plot percent non-native. This could suggest that either non-native plants altered arthropod species composition without changing diversity and richness, or that arthropod communities were similar amongst all plots regardless of percent non-native. The former could occur if a diverse and species-rich community of non-native arthropods replaced the pre-existing native community. It is well documented that non-native plants can be toxic to and reduce the abundance of native arthropod species [12,71]. Because most herbivorous insects are specialists that have coevolved with one or a few plant lineages, native arthropods are likely unequipped to combat the chemical and physical defenses of non-native plants [72,73]. Therefore, non-native insects, especially those that have coevolved with a specific non-native plant, may fill niches opened by the loss of native insects, leading to no net reductions in overall arthropod abundance.

Alternatively, it is possible that the introduction of non-native plants to our study area had little to no effect on the pre-existing arthropod community. Some native insects demonstrate plasticity in host plant preference and could prefer to use non-native vegetation [74,75]. Non-native plants can also be highly suitable for native insects and actually increase arthropod fitness [76]. In such cases, native insect abundance could even increase post-invasion, although this was not detected in our study. In addition, many plants introduced to North America are closely related to native species (e.g., the non-native *Lonicera japonica* is a congener of *L. sempervirens*, L., an eastern North American native), increasing the likelihood that an herbivore that specializes in a native plant can exploit a closely related plant that produces similar chemical compounds [73]. We did not classify arthropods into species; therefore, the mechanisms driving our results are not known.

A non-significant change in arthropod abundance, regardless of the insect community composition, logically results in few impacts on bat use of forest habitats. First, North American bats are generalist insectivores, feeding on a wide array of arthropods [59,77], suggesting that bats can tolerate small changes in arthropod availability. Thus, changes in insect diversity are likely inconsequential to bats, so long as there are ample numbers of prey items to make foraging activities beneficial. Furthermore, bats may not discriminate between native and non-native insects. Indeed, bats have been shown to consistently consume such non-native insects as brown marmorated stink bugs (*Halyomorpha halys*, Stål) [78], spotted lanternflies (*Lycorma delicatula*, White; McHale et al., unpublished data), and emerald ash borers (*Agrilus planipennis*, Waterhouse) [79]. Thus, bats may cope with decreased abundances of native prey by consuming non-native species, which often occur in high abundances [80,81]. It is important to note that we caught arthropods in every survey plot, indicating that even plots with the lowest recorded arthropod abundances may have had ample prey to sustain bat foraging.

We found that the percent of non-native vegetation was also significantly and positively correlated with ground cover and the number of midstory trees and shrubs, but not correlated with canopy cover. Such findings suggest that non-native plants may proliferate in and alter the vegetative structure of the ground and midstory layers. Because North American insectivorous bats primarily use the midstory and canopy strata of forests, changes in ground-level structure likely have few direct impacts on bat activity. On the other hand, the increased number of midstory trees and shrubs, and therefore midstory clutter, associated with a higher degree of vegetative invasion, could impact bat maneuverability and foraging.

In addition to percent non-native, the number of midstory trees and shrubs also had a significant, albeit negative, effect on overall bat activity. Such results indicate that bats were more active in plots with higher percent non-native and lower midstory clutter, which is interesting given our finding that percent non-native was positively associated with midstory trees and shrubs. Logically, bats should be less active in plots with high percent non-native due to the association with increased midstory clutter. However, our model results may indicate that bats were either (1) partitioning their time between sites with only high percent non-native or only a few midstory trees and shrubs; or (2) using sites that had high percent non-native and low midstory clutter. Regarding the latter, we did survey heavily invaded plots that had fewer than 25 midstory trees and shrubs. In such plots, it is possible that non-native vegetation was concentrated at the ground layer, resulting in a high calculated plot percent non-native despite there being few, mostly native, midstory trees and shrubs. Additionally, midstory clutter did not have a significant effect on terminal buzzes, which could suggest that the habitat structure affects bat locomotion more so than foraging. While foraging bats likely compromise their own habitat preferences for the habitat preferences of their prey, commuting bats may have fewer tradeoffs to consider and more often travel down paths of least resistance [82]. The discrepancy in the relationships between percent non-native, number of midstory trees and shrubs, and bat activity may also suggest that non-native plants may facilitate bat activity in ways that do not alter habitat structure, such as by improving habitat quality (e.g., providing roost sites). Further research should be conducted to determine the mechanisms by which non-native plants affect bats.

Although none of our measured predictors significantly affected total high-frequency passes, both percent non-native and the number of midstory trees and shrubs significantly affected total low-frequency passes, much like the model for total passes. This similarity is likely explained by the dominance within our dataset of bats that produce low-frequency pulses. The low-frequency echolocating big brown bat, responsible for over half of our recorded passes, is the most common bat in New Jersey [83]. We attribute the model results for total low-frequency passes, which comprised less than 15% of all passes recorded, to an inadequate sample size. There are several explanations for why we recorded so few low-frequency passes. First, white-nose syndrome has significantly reduced populations of previously common species that produce high-frequency pulses, such as the little brown and northern long-eared bats [84,85]. Alternatively, many areas within Jockey Hollow are dominated by non-native vegetation and have low densities of midstory trees and shrubs, both of which seem to facilitate the activity of bats with low-frequency pulses. Therefore, the noticeably low number of high-frequency passes recorded is not surprising but does prevent us from concluding if non-native plants or habitat structure affect bats at Jockey Hollow differently depending on their foraging preferences.

Implications for Forest Management

Managers often recommend removing non-native plants in both forested and urban landscapes because it can significantly increase insectivorous bat activity [43,86,87]. However, our results suggest that in the short term, non-native plants might not be detrimental to bats and may even benefit them by maintaining open flight space beneath the forest canopy. However, we are uncertain about how bats are impacted in the long term by non-native shrubs, especially in conjunction with deer browse. In the United Kingdom, the presence of a non-native rhododendron and red deer (*Cervus elaphus*, L.) negatively affected insectivorous bats that prefer to forage in open spaces [43]. Both non-native plants and deer browse can hinder the survival of native seedlings, which could eventually decrease plant richness and woody species recruitment [88,89]. As native trees die, these habitats may transition to having cluttered midstories and little to no canopy, which could negatively impact bats.

Our findings demonstrate that there are complex relationships between native and non-native species that warrant further study. Because bats seem to highly value habitat structure when choosing where to travel and forage, management activities that remove non-native plants while maintaining structural attributes preferred by bats are likely to sustain bat activity. Therefore, reducing midstory clutter and removing non-native shrubs could create travel corridors that facilitate foraging for bats with low-frequency pulses, the most common species detected at Jockey Hollow. Unfortunately, we collected insufficient data to draw conclusions about what structural habitat attributes significantly influenced high-frequency bat activity. Such an evidence base is needed to facilitate habitat-based bat conservation, particularly for bats impacted by other threats such as white-nose syndrome and wind energy infrastructure [90,91]. Because our model results were largely driven by big brown bats, we cannot advocate for a single management solution for increasing the activity of all bats. It is likely that a variety of management techniques that promote or maintain structurally heterogeneous habitats will support the greatest bat diversity. Further studies targeting bats with high-frequency pulses can reduce the uncertainty surrounding the impacts of non-native plants on bats of the highest conservation concern.

Author Contributions: Conceptualization, B.M. and K.K.; Methodology, K.K.; Data curation, M.M., E.D. and K.K.; Formal analysis, M.M.; Writing, M.M., E.D. and B.M.; Project administration, B.M.; Funding acquisition, B.M. All authors have read and agreed to the published version of the manuscript.

Funding: This research was funded by the National Park Service Cooperative Research and Training Grant [number P20AC00480].

Data Availability Statement: Raw data associated with this paper are stored within the Rutgers Libraries Data Portal, which maintains all Rutgers-produced datasets. Data can be freely accessed through this portal.

Acknowledgments: Special thanks to Park Biologist Bob Masson, Park Law Enforcement, for facilitating field surveys, and to Justin Martinez and Kelly Faller for assisting with data collection. We also thank Mark Ford, Samantha Hoff, Joy O'Keefe, and Susan Loeb for their constructive feedback on previous drafts.

Conflicts of Interest: The authors declare that they have no known competing financial interests or personal relationships that could have appeared to influence the work reported in this paper.

Appendix A

Table A1. Counts of arthropods collected in blacklight traps deployed at 23 plots in northern New Jersey, USA, in summer 2020.

Plot	Total Arthropod Abundance	Putative Prey Abundance	Shannon Diversity Index	Araneae	Blattodea	Coleoptera	Diptera	Hemiptera	Hymenoptera	Lepidoptera	Mantodea	Megaloptera	Neuroptera	Odonata	Opiliones	Orthoptera	Plecoptera	Siphonaptera	Trichoptera
AD2	753	739	1.34	4	0	91	58	0	18	373	0	0	3	0	4	2	1	0	199
AD3	1620	1462	1.54	1	0	80	676	15	22	365	0	0	130	0	0	0	27	0	304
AD4	702	694	1.33	0	0	104	17	6	28	296	0	1	0	0	0	0	7	0	243
AD9	1422	1373	1.55	11	4	195	46	130	62	688	0	0	2	0	0	0	32	0	252
AD10	1233	1218	1.35	2	4	589	18	65	29	365	0	1	2	0	3	0	3	0	152
AD11	610	561	1.59	4	3	122	17	27	11	278	0	0	0	0	32	3	7	0	106
AD12	100	96	1.50	0	0	10	9	1	11	50	0	0	0	0	4	0	0	0	15
AD13	507	501	0.76	2	0	5	3	5	10	85	0	0	0	0	0	4	0	0	393
AD15	1218	1145	1.45	0	0	314	10	28	16	493	0	1	37	0	0	0	35	0	284
AD16	275	275	1.04	0	0	16	43	1	7	185	0	0	0	0	0	0	0	0	23
AD17	789	784	1.45	0	0	275	27	44	23	285	0	0	0	2	1	2	0	0	130
AD18	641	636	1.64	4	0	187	67	59	29	202	0	0	1	0	0	0	0	0	92
AD19	1463	1452	1.05	0	0	654	13	55	52	678	0	0	1	1	0	1	4	4	0
AD20	1035	1029	1.11	0	0	159	20	11	50	674	0	0	6	0	0	0	0	0	115
AD21	1182	1174	1.02	0	6	361	41	80	2	690	2	0	0	0	0	0	0	0	0
AD22	1047	1039	1.28	0	0	124	36	14	24	366	0	0	4	0	0	0	4	0	475
AD23	1423	1387	1.56	1	10	469	55	58	89	528	0	0	2	1	4	3	15	0	188
AD24	852	704	1.63	1	0	292	10	28	9	177	0	0	23	0	0	0	124	0	188
AD25	453	453	1.66	0	0	54	36	70	41	150	0	0	0	0	0	0	0	0	102
AD26	174	172	0.80	0	0	18	1	1	1	134	0	0	0	0	2	0	0	0	17
AD27	1475	1438	1.17	0	0	128	19	11	9	701	0	0	11	0	0	1	25	0	570
AD28	897	893	1.39	0	0	191	17	25	42	384	0	0	1	0	0	0	3	0	234
AD29	603	600	1.26	0	0	58	24	0	30	252	0	0	0	0	0	1	2	0	236

Appendix B

Table A2. Counts of passes recorded by acoustic detectors deployed at 23 plots in northern New Jersey, USA, in summer 2020.

Plot	Terminal Buzzes	Total Passes	Low Freq. Unknown	High Freq. Unknown	*E. fuscus*	*L. borealis*	*L. cinereus*	*L. noctivagans*	*M. lucifugus/sodalis*	*M. septentrionalis*	*N. humeralis*	*P. subflavus*
AD2	9	140	9	9	51	68	0	1	2	0	0	0
AD3	3	53	22	0	20	0	9	2	0	0	0	0
AD4	4	46	6	7	23	5	0	0	0	0	5	0
AD9	3	54	14	0	39	1	0	0	0	0	0	0

Table A2. *Cont.*

Plot	Terminal Buzzes	Total Passes	Low Freq. Unknown	High Freq. Unknown	*E. fuscus*	*L. borealis*	*L. cinereus*	*L. noctivagans*	*M. lucifugus/sodalis*	*M. septentrionalis*	*N. humeralis*	*P. subflavus*
AD10	1	15	1	1	5	7	1	0	0	0	0	0
AD11	17	100	15	3	51	27	3	1	0	0	0	0
AD12	0	5	1	1	1	0	0	0	1	1	0	0
AD13	9	190	62	33	25	15	22	29	2	0	2	0
AD15	6	155	38	30	44	31	0	3	0	0	9	0
AD16	0	40	14	3	21	2	0	0	0	0	0	0
AD17	10	114	24	27	45	16	0	1	0	0	1	0
AD18	80	525	60	2	115	21	271	55	1	0	0	0
AD19	36	291	26	1	252	5	4	2	0	1	0	0
AD20	30	297	32	6	224	31	4	0	0	0	0	0
AD21	4	56	3	2	46	5	0	0	0	0	0	0
AD22	1	92	15	2	35	38	0	0	2	0	0	0
AD23	9	92	33	9	35	3	8	0	0	0	4	0
AD24	5	179	116	17	23	3	14	1	2	1	0	2
AD25	19	247	59	2	105	18	43	20	0	0	0	0
AD26	0	22	1	2	16	1	1	1	0	0	0	0
AD27	4	60	7	5	41	1	5	1	0	0	0	0
AD28	16	157	23	12	90	7	18	4	0	0	3	0
AD29	37	684	80	20	564	2	9	9	0	0	0	0
Total	303	3614	661	194	1871	307	412	130	10	3	24	2

Appendix C

Table A3. Top models (<2 ΔAICc scores from top model) for total low-frequency passes recorded in northern New Jersey, USA, in summer 2020.

Model	AICc	ΔAICc from Top Model
1. Midstory Trees and Shrubs + Percent Non-native	270.23	0.00
2. Percent Non-native	270.94	0.71
3. Canopy Cover + Midstory Trees and Shrubs + Percent Non-native	271.60	1.36

Table A4. Effect sizes for predictors in top models for total low-frequency passes recorded in northern New Jersey, USA, in summer 2020.

Predictor	Model 1	Model 2	Model 3
Midstory Trees and Shrubs	−0.66 ± 0.29 SE	—	−0.61 ± 0.28 SE
Percent Non-native	1.25 ± 0.39 SE	1.05 ± 0.38 SE	0.95 ± 0.39 SE
Canopy Cover	—	—	−0.29 ± 0.18 SE

Table A5. Top models (<2 ΔAICc scores from top model) for total passes recorded in northern New Jersey, USA, in summer 2020.

Model	AICc	ΔAICc from Top Model
1. Midstory Trees and Shrubs + Percent Non-native	279.58	0.00
2. Percent Non-native	280.21	0.62
3. Canopy Cover + Midstory Trees and Shrubs + Percent Non-native	281.45	1.87
4. Canopy Cover	281.50	1.91

Table A6. Effect sizes for predictors in top models for total passes recorded in northern New Jersey, USA, in summer 2020.

Predictor	Model 1	Model 2	Model 3	Model 4
Midstory Trees and Shrubs	−0.59 ± 0.36 SE	—	−0.56 ± 0.26 SE	—
Percent Non-native	1.03 ± 0.36 SE	0.84 ± 0.35 SE	0.79 ± 0.37 SE	—
Canopy Cover	—	—	−0.23 ± 0.17 SE	−0.41 ± 0.18 SE

Table A7. Top models (<2 ΔAICc scores from top model) for terminal buzzes recorded in northern New Jersey, USA, in summer 2020.

Model	AICc	ΔAICc from Top Model
1. Percent Non-native	162.52	0.00
2. Midstory Trees and Shrubs + Percent Non-native	164.35	1.83

Table A8. Effect sizes for predictors in top models for terminal buzzes recorded in northern New Jersey, USA, in summer 2020.

Predictor	Model 1	Model 2
Percent Non-native	1.54 ± 0.45 SE	1.65 ± 0.48 SE
Midstory Trees and Shrubs	—	−0.37 ± 0.34 SE

Table A9. Top models (<2 ΔAICc scores from top model) for total high-frequency passes recorded in northern New Jersey, USA, in summer 2020.

Model	AICc	ΔAICc from Top Model
1. Null model	196.73	0.00
2. Percent Non-native	198.18	1.45
3. Midstory Trees and Shrubs	198.32	1.58

Table A10. Effect sizes for predictors in top models for total high-frequency passes recorded in northern New Jersey, USA, in summer 2020.

Predictor	Model 1	Model 2	Model 3
Percent Non-native	—	−0.50 ± 0.41 SE	—
Midstory Trees and Shrubs	—	—	−0.32 ± 0.31 SE

References

1. Kettenring, K.M.; Adams, C.R. Lessons learned from invasive plant control experiments: A systematic review and meta-analysis. *J. Appl. Ecol.* **2011**, *48*, 970–979. [CrossRef]
2. Mack, R.N.; Simberloff, D.; Mark Lonsdale, W.; Evans, H.; Clout, M.; Bazzaz, F.A. Biotic invasions: Causes, epidemiology, global consequences, and control. *Ecol. Appl.* **2000**, *10*, 689–710. [CrossRef]
3. Fletcher, R.A.; Brooks, R.K.; Lakoba, V.T.; Sharma, G.; Heminger, A.R.; Dickinson, C.C.; Barney, J.N. Invasive plants negatively impact native, but not exotic, animals. *Glob. Change Biol.* **2019**, *25*, 3694–3705. [CrossRef] [PubMed]
4. Hall, L.S.; PKrausman, R.; Morrison, M.L. The habitat concept and a plea for standard terminology. *Wildl. Soc. Bull.* **1997**, *25*, 173–182.
5. Asner, G.P.; Hughes, R.F.; Vitousek, P.M.; Knapp, D.E.; Kennedy-Bowdoin, T.; Boardman, J.; Martin, R.E.; Eastwood, M.; Green, R.O. Invasive plants transform the three-dimensional structure of rain forests. *Proc. Natl. Acad. Sci. USA* **2008**, *105*, 4519–4523. [CrossRef]
6. Vitousek, P.M.; Walker, L.R.; Whiteaker, L.D.; Mueller-Dombois, D.; Matson, P.A. Biological invasion by Myrica faya alters ecosystem development in Hawaii. *Science* **1987**, *238*, 802–804. [CrossRef]
7. Drake, K.K.; Bowen, L.; Nussear, K.E.; Esque, T.C.; Berger, A.J.; Custer, N.A.; Waters, S.C.; Johnson, J.D.; Miles, A.K.; Lewison, R.L. Negative impacts of invasive plants on conservation of sensitive desert wildlife. *Ecosphere* **2016**, *7*, e01531. [CrossRef]

8. Dutra, H.P.; Barnett, K.; Reinhardt, J.R.; Marquis, R.J.; Orrock, J.L. Invasive plant species alters consumer behavior by providing refuge from predation. *Oecologia* **2011**, *166*, 649–657. [CrossRef]
9. McCary, M.A.; Mores, R.; Farfan, M.A.; Wise, D.H. Invasive plants have different effects on trophic structure of green and brown food webs in terrestrial ecosystems: A meta-analysis. *Ecol. Lett.* **2016**, *19*, 328–335. [CrossRef]
10. Manchester, S.J.; Bullock, J.M. The impacts of non-native species on UK biodiversity and the effectiveness of control. *J. Appl. Ecol.* **2000**, *37*, 845–864. [CrossRef]
11. Mangla, S.; Inderjit; Callaway, R.M. Exotic invasive plant accumulates native soil pathogens which inhibit native plants. *J. Ecol.* **2008**, *96*, 58–67. [CrossRef]
12. Bezemer, T.M.; Harvey, J.A.; Cronin, J.T. Response of native insect communities to invasive plants. *Annu. Rev. Entomol.* **2014**, *59*, 119–141. [CrossRef] [PubMed]
13. Collier, M.H.; Vankat, J.L.; Hughes, M.R. Diminished plant richness and abundance below Lonicera maackii, an invasive shrub. *Am. Midl. Nat.* **2002**, *147*, 60–71. [CrossRef]
14. Crooks, J.A. Characterizing ecosystem-level consequences of biological invasions: The role of ecosystem engineers. *Oikos* **2002**, *97*, 153–166. [CrossRef]
15. McKinney, A.M.; Goodell, K. Shading by invasive shrub reduces seed production and pollinator services in a native herb. *Biol. Invasions* **2010**, *12*, 2751–2763. [CrossRef]
16. Stewart, P.S.; Hill, R.A.; Stephens, P.A.; Whittingham, M.J.; Dawson, W. Impacts of invasive plants on animal behaviour. *Ecol. Lett.* **2021**, *24*, 891–907. [CrossRef] [PubMed]
17. Maynard-Bean, E.; Kaye, M. Invasive shrub removal benefits native plants in an eastern deciduous forest of North America. *Invasive Plant Sci. Manag.* **2019**, *12*, 3–10. [CrossRef]
18. Trammell, T.L.; D'Amico, V., III; Avolio, M.L.; Mitchell, J.C.; Moore, E. Temperate deciduous forests embedded across developed landscapes: Younger forests harbour invasive plants and urban forests maintain native plants. *J. Ecol.* **2020**, *108*, 2366–2375. [CrossRef]
19. Silaeva, T.; Andreychev, A.; Kiyaykina, O.; Balčiauskas, L. Taxonomic and ecological composition of forest stands inhabited by forest dormouse Dryomys nitedula (Rodentia: Gliridae) in the Middle Volga. *Biologia* **2021**, *76*, 1475–1482. [CrossRef]
20. Augustine, K.E.; Kingsolver, J.G. Biogeography and phenology of oviposition preference and larval performance of *Pieris virginiensis* butterflies on native and invasive host plants. *Biol. Invasions* **2018**, *20*, 413–422. [CrossRef]
21. Davis, S.L.; Cipollini, D. Do mothers always know best? Oviposition mistakes and resulting larval failure of *Pieris virginiensis* on *Alliaria petiolata*, a novel, toxic host. *Biol. Invasions* **2014**, *16*, 1941–1950. [CrossRef]
22. Magee, C.S.; Van Clef, M. Plant structure of hooded and Kentucky warbler breeding sites in New Jersey. *Cassinia* **2016**, *76*, 30–35.
23. Borgmann, K.L.; Rodewald, A.D. Nest predation in an urbanizing landscape: The role of exotic shrubs. *Ecol. Appl.* **2004**, *14*, 1757–1765. [CrossRef]
24. Carter, E.T.; Eads, B.C.; Ravesi, M.J.; Kingsbury, B.A. Exotic invasive plants alter thermal regimes: Implications for management using a case study of a native ectotherm. *Funct. Ecol.* **2015**, *29*, 683–693. [CrossRef]
25. Welch, J.N.; Leppanen, C. The threat of invasive species to bats: A review. *Mammal Rev.* **2017**, *47*, 277–290. [CrossRef]
26. Arnett, E.B.; Baerwald, E.F.; Mathews, F.; Rodrigues, L.; Rodríguez-Durán, A.; Rydell, J.; Villegas-Patraca, R.; Voigt, C.C. Impacts of wind energy development on bats: A global perspective. In *Bats in the Anthropocene: Conservation of Bats in a Changing World*; Springer: Berlin/Heidelberg, Germany, 2016; pp. 295–323.
27. Cheng, T.L.; Reichard, J.D.; Coleman, J.T.; Weller, T.J.; Thogmartin, W.E.; Reichert, B.E.; Bennett, A.B.; Broders, H.G.; Campbell, J.; Etchison, K. The scope and severity of white-nose syndrome on hibernating bats in North America. *Conserv. Biol.* **2021**, *35*, 1586–1597. [CrossRef]
28. Brigham, R.M. *Bats in Forests: What We Know and What We Need to Learn. Bats in Forests: Conservation and Management*; Johns Hopkins University Press: Baltimore, MD, USA, 2007; pp. 1–15.
29. Harvey, M.J.; Altenbach, J.S.; Best, T.L. *Bats of the United States and Canada*; JHU Press: Baltimore, MD, USA, 2011.
30. Wunder, L.; Carey, A.B. Use of the forest canopy by bats. *Northwest Sci.* **1996**, *70*, 79–85.
31. Bender, M.J.; Perea, S.; Castleberry, S.B.; Miller, D.A.; Wigley, T.B. Influence of insect abundance and vegetation structure on site-occupancy of bats in managed pine forests. *For. Ecol. Manag.* **2021**, *482*, 118839. [CrossRef]
32. Froidevaux, J.S.; Barbaro, L.; Vinet, O.; Larrieu, L.; Bas, Y.; Molina, J.; Calatayud, F.; Brin, A. Bat responses to changes in forest composition and prey abundance depend on landscape matrix and stand structure. *Sci. Rep.* **2021**, *11*, 10586. [CrossRef]
33. Brigham, R.; Grindal, S.; Firman, M.; Morissette, J. The influence of structural clutter on activity patterns of insectivorous bats. *Can. J. Zool.* **1997**, *75*, 131–136. [CrossRef]
34. Hutson, A.M.; Hutson, A.M.; Mickleburgh, S.P.; Racey, P. *Microchiropteran Bats: Global Status Survey and Conservation Action Plan*; IUCN: Gland, Switzerland, 2001; Volume 56.
35. Frick, W.F. Acoustic monitoring of bats, considerations of options for long-term monitoring. *Therya* **2013**, *4*, 69–70. [CrossRef]
36. Russo, D.; Voigt, C.C. The use of automated identification of bat echolocation calls in acoustic monitoring: A cautionary note for a sound analysis. *Ecol. Indic.* **2016**, *66*, 598–602. [CrossRef]
37. Denzinger, A.; Schnitzler, H.-U. Bat guilds, a concept to classify the highly diverse foraging and echolocation behaviors of microchiropteran bats. *Front. Physiol.* **2013**, *4*, 164. [CrossRef] [PubMed]
38. Jones, G.; Teeling, E.C. The evolution of echolocation in bats. *Trends Ecol. Evol.* **2006**, *21*, 149–156. [CrossRef] [PubMed]

39. Loeb, S.C.; Rodhouse, T.J.; Ellison, L.E.; Lausen, C.L.; Reichard, J.D.; Irvine, K.M.; Ingersoll, T.E.; Coleman, J.T.; Thogmartin, W.E.; Sauer, J.R. *A Plan for the North American Bat Monitoring Program (NABat)*; General Technical Report SRS-208; US Department of Agriculture Forest Service, Southern Research Station: Asheville, NC, USA, 2015; Volume 208, pp. 1–100.
40. Schnitzler, H.-U.; Kalko, E.K. Echolocation by insect-eating bats: We define four distinct functional groups of bats and find differences in signal structure that correlate with the typical echolocation tasks faced by each group. *Bioscience* **2001**, *51*, 557–569. [CrossRef]
41. Drake, E.; Vonhof, M.; Maslo, B. Bat use of golf courses depends on surrounding landscape context. *Urban Ecosyst.* **2023**, *26*, 1469–1482. [CrossRef]
42. Welch, J.N.; Fordyce, J.A.; Simberloff, D.S. Indirect impacts of invaders: A case study of the Pacific sheath-tailed bat (*Emballonura semicaudata rotensis*). *Biol. Conserv.* **2016**, *201*, 146–151. [CrossRef]
43. Beason, R.D.; Riesch, R.; Koricheva, J. Tidying up the cluttered understorey: Foraging strategy mediates bat activity responses to invasive rhododendron. *For. Ecol. Manag.* **2020**, *475*, 118392. [CrossRef]
44. Aldridge, H.; Rautenbach, I. Morphology, echolocation and resource partitioning in insectivorous bats. *J. Anim. Ecol.* **1987**, *56*, 763–778. [CrossRef]
45. Wund, M.A. Variation in the echolocation calls of little brown bats (*Myotis lucifugus*) in response to different habitats. *Am. Midl. Nat.* **2006**, *156*, 99–108. [CrossRef]
46. Bellamy, C.; Scott, C.; Altringham, J. Multiscale, presence-only habitat suitability models: Fine-resolution maps for eight bat species. *J. Appl. Ecol.* **2013**, *50*, 892–901. [CrossRef]
47. Wordley, C.F.; Sankaran, M.; Mudappa, D.; Altringham, J.D. Landscape scale habitat suitability modelling of bats in the Western Ghats of India: Bats like something in their tea. *Biol. Conserv.* **2015**, *191*, 529–536. [CrossRef]
48. Sneddon, L.; Zaremba, R.; Largay, E.; Podniesinski, G.; Perles, S.; Thompson, J. *Vegetation Classification and Mapping of Morristown National Historical Park: New Jersey*; Technical Report NPS/NER/NRTR—2008/116; National Park Service: Philadelphia, PA, USA, 2008.
49. Wagner, R.; Cole, C.A.; Gorenflo, L.; Orland, B.; Tamminga, K. *Morristown National Historical Park Natural Resource Condition Assessment: New Jersey*; Natural Resource Report NPS/NETN/NRR—2014/869; National Park Service: Philadelphia, PA, USA, 2014.
50. Gilbert, A.T.; O'Connell, A.F.; Annand, E.M.; Talancy, N.W.; Sauer, J.R.; Nichols, J.D. *An Inventory of Terrestrial Mammals at National Parks in the Northeast Temperate Network and Sagamore Hill National Historic Site*; 2007-5245; U.S. Geological Survey: Reston, VA, USA, 2008. Available online: https://irma.nps.gov/DataStore/Reference/Profile/660794 (accessed on 27 January 2023).
51. Rogers, D.S.; Belk, M.C.; González, M.W.; Coleman, B.L. Patterns of habitat use by bats along a riparian corridor in northern Utah. *Southwest. Nat.* **2006**, *51*, 52–58. [CrossRef]
52. Ketzler, L.P.; Comer, C.E.; Twedt, D.J. Nocturnal insect availability in bottomland hardwood forests managed for wildlife in the Mississippi Alluvial Valley. *For. Ecol. Manag.* **2017**, *391*, 127–134. [CrossRef]
53. Szewczak, J.; Szewczak, N. SonoBat. 2017. Available online: https://sonobat.com/ (accessed on 29 November 2023).
54. Szewczak, J.M.; Corcoran, A.; Kennedy, J.-P.; Weller, T.J.; Ormsbee, P.C. *Echolocation Call Characteristics of Eastern US Bats—SonoBat v.4x*; Humboldt State University Bat Lab: Arcata, CA, USA, 2017.
55. Szewczak, J.M.; Harris, L.S. *A Field Test of Two Acoustic Classification Systems to Discriminate Indiana Bats*; Northeast Bat Working Group: Albany, NY, USA, 2013.
56. Dodd, L.E.; Lacki, M.J.; Britzke, E.R.; Buehler, D.A.; Keyser, P.D.; Larkin, J.L.; Rodewald, A.D.; Wigley, T.B.; Wood, P.B.; Rieske, L.K. Forest structure affects trophic linkages: How silvicultural disturbance impacts bats and their insect prey. *For. Ecol. Manag.* **2012**, *267*, 262–270. [CrossRef]
57. Froidevaux, J.S.; Fialas, P.C.; Jones, G. Catching insects while recording bats: Impacts of light trapping on acoustic sampling. *Remote Sens. Ecol. Conserv.* **2018**, *4*, 240–247. [CrossRef]
58. Oksanen, J.; Simpson, G.L.; Blanchet, F.G.; Kindt, R.; Legendre, P.; Minchin, P.R.; O'Hara, R.B.; Solymos, P.; Stevens, M.H.H.; Szoecs, E.; et al. _vegan: Community Ecology Package_, R Package Version 2.6-4. 2022. Available online: https://www.researchgate.net/publication/360782912_vegan_community_ecology_package_version_26-2_April_2022 (accessed on 14 December 2023).
59. Maslo, B.; Mau, R.L.; Kerwin, K.; McDonough, R.; McHale, E.; Foster, J.T. Bats provide a critical ecosystem service by consuming a large diversity of agricultural pest insects. *Agric. Ecosyst. Environ.* **2022**, *324*, 107722. [CrossRef]
60. R Core Team. *R: A Language and Environment for Statistical Computing*; R Core Team: Vienna, Austria, 2022.
61. Wickham, H.; François, R.; Henry, L.; Müller, K.; Vaughan, D. _dplyr: A Grammar of Data Manipulation_, R Package Version 1.1.0. 2023. Available online: https://www.researchgate.net/publication/275646200_dplyr_A_Grammar_of_Data_Manipulation (accessed on 14 December 2023).
62. Wickham, H. *ggplot2: Elegant Graphics for Data Analysis*; Springer: New York, NY, USA, 2016.
63. Canty, A.; Ripley, B. Boot: Bootstrap R (S-Plus) Functions, R Package Version 1.3-18; Computer Software. 2016. Available online: https://www.researchgate.net/publication/311467889_boot_Bootstrap_R_S-Plus_Functions (accessed on 27 November 2023).
64. Bates, D.; Mächler, M.; Bolker, B.; Walker, S. Fitting linear mixed-effects models using lme4. *arXiv* **2014**, arXiv:1406.5823.
65. Wang, C.; Abbar, S.; Pan, X.; Ranabhat, S.; Cooper, R. Diversity and prevalence of nuisance arthropods detected by sticky traps in apartments in New Jersey. *J. Econ. Entomol.* **2023**, *116*, 1317–1320. [CrossRef]
66. Venables, W.N.; Ripley, B.D. *Modern Applied Statistics with S-PLUS*; Springer Science & Business Media: Berlin/Heidelberg, Germany, 2013.

67. Barton, K. _MuMIn: Multi-Model Inference_, R Package Version 1.47.1. 2022. Available online: https://www.researchgate.net/publication/305346822_MuMIn_Multi-model_inference (accessed on 27 November 2023).
68. Burnham, K.; Anderson, D. *Model Selection and Multimodel Inference: A Practical Information-Theoretic Approach*, 2nd ed.; Springer, New York, NY, USA, 2002.
69. Olckers, T.; Hulley, P. Impoverished insect herbivore faunas on the exotic bugweed Solanum mauritianum Scop. relative to indigenous Solanum species in Natal/KwaZulu and the Transkei. *J. Entomol. Soc. S. Afr.* **1991**, *54*, 39–50.
70. Samways, M.J.; Caldwell, P.; Osborn, R. Ground-living invertebrate assemblages in native, planted and invasive vegetation in South Africa. *Agric. Ecosyst. Environ.* **1996**, *59*, 19–32. [CrossRef]
71. Sunny, A.; Diwakar, S.; Sharma, G.P. Native insects and invasive plants encounters. *Arthropod-Plant Interact.* **2015**, *9*, 323–331. [CrossRef]
72. Bernays, E.; Graham, M. On the evolution of host specificity in phytophagous arthropods. *Ecology* **1988**, *69*, 886–892. [CrossRef]
73. Tallamy, D.W. Do alien plants reduce insect biomass? *Conserv. Biol.* **2004**, *18*, 1689–1692. [CrossRef]
74. Fielding, D.J.; Conn, J.S. Feeding preference for and impact on an invasive weed (*Crepis tectorum*) by a native, generalist insect herbivore, *Melanoplus borealis* (Orthoptera: Acrididae). *Ann. Entomol. Soc. Am.* **2011**, *104*, 1303–1308. [CrossRef]
75. Parker, J.D.; Hay, M.E. Biotic resistance to plant invasions? Native herbivores prefer non-native plants. *Ecol. Lett.* **2005**, *8*, 959–967. [CrossRef]
76. Harvey, J.A.; Biere, A.; Fortuna, T.; Vet, L.E.; Engelkes, T.; Morriën, E.; Gols, R.; Verhoeven, K.; Vogel, H.; Macel, M. Ecological fits, mis-fits and lotteries involving insect herbivores on the invasive plant, *Bunias orientalis*. *Biol. Invasions* **2010**, *12*, 3045–3059. [CrossRef]
77. Clare, E.L.; Fraser, E.E.; Braid, H.E.; Fenton, M.B.; Hebert, P.D. Species on the menu of a generalist predator, the eastern red bat (*Lasiurus borealis*): Using a molecular approach to detect arthropod prey. *Mol. Ecol.* **2009**, *18*, 2532–2542. [CrossRef] [PubMed]
78. Maslo, B.; Valentin, R.; Leu, K.; Kerwin, K.; Hamilton, G.C.; Bevan, A.; Fefferman, N.H.; Fonseca, D.M. Chirosurveillance: The use of native bats to detect invasive agricultural pests. *PLoS ONE* **2017**, *12*, e0173321. [CrossRef]
79. Münzer, O.M.; Schaetz, B.; Kurta, A. Consumption of insect pests by the evening bat (*Nycticeius humeralis*) in southeastern Michigan. *Mich. Entomol. Soc.* **2016**, *49*, 36.
80. Brockerhoff, E.; Liebhold, A. Ecology of forest insect invasions. *Biol. Invasions* **2017**, *19*, 3141–3159. [CrossRef]
81. Murman, K.; Setliff, G.P.; Pugh, C.V.; Toolan, M.J.; Canlas, I.; Cannon, S.; Abreu, L.; Fetchen, M.; Zhang, L.; Warden, M.L. Distribution, survival, and development of spotted lanternfly on host plants found in North America. *Environ. Entomol.* **2020**, *49*, 1270–1281. [CrossRef] [PubMed]
82. Verboom, B. *The Use of Edge Habitats by Commuting and Foraging Bats*; Wageningen University and Research: Wageningen, The Netherlands, 1998.
83. NJ Department of Health. *What You Should Know About Bats...*; NJ Department of Health: Trenton, NJ, USA, 2014.
84. Maslo, B.; Valent, M.; Gumbs, J.F.; Frick, W.F. Conservation implications of ameliorating survival of little brown bats with white-nose syndrome. *Ecol. Appl.* **2015**, *25*, 1832–1840. [CrossRef] [PubMed]
85. O'Keefe, J.M.; Pettit, J.L.; Loeb, S.C.; Stiver, W.H. White-nose syndrome dramatically altered the summer bat assemblage in a temperate Southern Appalachian forest. *Mamm. Biol.* **2019**, *98*, 146–153. [CrossRef]
86. Lintott, P.R.; Bunnefeld, N.; Park, K.J. Opportunities for improving the foraging potential of urban waterways for bats. *Biol. Conserv.* **2015**, *191*, 224–233. [CrossRef]
87. Smith, D.A.; Gehrt, S.D. Bat response to woodland restoration within urban forest fragments. *Restor. Ecol.* **2010**, *18*, 914–923. [CrossRef]
88. DiTommaso, A.; Morris, S.H.; Parker, J.D.; Cone, C.L.; Agrawal, A.A. Deer browsing delays succession by altering aboveground vegetation and belowground seed banks. *PLoS ONE* **2014**, *9*, e91155. [CrossRef]
89. Stinson, K.A.; Campbell, S.A.; Powell, J.R.; Wolfe, B.E.; Callaway, R.M.; Thelen, G.C.; Hallett, S.G.; Prati, D.; Klironomos, J.N. Invasive plant suppresses the growth of native tree seedlings by disrupting belowground mutualisms. *PLoS Biol.* **2006**, *4*, e140. [CrossRef]
90. Frick, W.F.; Baerwald, E.F.; Pollock, J.F.; Barclay, R.M.; Szymanski, J.; Weller, T.; Russell, A.; Loeb, S.; Medellin, R.; McGuire, L. Fatalities at wind turbines may threaten population viability of a migratory bat. *Biol. Conserv.* **2017**, *209*, 172–177. [CrossRef]
91. Hoyt, J.R.; Kilpatrick, A.M.; Langwig, K.E. Ecology and impacts of white-nose syndrome on bats. *Nat. Rev. Microbiol.* **2021**, *19*, 196–210. [CrossRef]

MDPI AG
Grosspeteranlage 5
4052 Basel
Switzerland
Tel.: +41 61 683 77 34

Forests Editorial Office
E-mail: forests@mdpi.com
www.mdpi.com/journal/forests

www.ingramcontent.com/pod-product-compliance
Lightning Source LLC
Chambersburg PA
CBHW041401210926
51850CB00013B/3146